# Recent Advances in Rice Research

*Edited by Mahmood-ur-Rahman Ansari*

Published in London, United Kingdom

IntechOpen

*Supporting open minds since 2005*

Recent Advances in Rice Research
http://dx.doi.org/10.5772/intechopen.87472
Edited by Mahmood-ur-Rahman Ansari

Contributors
Mohammad Rondhi, Ahmad Fatikhul Khasan, Anik Suwandari, Rokhani, Yasuhiro Mori, Takumi Kondo, Ahmad Asrofi, Ramlakhan Verma, Debarchana Jena, Diptibala Rout, Vineeta Singh, Manish Kumar Ahlavat, Pandurang Arsode, Jawahar Lal Katara, Sanghamitra Samantaray, Prakash Singh, Samuel Oppong Abebrese, Alex Yeboah, Nor'Aishah Hasan, Mohd Rafii Yusoff, Abdul Rahim Harun, Faiz Ahmad, Satyen Mondal, Jamil Hasan, Priya Lal Biswas, Emam Ahmed, Tuhin Halder, Amina Khatun, Muhammad Nasim, Evangelina S. Ella, Md. Panna Ali, Tofazzal Islam, Endang M. Septiningsih, Leka Lutpiatina, Danbaba Nahemiah, Iro Nkama, Paul Yahaya Idakwo, Mamudu Halidu Badau, Aliyu Umar, Arif Hasan Khan Robin, Afsana Hannan, Md. Najmol Hoque, Lutful Hassan, Prinya Wongsa, Zeti-Azura Mohamed-Hussein, Rabiatul-Adawiah Zainal-Abidin, Mahmood-Ur- Rahman Ansari, Munazza Ijaz, Roshina Shahzadi, Shazia Anwer Bukhari, Akmaral U. Umirbekovna Issayeva, Nnaemeka Success Esiobu



First published in London, United Kingdom, 2021 by IntechOpen
IntechOpen is the global imprint of INTECHOPEN LIMITED, registered in England and Wales, registration number: 11086078, 5 Princes Gate Court, London, SW7 2QJ, United Kingdom
Printed in Croatia

British Library Cataloguing-in-Publication Data
A catalogue record for this book is available from the British Library

Additional hard and PDF copies can be obtained from orders@intechopen.com

Recent Advances in Rice Research
Edited by Mahmood-ur-Rahman Ansari
p. cm.
Print ISBN 978-1-83881-031-3
Online ISBN 978-1-83881-032-0
eBook (PDF) ISBN 978-1-83881-036-8

For EU product safety concerns:
IN TECH d.o.o., Prolaz Marije Krucifikse Kozulić 3, 51000 Rijeka, Croatia, info@intechopen.com or visit our website at intechopen.com.

# Meet the editor

Dr. Mahmood-ur-Rahman Ansari is an Associate Professor of Molecular Biology at the Department of Bioinformatics and Biotechnology, GC University – Faisalabad, Pakistan. He received his BSc (Hons) in Plant Breeding and Genetics from the University of Agriculture, Faisalabad, Pakistan in 2003. He received his MPhil and Ph.D. in Molecular Biology from the National Centre of Excellence in Molecular Biology, Lahore, Pakistan in 2006 and 2011 respectively. He has published over 60 papers in international peer-reviewed journals in the field of Molecular Biology, Biotechnology, and Bioinformatics. He has also published over 10 book chapters and edited 3 books. He is involved in research projects to understand the molecular mechanisms of stress tolerance in plants. He has been involved in genetic modification of rice and cotton, their greenhouse and field testing as well as biosafety studies. He is a member of various national and international professional societies.

# Contents

# Preface

Rice is the most important food crop in the world. It is a staple food for more than 40% of the world's population. It feeds more than two billion people in developing countries. It has a major share in the global agricultural economy; however, its production and yield are compromised due to various biotic, abiotic, and climatic factors. To meet the food security standards, modern and sophisticated technologies have been used for the improvement of rice production. Starting from the conventional breeding strategies to the modern technologies, there is a long history of rice research. In this book, the milestones of rice research are discussed with recent and innovative results. The book deals with rice biology, breeding, genomics, economics, etc. Different sections of the book discuss diverse aspects of "Recent Advances in Rice Research". This book contains five sections including an Introductory Chapter in the first section that deals with an overview of recent technologies for the improvement of rice.

Section 2 describes the overview of the current status and innovations in hybrid rice research worldwide. Section 3 discusses the latest information regarding various metabolites produced and evaluated in rice using recent technologies. It contains information about phenolic compounds and flavonoids produced in rice. This section also discusses rice biology including marker-assisted breeding and pathogens transmitted through contaminated rice. Section 4 highlights the role of different stress conditions that limit crop production. This section also helps the readers to understand the mechanism of the root system and its role in abiotic stress tolerance. Section 5 describes very important aspects of rice including economics, value addition, and product development. This section contains information regarding the potential impacts of COVID-19 on rice yield and production, husk value addition, formulation of the sheet mask, and its evaluation.

The information available in this book will greatly help research students, scientists, academicians, and the general public as well as other stakeholders. In the end, I thank IntechOpen for giving me an opportunity to edit this book. I am also thankful to Ms. Sandra Maljavac, Author Service Manager, for her valuable help throughout the editing process. I must thank my students, Ms. Munazza Ijaz and Ms. Roshina Shahzadi for their assistance in the preparation of the "Introductory Chapter". I am especially thankful to all the authors for their valuable contributions in this book.

**Mahmood-ur-Rahman Ansari, PhD**
Department of Bioinformatics and Biotechnology,
GC University – Faisalabad,
Pakistan

## Section 1

# Introduction

Chapter 1

# Introductory Chapter: Recent Advances in Rice Biotechnology for Abiotic Stress Tolerance

*Munazza Ijaz, Roshina Shahzadi, Akmaral U. Issayeva, Shazia Anwer Bukhari and Mahmood-ur-Rahman*

## 1. Introduction

The world population is rising gradually, and it would be approximately 9.1 billion in 2050 but the production of agriculture is not escalating with this speed. The global production of agriculture should be increased up to 100-110 percent till 2050 to feed the whole population. Moreover, agricultural production and crop growth are negatively affected by abiotic stresses. So, it is a major hurdle in the way of increasing food production world-wide. The drought, heat, cold, salinity, ultra violet radiations and heavy metal toxicity are the major abiotic stresses that affect the crop yield. The significant source of calories for human beings is cereals. The maize, wheat and rice produce 10%, 17% and 23% respectively. *Oryza sativa* (rice) serve as a staple food and a famous cost-effective cereal. Acceptance of green revolution cultivars lead towards an essential revolution in the production of rice [1]. First green revolution came into being by producing high-yielding semi dwarf rice and wheat cultivars via plant breeding and providing high doses of nitrogen fertilizers. It helped in preventing the famine in semi-arid regions of Southern Asia [2].

The first transgenic rice was developed almost two decades before that initiated the high-throughput transformation protocols for the development of high yielding cultivars and it laid the foundation of transgenic rice biotechnology [3]. After the genome sequencing of rice, the development of transgenic rice varieties for better stress tolerance, high yield and better nutritional qualities became relatively easy and gained more importance. Similarly, the donor genes from the other species (like bacteria, fungi, viruses, animals and insects) can be inserted in the rice for obtaining the improved cultivars for making agriculture more sustainable. Improvement of some other significant traits (like photosynthetic rate, aroma and nutrition of rice grains) with the good stress tolerance has gained attention in past few years [2]. Some examples of these traits will be discussed in the next sections comprehensively.

The transgenic or genetically improved crops are an integral part of the agriculture industry in the modern world. The transgenic cereal crops have been commercialized and cultivated by more than twenty-eight countries. In 1996, only 1.7 million hectares of land have been cultivated with the transgenic crop that was increased up to 170 million hectares till 2012 and still it is increasing. Continuous efforts are made to bring the rice yield maximum for developing the large market of it globally [2]. Therefore, it is necessary to develop more cultivars of rice by using either old plant breeding techniques or new targeted genome editing techniques like zinc finger proteins (ZFNs), transcription activator-like nucleases (TALENs) and clustered regularly interspaced short palindromic repeats associated with Cas9 protein (CRISPR-Cas9) to develop cultivar of rice for improved agronomical traits.

## 2. Rice biotechnology for abiotic stress tolerance

Rice is an economically important crop and serious efforts are needed to increase its production. The production of rice can be increased in various ways like use of the available agricultural land at its maximum, utilization of marginal lands for growing the rice crop and using the lands effected by various abiotic stresses (e.g. drought stress, heat stress, cold stress, salt stress, UV radiation stress and metal toxicity) by growing the transgenic rice cultivars that are developed with the help of biotechnology. This is necessary because the agricultural land is decreasing and demand for food is increasing with the increasing world population and it is becoming very difficult to meet the food demands of this exponentially growing population with the same efficiency of the present day cultivars of rice [4].

The production of rice is greatly influenced by various abiotic stresses and it can be compensated by the various genes and regulatory networks present in different stress tolerant cultivars of rice and these genes and networks can be taken from the other plant species that can be used for developing new cultivars of rice (**Figure 1**). Abiotic stresses can induce the expression of many genes. The complex transcriptional networks are responsible for regulating the induction of stress tolerant genes. The important genes that are involved in transcriptional networks are studied by molecular techniques and these genes are useful in developing abiotic stress tolerant transgenic rice [5]. So, various biotechnology methods are used to develop rice cultivars that can combat with abiotic stresses (**Table 1**).

**Figure 1.**
*Abiotic stresses (e.g. drought, cold, heat, salinity, UV radiations, heavy metals) follow a same mechanism to survive and retain their growth and development. It has four main steps that are (a) avoidance: avoiding the contact with stresses, (b) escape: changing the life-cycle, (c) recovery: vegetative growth potency and (d) tolerance: nullifying the impacts of stresses.*

*DOI: http://dx.doi.org/10.5772/intechopen.94036*

| S. No. | Transgene(s) | Improved Trait(s) | References |
|---|---|---|---|
| 1. | *OsHsp101, AtHsp101* | Heat stress tolerance enhanced | [6] |
| 2. | *OsMAPK44* | Drought and salinity stress tolerance enhanced | [7] |
| 3. | *OsPIP1;3* | Cold stress tolerance ability enhanced | [8] |
| 4. | *Choline mono-oxygenase* | Heat and salinity stress tolerance enhanced | [9] |
| 5. | *AVP1, SsNHX1* | Improved ROS and salinity stress tolerance | [10] |
| 6. | *OsSBPase* | Improved photosynthetic efficiency and heat tolerance ability | [11] |
| 7. | *HvCBF4* | Salinity, drought and cold tolerance capacity enhanced | [12] |
| 8. | *SUB1A* | Enhanced submergence tolerance | [13] |
| 9. | *OsDREB1F* | Salinity, drought and cold tolerance capacity enhanced | [14] |
| 10. | *STAR1, STAR2* | Enhanced tolerance to Aluminum toxicity | [15] |
| 11. | *ZFP245* | Improved ROS, cold and drought stress tolerance | [16] |
| 12. | *P5CSF129A* | Salinity stress tolerance enhanced | [17] |
| 13. | *Isoflavone reductase* | Salinity stress tolerance enhanced | [18] |
| 14. | *OsHMA3* | Enhanced drought and submergence stress tolerance | [19] |
| 15. | *OsMAPK2* | Tolerance to phosphate deficiency | [20] |
| 16. | *OsNAC5* | Salinity, drought and cold tolerance capacity enhanced | [21] |
| 17. | *OsLEA3-2* | Salinity and drought tolerance capacity enhanced | [22] |
| 18. | *OsMYB55* | Heat stress tolerance enhanced | [23] |
| 19. | *PCK, PPDK* | Improved ROS stress tolerance | [24] |
| 20. | *OsETOL1* | Enhanced submergence stress tolerance | [25] |
| 21. | *OsMYB48-1* | Salinity and drought tolerance capacity enhanced | [26] |
| 22. | *VrDREB2A* | Increased tolerance to salinity and drought stress | [27] |
| 23. | *TaMYB3R1* | Enhanced drought and salt stress tolerance | [28] |
| 24. | *CaPUB1* | Enhanced cold stress tolerance | [29] |
| 25. | *OsLEA4* | Enhanced drought, salt and heavy metal stress tolerance | [30] |
| 26. | *OsNAC2* | Enhanced drought and salt stress tolerance | [31] |
| 27. | *OsGS* | Improved ROS and drought stress tolerance | [32] |
| 28. | *TsPIP1;1* | Enhanced salinity stress tolerance | [33] |
| 29. | *RhMYB96* | Enhanced salt tolerance | [34] |

**Table 1.**
*Some transgenes inserted for improving the abiotic stresses.*

## 2.1 Drought stress

Drought is affecting agricultural land worldwide from few past years. Many molecular, physiological and metabolic changes occur in plants due to drought stress that damages their growth and development [35]. During drought stress, plants respond variously and express change in their physiology and morphology.

Rice drought resistance is achieved by four procedures that are (a) avoidance: avoiding the contact with stress, (b) escape: changing life-cycle, (c) recovery: vegetative growth potency and (d) tolerance: nullifying the impacts of stress. These procedures are fulfilled by variable mechanisms like reduced leaf area, leaf rolling, senescence of older leaves, increased root proliferation, dense root system, scavenging reactive oxygen species (ROS), early flowering, osmotic adjustment, stomatal closure that minimize the water loss, changes in elasticity of cell wall and maximum uptake of deep water allow plants to survive during extended periods of drought and even allow them to reproduce in limited water supply by maintaining physiological activities [36]. Osmotic potential is kept lower inside the plant cells than outside the cells by an important mechanism known as osmotic adjustment. So, it allows the plants to retain its turgidity and prevents the water loss. Drought stress tolerance in plants can be achieved by the accumulation of inorganic and organic substances like proline, potassium ions, glucose and sucrose.

This mechanism participates in osmotic adjustment and maintains the turgor pressure. Plants exposed to drought stress produce high level of ROS that are extremely toxic and damages the DNA, carbohydrates, proteins and lipid. In drought tolerant plants, non-enzymatic antioxidants and antioxidant enzymes are produced that protect plants from the deleterious effects of ROS. Currently, the production system of rice depends upon the excessive supply of water and therefore, it is more susceptible to drought stress. The drought stress is the most significant restrictive factor in the production of rice and it is becoming a very severe problem. The varieties of rice favored by farmers in the tropical and sub-tropical regions are vulnerable to drought stress [35]. Rice is a significant economic crop of Asia that is mainly cultivated in lowland regions, where agriculture depends on the seasonal rainfall. Thus, rice crop in these areas is very susceptible to drought stress but this problem can be rectified by introducing some drought tolerant genes in the economically important cultivars of rice. For example, jasmine rice or 'Khao Dawk Mali 105' (KDML105) is a world-famous cultivar of rice exported from Thailand, had suffered from drought stress because of limited irrigation and unpredictable rainfalls. Drought tolerance can be achieved after keeping genetic background of KDML105 conserved by chromosome-segment substitution lines (CSSL) which have drought-tolerant quantitative-trait loci (DT-QTL) obtained from the back crossing between drought tolerant donor IR58586-F2-CA-143 (DH212) and KDML105. For more understanding, the drought tolerance related physiological responses consider another CSSL named as CSSL1-16. This line has properties like high proline, high water status, good membrane stability and great osmotic adjustment. Furthermore, it can recommence growth after recovery from stress. So, it can be used as a potential candidate for developing further drought tolerant cultivars while keeping their own genetic background preserved [37].

Another example of producing drought tolerant rice plants is introduction of pea DNA Helicase-47 (PDH47) from the plant *Pisum sativum* by using a constitutive promoter of cauliflower mosaic virus 35S- CaMV had been introduced in the ASD16 a cultivar of Indica rice via *Agrobacterium tumefaciens*. The transcripts of PDH47 are upregulated during drought stress that is correlated with increased water status, hydrogen peroxide accumulation and proline accumulation. It also regulates many stress responsive genes present endogenously during drought stress in transgenic rice [38]. There are many more genes that are expressed during the drought stress in different plants and are also potential candidates for producing transgenic rice that are drought tolerant.

## 2.2 Heat stress

Heat stress is a major limiting factor in the production of crops across the world because of global warming. A negative correlation is present between the increased temperature and crop yield especially in case of rice, wheat, barley and maize [39]. The heat stress can damage rice plants severely by inhibiting the metabolic activities, seed setting, plant growth and pollen fertility; therefore production of rice is reduced [40]. Excessive heat can also decrease the photosynthetic ability of plant, reduce water use efficiency, seed weight and grain mass and shorten the leaf area. Heat stress can damage both at vegetative and reproductive stage from sprouting to maturity. But, flowering and booting are the two more critical stages that can cause complete sterility in rice cultivars [41]. Heat tolerance belongs to plants that can lessen the effects of stress and give enough economic yields even at high temperature. Like other plant species, rice also has variations in germplasm to combat with heat stress. Tolerance is achieved by adjusting different molecular, morphological and physiological traits in rice cultivars. High temperature enhances the expression of stress tolerating genes and increases the metabolite production which is beneficial for achieving stress tolerance in plants [42]. During heat stress plants adopt multiple mechanisms like avoidance, survival and escape. These mechanisms impose avoidance for short term and develop resistance for long term survival. At the cellular level, effects of stress can be neutralized by certain factors and mechanisms; such as transcriptional control, antioxidant defense, osmolytes, late embryogenesis abundant (LEA) proteins and the factors that participate in signaling cascades. In hot environments, yield is reduced because of early maturity comes under the domain of avoidance strategies while suffering from heat stress [41].

Several morphological markers like long anthers, high pollen fertility, long basal pores and large basal dehiscence can be used for determining heat tolerant rice. As well as grain yield, number of spikelet, weight of thousand grains and seed setting percentage can be used for screening rice germplasm against high temperature stress; a great reduction is found in all these factors during heat stress. So, these parameters can be used to screen heat tolerant rice. Opening of the rice spikelet in early morning can be another beneficial criteria for the selection of heat tolerant rice cultivars [41]. A thermo-tolerant wild cultivar known as *Oryza meridionalis* can maintain high photosynthesis rate at high temperature. This high rate of photosynthesis is due to good stability and activity of Rubisco and it can be used as a significant physiological marker to determine the heat tolerance in rice [43]. The chlorophyll content and electrolyte leakage from leaves and roots can be increased during heat stress; it can also be used as a marker to analyze the heat tolerance. The fluid content of the membrane, lipid molecule present in membranes, cooling effect of transpiration and leaf position are some other beneficial physiological marker to determine the heat tolerance in rice.

Plant breeding efficacy has been increased by using marker assisted selection (MAS). Numerous molecular markers are used in MAS, but, single nucleotide polymorphism (SNPs) and simple sequence repeats (SSRs) are extensively used. This procedure is used to gather the information about the genes involved in providing resistance against biotic and abiotic stresses [44]. Still, there are very limited numbers of rice cultivars that are engineered to tolerate heat stress. Because the SNPs are present more than the SSRs in the rice genome, so SNPs are more prone to use for developing heat tolerance. A few SNPs related with heat tolerance have been known. Due to the very complex nature of this trait, each marker can contribute very less towards the variance. Therefore, several markers that are linked with various quantitative trait loci (QTLs) are used to improve a cultivar for heat tolerance [45]. Many genes are responsible for heat tolerance at various stages; for example, ZFP is a gene that is related with heat tolerance at seedling stage [46] and OsWRKY11 is another gene that is responsible for heat tolerance in rice [47]. The heat tolerant genes are needed to be identified in rice cultivars and the genes from the other plant species can also be transferred to the rice cultivars to improve the heat tolerance. Other approaches like use of mutagens and a recent technology like CRISPR-Cas9 should be applied for further development of heat tolerant rice cultivars.

### 2.3 Cold stress

The cold stress is a significant environmental factor that affects the development and growth of rice crop. At the stage of seedling development, sudden decrease in temperature can influence the development of chlorophyll [48]. The damage of rice seedling due to cold stress ultimately decreases the grain yield. So, the cold stress is a major limitation which can be overcome by using cold tolerant rice varieties [49]. Because rice crop is evolved in tropical region, so it has limited adaptability to chilling stress. Improving rice varieties to make them cold tolerant, enable the cultivation of rice in northerly latitudes. Different signal transduction pathways and genetic networks are involved in controlling chilling tolerance [50]. In japonica rice, chilling tolerance is achieved by interactions between rice G-protein α-subunit 1 (RGA1) and chilling tolerance divergence 1 (COLD1) followed by the calcium signaling initiated in the response of downstream network of stress response belongs to C repeat binding factor (CBF) that is a transcription factor [51]. Still, there is limited information available about the stress response and adaptation. The plant show response to abnormal surrounding temperature by changing their gene expression and adapt the suitable architecture due to the developmental plasticity. In SAM (shoot apical meristem) intrinsic signals can be disrupted by cold stress and stress tolerance is enhanced by setting the dormancy cycling at the SAM [52]. For the survival mechanism against cold stress, protection of niche forms of root stem cell is a sacrifice [53]. The differentiated cells are properly organized, and meristematic activity is maintained by this rehabilitated development in response to the cold temperature. During cold stress, many specific genes like OsMYB3R-2 are activated by the various transcription factors to maintain the mitotic cell and their cold tolerance. The survival and growth are enhanced by keeping maintained the cellular activity and cell behavior during and after the cold stress.

The formation of axillary bud following by its outgrowth in axils of primordia of leaf controls the shoot branching. The axillary bud initiation is controlled by MOC1 (Monoclum 1) and LAX 1 (Lax Panicle 1). The signaling pathways involved in the axillary bud outgrowth formation and biosynthesis of strigolactone are important for tillering in rice crop and are controlled by the dwarf genes, that are dwarf 53 (D53), dwarf 27 (D27), dwarf 17 (D17), dwarf 14 (D14), dwarf 10 (D10) and dwarf 3 (D3) [54]. In rice the outgrowth of axillary buds is repressed by the OsTB1 that acts in the downstream region of the dwarf genes. The interaction between the OsMADS57 and OsTB1 is responsible for reducing the inhibitory effect on the D14, a gene responsible for the organogenesis of rice tillers by producing a receptor for the hormone strigolactone [55], and it enables D14 to regulate the initiation of axillary bud [56]. The plant endogenous conditions and environmental situations control the development of axillary buds that are basically indeterminate structures [57]. To adapt the cold environment a network has been identified having core gene OsMADS57. The rice tiller growth is maintained by the overexpression of OsMADS57 during cold stress. OsMADS57 directly binds with the promoter of the

OsWRKY94 gene and enhances its expression in response to chilling stress while retarding its expression during normal temperature. Moreover, during normal temperature and cold stress, OsTB1 directly target and suppress the OsWRKY94 gene and under the cold stress OsMADS57 is responsible for promoting the transcription of D14 and suppress the tillering while during normal conditions expression of D14 was repressed to enhance the tillering. It shows that OsTB1 and OsMADS57 equally contribute to develop the cold tolerance in rice by targeting OsWRKY94 [52]. So, they can be used as potential candidates for developing cold tolerant rice cultivars.

Cold tolerance can also be achieved by the biogenesis of chloroplast in rice crop. In plant development and growth, chloroplast plays an important role and its development is affected by the chilling temperature. The genes and regulators that are involved in the biogenesis of chloroplast are identified and characterized. The mutant of WSL5 (white strip leaf 5) in rice has been characterized. The white stripped leaves are developed by this mutant during early leaf development stage and show albino phenotype during chilling stress. A unique chloroplast-targeted-pentatricopeptide repeat protein is encoded by the WSL5; the molecular and genetic analysis has revealed it. The RNA sequence analysis revealed that in the mutant, expression of the nuclear-encoded-photosynthetic genes was significantly repressed and the genes responsible for the chloroplast formation were also changed significantly. WSL5 gene is required for the development of chloroplast during cold stress [58]. A transcription factor OsMYB3R-2 from rice having a DNA binding domain is involved in enhancing the cold tolerance [59]. Another transcription factor MYBS3 of rice having a DNA binding repeat also play a significant role in adapting the cold environment [60].

The progress in developing the genetically modified plants by introducing and overexpressing the novel genes appears to be a very good practically possible option to speed up the process of breeding in plants. Instinctively, the faster way of inserting the genes with beneficial traits is genetic engineering instead of molecular or conventional breeding; and it would be the only possible option if the gene of interest belongs to some other species of plants, distant relatives, or from any non-plant source. The utilization of gene knockout strategies and genomic approaches are developing to enhance the efforts to measure thoroughly and make it easy to completely understand the complicated quantitative traits like tolerance to different temperature fluctuations and extremes. Many relevant genes have been identified by using molecular and genetic techniques to understand the plant response for the cold stress and these genes can be used further for developing cold tolerant varieties [61].

### 2.4 Salt stress

Rice crop is susceptible to salt stress and 1/3 of total world agricultural land is affected with it. The salinity of both water and soil has negative impact on the production of rice. The increasing level of sodium ions in the agricultural lands is becoming a severe threat for the agriculture worldwide. The plants suffer an osmotic stress due to the accumulation of salts at the outer side of the roots and suffer an ionic stress due to the accumulation of salts at the inner side of the plants [4]. The increase in food supply must be equal to the rate of increasing population and this demand can only be satisfied via utilizing all the available resources of land at their maximum potential. So, it is also necessary to use the saline soils at their fullest production potential. Different methods like agronomic adjustments, reclamation and different biological amendments are carried out in combination for increasing the production of saline soils. The use of salinity tolerant genetically improved crop varieties is the most suitable option for the sustainable crop production in these areas [62]. The genetic diversity of crops regarding salinity tolerance must be evaluated for developing the salt tolerant crop varieties. The molecular mapping approaches have made it possible to identify the genomic regions responsible for the salt tolerance and assessment of the genetic diversity of different crops and varieties is becoming very easy [63]. The chromosomal regions (QTLs) responsible for the tolerance of salt stress in rice crop can be identified through various molecular mapping approaches.

The salt stress badly effects the physiological, morphological and biochemical features of the rice. It has the negative impact on plant height, shoot dry weight, total tillers, total dry matter and root dry weight. The various physiological attributes that are affected by the salt stress are senescence, calcium ion uptake, sodium ion uptake potassium ion uptake, total cations uptake, osmotic potential, transcription efficiency and relative growth rate [64]. The biochemical features of rice that are effected by the salt stress are proline content, anthocyanins, peroxidase (POX) activity, calcium content, sodium content, potassium content, chlorophyll content and hydrogen peroxide content [4]. Various levels of salinity tolerance are observed at the whole plant level and leaves of rice [65]. Likewise, the rice plant behavior against the salinity stress may vary at reproductive and vegetative phase and this may not be related to the net relative tolerance of the plant. It is mandatory to know the stage that is more susceptible to the salt stress because it is important for comparing the performance of different cultivars during stress. The process of photosynthesis is necessary for the good vegetative and reproductive growth. In the leaf tissues, increased sodium concentration adversely affects the essential cellular metabolism and net photosynthesis. In the process of photosynthesis, chlorophyll content is significantly important but during salinity stress, there is no relationship found between the photosynthesis and chlorophyll content because net photosynthesis rate is decreased by the sodium ion concentration which does not have any connection with the chlorophyll content. It shows disturbance in some other cellular processes of photosynthesis due to salinity stress. The carbon dioxide fixation and stomatal aperture are affected by the sodium ion accumulation in the leaf at the same time, so it can be a reason for the decrease in photosynthesis during salt stress [4]. Different mechanisms have been evolved in rice plant to cope with salt stress conditions. An example of this type of mechanism is compartmenting of salts within the plant.

Numerous genes and QTLs are activated during the salt stress, which could be determined by the different molecular mapping approaches. Many types of molecular marker are present to identify different QTLs. SNP (Single nucleotide polymorphism), SSLP (simple sequence length polymorphism), RFLP (restriction fragment length polymorphism), AFLP (Amplified fragment length polymorphism), STS (sequence tagged sites) and SSRs (simple sequence repeats) are various markers of DNA that are used for genotyping in the studies of molecular mapping [66]. For example: the QTLs of pollen fertility, sodium ion concentration and calcium, sodium and potassium accumulation have been identified in F2 population of rice by using SSR marker [67]. The QTLs for yield related traits and morphological traits are identified by F2 population of rice as plant material and SSR as DNA marker [66] and the QTLs of potassium and sodium uptake for increasing the salinity tolerance in rice are determined by the AFLP, RFLP and SSR [68]. QTLs of salt stress responsive genes have been identified by the SNP in the rice crop [69]. The rice germplasm and these identified QTLs are found salt stress tolerant and are useful for three main reasons: (1) salt tolerance can be understood by the molecular genetics in rice, (2) rice germplasm that is tolerant to salt stress can be introduced to make them salt tolerant and (3) for the screening of rice germplasm, identified QTLs are used against salinity stress [4]. Moreover, new genes responsible for the salt tolerance can be identified and incorporated in the rice cultivars to make them salt tolerant.

### 2.5 UV radiation stress

The ultraviolet (UV) radiations are present in the region of solar electromagnetic spectrum that has the wavelength (λ) from 200 nm to 400 nm. The UV radiations have shorter wavelength as compared to photosynthetically active radiation that has the wavelength ranges from 700 nm to 400 nm. The UV radiations are composed of three different types that are UV-A, UV-B and UV-C. The UV-C radiations have a very smaller wavelength that is 200-280 nm and it emits photons with high energy which are absorbed totally by ozone layer and are not be able to reach the Earth surface [70]. The UV-A radiations ranges from 315 nm to 415 nm and are more constant. The UV-A radiations causes a little harm to the plants, so the major source of damage are UV-B radiations that have wavelength between 280 and 315 nm [71]. UV-B can cause mutations, reduction in photosynthetic activity, reduction in chlorophyll content, lower electron transfer rate, damage genetic material, decrease the biomass, reduce leaf size, lessen leaf number and eventually decrease the plant productivity. At the cell level, it initiates the oxidative stress through enhancing the level of ROS that ultimately damages the DNA, lipids and proteins; so, the integrity and functionality of cell membrane and enzymes is compromised. The light for photosynthesis can be maximally captured in higher plant by the exposure with UV-B. It is important for increasing the secondary metabolites, enzymatic and non-enzymatic antioxidants, bioactive compounds and cyto-solutes (sugars, glycine, betane, proline) for the survival of plants. The adverse effects of the radiation can be mitigated by the growth regulators present in the plants. The specific signaling pathways are present in plants that are involved in regulating protective gene expression responses against UV-B are vital for survival of plants in sunlight. The identification of the genes responsible for the UV-B radiation, survival is necessary to develop the crop varieties resistant to UV radiation stress [72].

### 2.6 Heavy metal toxicity

Many solutes are present in the rhizosphere required for the growth and development of plants. Plants uptake these solutes by roots and distribute them in the whole plant body. The successful plant life is ensured by up taking the other components with water from the rhizospheric soil by roots. The developmental plasticity and physiological activity in the plant roots is carried out by the water uptake with soluble elements. The distribution and uptake of these inorganic materials inside the plants is an intrinsic property of energy and material fluidity. In plant cells, a plethora of physiological and structural functions is supported by these essential ions but if these essential ions are present in the non-physiological concentrations, they can turn out as limiting factors. The cellular homeostasis is affected by their availability to plants, Inequalities in comparative abundance of these elements in soil and their rate of uptake. The defense system and adaption of plants is dependent on the developmental and physiological changes triggered by ion toxicity; but it can cause the permanent damage to the plant. In the rhizospheric soil, the ions of heavy metals are also present that can be absorbed by the roots with the water and nutrients and can be incorporated into the plant tissues. The zinc, iron, manganese, copper, aluminum, chromium, cadmium, cobalt, lead, arsenic, nickel and molybdenum are some toxic metals for plants [73].

The concentration of metal ions is excessive in polluted areas and plants growing in those areas suffer from metal toxicity. Some soils have high level of heavy metals naturally like serpentine soils and the mining areas also have high heavy metal content due to activities of human beings. The environmental pollutants including high concentrations of heavy metals are becoming a major challenge for all organisms (plants, animals, microbes, etc.) across the globe. The ionic homeostasis is achieved in plants by maintaining the levels of essential and non-essential ions. Furthermore, this homeostasis in cells and tissues must be maintained in an organ specific manner. The risk of heavy metals can be mitigated by minimizing their uptake, avoiding their exposure and intracellular sequestering in the plants. The cellular structures, membranes structures, basic metabolism and transport processes are affected, if plant is not able to combat with the harmful effects and internalization of heavy metals. The stress reactions are triggered by the high levels of non-toxic and toxic metal ions that disturb the development at molecular, structural and physiological level. The defense mechanism is started in the plants upon the beginning of heavy metal toxicity [74]. The oxidative stress is caused by the heavy metal toxicity, so cells and sub-cellular compartments are protected by plant defense system. The plant growth and development are continued if the heavy metal ions are present in low or moderate concentrations. But, plants cannot survive during the long-term exposure of high level of heavy metal toxicity. The metal tolerance is a unique feature of some plant species in stress biology. The compartmentalization and of heavy metals, restriction of heavy metals in apoplast, extracellular attenuation and chelation of heavy metals and exclusion strategies are some mechanisms of achieving stress tolerance in the plants.

The most practical approach adopted by the plants to combat with the heavy metal toxicity is hyper accumulation and hyper tolerance of heavy metals inside the plant without effecting its viability [75]. The hyper accumulation of metal ions protects the plants from pathogens and herbivores in some conditions (Hörger et al. 2013). The contamination of heavy metals is present in the wide range of terrestrial habitats. In food chain, plants have vital importance, so heavy metal accumulation in plants causes severe contamination in the food chain. So, it is significantly important to know that how plants combat with the heavy metal exposure. The knowledge of these mechanisms is used in developing the practical biological methods to develop the tolerance against heavy metal toxicity [74]. Many promising and existing biotechnological approaches are used are used for this purpose [75]. Before concluding the major efforts used for making plant to heavy metal stress, it is necessary to understand the key factors involved in adjusting the heavy metal tolerance. The total cell protein study is called proteomics; it is a most valuable tool for understanding the basic procedures of development and stress responses in plants [73]. The sensing, transport and uptake of heavy metal, their sequestration in the vacuole and formation of phyto-chelating-heavy metal complex are some main procedures for influencing the plant susceptibility for the metal ions. These processes are controlled at transcriptional and post-transcriptional levels and the secondary process responsible for maintaining the plant during metal toxicity includes the defense reaction derived by the production of antioxidants to preserve the membrane lipid integrity. Diverse genes are responsible for controlling these processes. The heavy metal tolerance of the plants is achieved by the various biotechnological, molecular and genetic tools [74].

## 3. Rice biotechnology under climate change conditions

Due to climate change, the food availability is affected in 21st century to feed the population which is increasing day by day. Globally, climate change have bad effects on the human health, food security and water resources [76]. About 30-60% of rice income is decreased per year because of different types of stresses [77]. The rice fields are susceptible due to climate change, expansion of urban area and ruined of cultivation field. Due to some cases, 100% rice production is destroy [78]. The rice yield is reduced due to increase in temperature by accumulation of carbon dioxide in environment. The production and quality of rice (*Oryza sativa* L.) is badly affected by climate change which is mainly abiotic stresses (heat, flood, salinity, cold and drought) and biotic stresses (pest, weeds and pathogens). Climate change cause the formation of unripe grains, white grains, and fractured grains. The grain damage is reduced by use of thermo-stable verities which are formed through conventional breeding and genetic engineering. Thermo tolerant 1 (TT1) is the most important (QTL) present in *Oryza glaberrima*, it codes α2 which is the subunit of 26S proteasome and is involved in destruction of ubiquitinated protein.

The removal of cytotoxic degraded protein is done by OgTT1 (*Oryza glaberrima* thermo tolerant 1) maintain the cell from high temperature. Over expression of TT1 increase the ability of heat tolerant stress in rice. It helps to grow the rice in global warming [79]. To protect the food, production of heat tolerant varieties which can be grown in high temperature is a major challenge. In *Arabidopsis thaliana*, Dehydration responsive element binding protein 2A (DREB2A) is regulated by DNA polymerase subunit B3-1 (DPB3-1), is a transcriptional controller which regulate the growth of plant and increase the thermostability [80]. If temperature is round about 35°C in chamber trial at the time of flowering, then spikelet infertility occurs due to heat stress. When the temperature in Japan cities (Tokai, Kanto) was reached at 40°C in 2007, it induced extensive unproductiveness. The increase in world temperature is expected, that is why we need to develop techniques to protect the grain quality and production from heat damage [81]. During the ripening of rice, the high temperature causes the poor-quality grain production. Late transplanting and thermo-stable varieties are used in preventive measure. Deep-flood irrigation is a process which falling the rate of milky white seeds when the temperature is high and low. Adequate quantity of nitrogen is necessary to protect the grain from disease at reproductive stage [81]. In future, if water availability is less, yield of rice will badly be affected.

Due to climate change, the crop rotation time can be reduced. There are many positive and negative effects of climate change on crop, so scientist try to develop the verities which can be cultivated in harsh conditions to provide enough food to global population [82]. Galactinol synthase 2 gene from Arabidopsis is introduced in two rice verities and have ability to produce the good quality in waterless field environment. The biotechnologically advanced verities have greater intensity of galactinol as compared to non-transgenic verities. The drought resistance rice has greater ability of photosynthesis, recovering and plant growth. It is clear that AtGolS2 gene is beneficial which provide higher quantity of rice in drought stress [83]. Rice varieties which are better grown in anaerobic environment produce less yield in aerobic situation. To accurately measure the expression of different genes in a condition, it is necessary to find out the genes which have same output in aerobic and anaerobic systems [84]. Transcriptional reaction and ecological variation is affected by macroclimate and meadow type, then we identify the genes which are involved in production and photosynthesis [82].

## 4. Future prospects

In modern-day agriculture, biotechnology has a wide range of applications. It helps in understanding the genetics of important traits as well as in developing new cultivars. The transgenic crop deployment and the marker assisted breeding are the two main areas. The remarkable progress has been made by the marker assisted rice breeding. Modern tools of biotechnology help in studying the functions of genes and mechanism of resistance through improved varieties via MAS. Exploitation of tissue culture techniques used as a vital mean or work as an assistant to other procedures together with recombinant DNA methods is at its frontline in plant improvement and modification for agriculture. The targeted genome editing methods like ZFNs, TALENs, CRISPR/Cas9 are implanted for the modification of non-operational alleles of the useful genes or to repress or knock out the non-useful alleles in plants without making the plant transgenic. The techniques and tools of biotechnology are a very sophisticated option for the rice breeders to develop the useful crop varieties.

## Acknowledgements

This work was supported by funds from Higher Education Commission of Pakistan.

## Author details

Munazza Ijaz[1], Roshina Shahzadi[1], Akmaral U. Issayeva[2], Shazia Anwer Bukhari[3] and Mahmood-ur-Rahman[1]*

1 Department of Bioinformatics and Biotechnology, GC University – Faisalabad, Faisalabad, Pakistan

2 Ecology and Biology Research Institute, Shymkent University, Shymkent, Kazakhstan

3 Department of Biochemistry, GC University – Faisalabad, Faisalabad, Pakistan

*Address all correspondence to: mahmoodansari@gcuf.edu.pk

## References

[1] Wani SH, Sah SK. Biotechnology and abiotic stress tolerance in rice. J. Rice Res. 2014;**2**:e105

[2] Clement WK, Wong MY, Jugah K, Maziah M. Producing Transgenic Rice with Improved Traits and Yield–How Far Have We Come? *Pertanika J. Schol. Res. Rev*. 2017;3(3).

[3] Hiei Y, Ohta S, Komari T, Kumashiro T. Efficient transformation of rice (*Oryza sativa* L.) mediated by Agrobacterium and sequence analysis of the boundaries of the T-DNA. The Plant Journal. 1994;**6**(2):271-282

[4] Saeed M. Abiotic stress tolerance in Rice (*Oryza sativa* L.): a genomics perspective of salinity tolerance. *Developments*. 2018;5:181.

[5] Todaka D, Shinozaki K, Yamaguchi-Shinozaki K. Recent advances in the dissection of drought-stress regulatory networks and strategies for development of drought-tolerant transgenic rice plants. Frontiers in Plant Science. 2015;**6**:84

[6] Agarwal M, Sahi C, Katiyar-Agarwal S, Agarwal S, Young T, Gallie DR, et al. Molecular characterization of rice hsp101: Complementation of yeast hsp104 mutation by disaggregation of protein granules and differential expression in indica and japonica rice types. Plant Molecular Biology. 2003;**51**(4):543-553

[7] Jeong MJ, Lee SK, Kim BG, Kwon TR, Cho WS, Park YT, et al. A rice (*Oryza sativa* L.) MAP kinase gene, OsMAPK44, is involved in response to abiotic stresses. Plant Cell, Tiss Org Cult. 2006;**85**(2):151-160

[8] Lian HL, Yu X, Ye Q, Ding XS, Kitagawa Y, Kwak SS, et al. The role of aquaporin RWC3 in drought avoidance in rice. Plant & Cell Physiology. 2004;**45**(4):481-489

[9] Shirasawa K, Takabe T, Takabe T, Kishitani S. Accumulation of glycinebetaine in rice plants that overexpress choline monooxygenase from spinach and evaluation of their tolerance to abiotic stress. Annals of Botany. 2006;**98**(3):565-571

[10] Zhao FY, Zhang XJ, Li PH, Zhao YX, Zhang H. Co-expression of the Suaeda salsa SsNHX1 and Arabidopsis AVP1 confer greater salt tolerance to transgenic rice than the single SsNHX1. Molecular Breeding. 2006;**17**(4):341-353

[11] Feng L, Wang K, Li Y, Tan Y, Kong J, Li H, et al. Overexpression of SBPase enhances photosynthesis against high temperature stress in transgenic rice plants. Plant Cell Reports. 2007;**26**(9):1635-1646

[12] Oh SJ, Kwon CW, Choi DW, Song SI, Kim JK. Expression of barley HvCBF4 enhances tolerance to abiotic stress in transgenic rice. Plant Biotechnology Journal. 2007;**5**(5):646-656

[13] Fukao T, Bailey-Serres J. Submergence tolerance conferred by Sub1A is mediated by SLR1 and SLRL1 restriction of gibberellin responses in rice. Proc. Natl. Acad. Sci. USA. 2008;**105**(43):16814-16819

[14] Wang Q, Guan Y, Wu Y, Chen H, Chen F, Chu C. Overexpression of a rice OsDREB1F gene increases salt, drought, and low temperature tolerance in both Arabidopsis and rice. Plant Molecular Biology. 2008;**67**(6):589-602

[15] Huang CF, Yamaji N, Mitani N, Yano M, Nagamura Y, Ma JF. A bacterial-type ABC transporter is involved in aluminum tolerance in rice. The Plant Cell. 2009;**21**(2):655-667

[16] Huang J, Sun SJ, Xu DQ, Yang X, Bao YM, Wang ZF, et al. Increased tolerance of rice to cold, drought and oxidative stresses mediated by the overexpression of a gene that encodes the zinc finger protein ZFP245. Biochemical and Biophysical Research Communications. 2009;**389**(3):556-561

[17] Kumar V, Shriram V, Kishor PK, Jawali N, Shitole MG. Enhanced proline accumulation and salt stress tolerance of transgenic indica rice by over-expressing P5CSF129A gene. Plant Biotechnol. Rep. 2010;**4**(1):37-48

[18] Kim SG, Kim ST, Wang Y, Kim SK, Lee CH, Kim KK, et al. Overexpression of rice isoflavone reductase-like gene (OsIRL) confers tolerance to reactive oxygen species. Physiologia Plantarum. 2010;**138**(1):1-9

[19] Ueno D, Yamaji N, Kono I, Huang CF, Ando T, Yano M, et al. Gene limiting cadmium accumulation in rice. Proc. Natl. Acad. Sci. USA. 2010;**107**(38):16500-16505

[20] Gaxiola RA, Edwards M, Elser JJ. A transgenic approach to enhance phosphorus use efficiency in crops as part of a comprehensive strategy for sustainable agriculture. Chemosphere. 2011;**84**(6):840-845

[21] Song SY, Chen Y, Chen J, Dai XY, Zhang WH. Physiological mechanisms underlying OsNAC5-dependent tolerance of rice plants to abiotic stress. Planta. 2011;**234**(2):331-345

[22] Duan J, Cai W. OsLEA3-2, an abiotic stress induced gene of rice plays a key role in salt and drought tolerance. PLoS One. 2012;**7**(9):e45117

[23] El-Kereamy A, Bi YM, Ranathunge K, Beatty PH, Good AG, Rothstein SJ. The rice R2R3-MYB transcription factor OsMYB55 is involved in the tolerance to high temperature and modulates amino acid metabolism. PLoS One. 2012;**7**(12):e52030

[24] Gu JF, Qiu M, Yang JC. Enhanced tolerance to drought in transgenic rice plants overexpressing C4 photosynthesis enzymes. Crop J. 2013;**1**(2):105-114

[25] Du H, Wu N, Cui F, You L, Li X, Xiong L. A homolog of ETHYLENE OVERPRODUCER, O s ETOL 1, differentially modulates drought and submergence tolerance in rice. The Plant Journal. 2014;**78**(5):834-849

[26] Xiong H, Li J, Liu P, Duan J, Zhao Y, Guo X, et al. Overexpression of OsMYB48-1, a novel MYB-related transcription factor, enhances drought and salinity tolerance in rice. PLoS One. 2014;**9**(3):e92913

[27] Chen H, Liu L, Wang L, Wang S, Cheng X. VrDREB2A, a DREB-binding transcription factor from Vigna radiata, increased drought and high-salt tolerance in transgenic Arabidopsis thaliana. Journal of Plant Research. 2016;**129**(2):263-273

[28] Cai H, Tian S, Dong H, Guo C. Pleiotropic effects of TaMYB3R1 on plant development and response to osmotic stress in transgenic Arabidopsis. Gene. 2015;**558**(2):227-234

[29] Min HJ, Jung YJ, Kang BG, Kim WT. CaPUB1, a hot pepper U-box E3 ubiquitin ligase, confers enhanced cold stress tolerance and decreased drought stress tolerance in transgenic rice (*Oryza sativa* L.). *Mol. Cells*. 2016;39(3):250.

[30] Hu T, Zhu S, Tan L, Qi W, He S, Wang G. Overexpression of OsLEA4 enhances drought, high salt and heavy metal stress tolerance in transgenic rice (Oryza sativa L.). *Environ. Exp*. Bot. 2016;**123**:68-77

[31] Shen J, Lv B, Luo L, He J, Mao C, Xi D, et al. The NAC-type transcription

factor OsNAC2 regulates ABA-dependent genes and abiotic stress tolerance in rice. Scientific Reports. 2017;7:40641

[32] Park SI, Kim YS, Kim JJ, Mok JE, Kim YH, Park HM, et al. Improved stress tolerance and productivity in transgenic rice plants constitutively expressing the Oryza sativa glutathione synthetase OsGS under paddy field conditions. Journal of Plant Physiology. 2017;**215**:39-47

[33] Li W, Qiang XJ, Han XR, Jiang LL, Zhang SH, Han J, et al. Ectopic expression of a Thellungiella salsuginea aquaporin gene, TsPIP1; 1, increased the salt tolerance of Rice. International Journal of Molecular Sciences. 2018;**19**(8):2229

[34] Jiang X, Li S, Ding A, Zhang Z, Hao Q, Wang K, et al. The novel rose MYB transcription factor RhMYB96 enhances salt tolerance in transgenic Arabidopsis. Plant Mol. Biol. Rep. 2018;**36**(3):406-417

[35] Zu X, Lu Y, Wang Q, Chu P, Miao W, Wang H, et al. A new method for evaluating the drought tolerance of upland rice cultivars. Crop J. 2017;**5**(6):488-498

[36] Saha P, Sade N, Arzani A, Wilhelmi MD, Coe KM, Li B, et al. Effects of abiotic stress on physiological plasticity and water use of *Setaria viridis* L. Plant Science. 2016;**251**:128-138

[37] Larkunthod P, Nounjan N, Siangliw JL, Toojinda T, Sanitchon J, Jongdee B, et al. Physiological responses under drought stress of improved drought-tolerant rice lines and their parents. Not. Bot. Horti. Agrobo. 2018;**46**(2):679-687

[38] Singha DL, Tuteja N, Boro D, Hazarika GN, Singh S. Heterologous expression of PDH47 confers drought tolerance in indica rice. *Plant Cell, Tiss. Org.* Cult. 2017;**130**(3):577-589

[39] Zhang CX, Feng BH, Chen TT, Zhang XF, Tao LX, Fu GF. Sugars, antioxidant enzymes and IAA mediate salicylic acid to prevent rice spikelet degeneration caused by heat stress. Plant Growth Regulation. 2017;**83**(2):313-323

[40] Zafar SA, Hameed A, Khan AS, Ashraf M. Heat shock induced morpho-physiological response in indica rice (Oryza sativa L.) at early seedling stage. *Pak. J.* Bot. 2017;**49**(2):453-463

[41] Zafar SA, Hameed A, Nawaz MA, Wei MA, Noor MA, Hussain M. Mechanisms and molecular approaches for heat tolerance in rice (Oryza sativa L.) under climate change scenario. Journal of Integrative Agriculture. 2018;**17**(4):726-738

[42] Hasanuzzaman M, Nahar K, Alam M, Roychowdhury R, Fujita M. Physiological, biochemical, and molecular mechanisms of heat stress tolerance in plants. International Journal of Molecular Sciences. 2013;**14**(5):9643-9684

[43] Scafaro AP, Yamori W, Carmo-Silva AE, Salvucci ME, von Caemmerer S, Atwell BJ. Rubisco activity is associated with photosynthetic thermotolerance in a wild rice (Oryza meridionalis). Physiologia Plantarum. 2012;**146**(1):99-109

[44] Shamsudin NA, Swamy BM, Ratnam W, Cruz MT, Raman A, Kumar A. Marker assisted pyramiding of drought yield QTLs into a popular Malaysian rice cultivar, MR219. BMC Genetics. 2016;**17**(1):30

[45] Ye C, Tenorio FA, Argayoso MA, Laza MA, Koh HJ, Redoña ED, et al. Identifying and confirming quantitative trait loci associated with heat tolerance

at flowering stage in different rice populations. BMC Genetics. 2015;**16**(1):41

[46] Wei H, Liu J, Wang Y, Huang N, Zhang X, Wang L, et al. A dominant major locus in chromosome 9 of rice (Oryza sativa L.) confers tolerance to 48 C high temperature at seedling stage. *J.* Hered. 2013;**104**(2):287-294

[47] Wu X, Shiroto Y, Kishitani S, Ito Y, Toriyama K. Enhanced heat and drought tolerance in transgenic rice seedlings overexpressing OsWRKY11 under the control of HSP101 promoter. Plant Cell Reports. 2009;**28**(1):21-30

[48] Kusumi K, Iba K. Establishment of the chloroplast genetic system in rice during early leaf development and at low temperatures. Frontiers in Plant Science. 2014;**5**:386

[49] Zhao J, Zhang S, Dong J, Yang T, Mao X, Liu Q, et al. A novel functional gene associated with cold tolerance at the seedling stage in rice. Plant Biotechnology Journal. 2017;**15**(9):1141-1148

[50] Zhao C, Lang Z, Zhu JK. Cold responsive gene transcription becomes more complex. Trends in Plant Science. 2015;**20**(8):466-468

[51] Zhu JK. Abiotic stress signaling and responses in plants. Cell. 2016;**167**(2):313-324

[52] Chen L, Zhao Y, Xu S, Zhang Z, Xu Y, Zhang J, et al. Os MADS 57 together with Os TB 1 coordinates transcription of its target Os WRKY 94 and D14 to switch its organogenesis to defense for cold adaptation in rice. The New Phytologist. 2018;**218**(1):219-231

[53] Hong JH, Savina M, Du J, Devendran A, Ramakanth KK, Tian X, et al. A sacrifice-for-survival mechanism protects root stem cell niche from chilling stress. Cell. 2017;**170**(1):102-113

[54] Jiang L, Liu X, Xiong G, Liu H, Chen F, Wang L, et al. DWARF 53 acts as a repressor of strigolactone signalling in rice. Nature. 2013;**504**(7480):401-405

[55] Yao R, Ming Z, Yan L, Li S, Wang F, Ma S, et al. DWARF14 is a non-canonical hormone receptor for strigolactone. Nature. 2016;**536**(7617):469-473

[56] Guo S, Xu Y, Liu H, Mao Z, Zhang C, Ma Y, et al. The interaction between OsMADS57 and OsTB1 modulates rice tillering via DWARF14. Nature Communications. 2013;**4**(1):1-2

[57] Janssen BJ, Drummond RS, Snowden KC. Regulation of axillary shoot development. Current Opinion in Plant Biology. 2014;**17**:28-35

[58] Liu X, Lan J, Huang Y, Cao P, Zhou C, Ren Y, et al. WSL5, a pentatricopeptide repeat protein, is essential for chloroplast biogenesis in rice under cold stress. Journal of Experimental Botany. 2018;**69**(16):3949-3961

[59] Ma Q, Dai X, Xu Y, Guo J, Liu Y, Chen N, et al. Enhanced tolerance to chilling stress in OsMYB3R-2 transgenic rice is mediated by alteration in cell cycle and ectopic expression of stress genes. Plant Physiology. 2009;**150**(1):244-256

[60] Su CF, Wang YC, Hsieh TH, Lu CA, Tseng TH, Yu SM. A novel MYBS3-dependent pathway confers cold tolerance in rice. Plant Physiology. 2010;**153**(1):145-158

[61] Sanghera GS, Wani SH, Hussain W, Singh NB. Engineering cold stress tolerance in crop plants. Current Genomics. 2011;**12**(1):30

[62] Singh R, Singh Y, Xalaxo S, Verulkar S, Yadav N, Singh S, et al. From QTL to variety-harnessing the benefits of QTLs for drought, flood and salt tolerance in mega rice varieties of India through a multi-institutional network. Plant Science. 2016;**242**:278-287

[63] Khan MS, Saeed M, Iqbal J. Association mapping validates previously identified quantitative trait loci for salt tolerance in rice (*Oryza sativa* L.). *Mol. Breed.* 2016;36(12):172.

[64] Negrão S, Schmöckel SM, Tester M. Evaluating physiological responses of plants to salinity stress. Annals of Botany. 2017;**119**(1):1-1

[65] Sytar O, Brestic M, Zivcak M, Olsovska K, Kovar M, Shao H, et al. Applying hyperspectral imaging to explore natural plant diversity towards improving salt stress tolerance. Sci. Total Environ. 2017;**578**:90-99

[66] Khan MS, Saeed M, Iqbal J. Quantitative trait locus mapping for salt tolerance at maturity stage in indica rice using replicated F 2 population. Braz. J. Bot. 2016;**39**(2):641-650

[67] Khan MS, Saeed M, Iqbal J. Identification of quantitative trait loci for Na ı, K ı and Ca ı ı accumulation traits in rice grown under saline conditions using F 2 mapping population. Braz. J. Bot. 2015;**38**(3):555-565

[68] KoyamaML, LevesleyA, KoebnerRM, Flowers TJ, Yeo AR. Quantitative trait loci for component physiological traits determining salt tolerance in rice. Plant Physiology. 2001;**125**(1):406-422

[69] Kumar V, Singh A, Mithra SA, Krishnamurthy SL, Parida SK, Jain S, et al. Genome-wide association mapping of salinity tolerance in rice (Oryza sativa). DNA Research. 2015;**22**(2):133-145

[70] Vishwakarma K, Upadhyay N, Kumar N, Yadav G, Singh J, Mishra RK, et al. Abscisic acid signaling and abiotic stress tolerance in plants: A review on current knowledge and future prospects. Frontiers in Plant Science. 2017;**8**:161

[71] Singh SS, KUMAR P, RA AK. Ultraviolet radiation stress: Molecular and physiological adaptations in trees. In: Abiotic stress tolerance in plants 2006 (pp. 91-110). Springer. Dordrecht.

[72] Gadi BR. Effect of UV - B radiation on plants. IJSRST. 2018;**4**:255-260

[73] Hossain Z, Komatsu S. Contribution of proteomic studies towards understanding plant heavy metal stress response. Front. Plant Sc. 2013;**3**:310

[74] Ovečka M, Takáč T. Managing heavy metal toxicity stress in plants: Biological and biotechnological tools. Biotechnology Advances. 2014;**32**(1):73-86

[75] DalCorso G, Manara A, Furini A. An overview of heavy metal challenge in plants: From roots to shoots. Metallomics. 2013;**5**(9):1117-1132

[76] Kang Y, Khan S, Ma X. Climate change impacts on crop yield, crop water productivity and food security–A review. Progress in Natural Science. 2009;**19**(12):1665-1674

[77] Sharma I, Kaur N, Pati PK. Brassinosteroids: A promising option in deciphering remedial strategies for abiotic stress tolerance in rice. Frontiers in Plant Science. 2017;**8**:2151

[78] Mottaleb KA, Rejesus RM, Murty MV, Mohanty S, Li T. Benefits of the development and dissemination of climate-smart rice: Ex ante impact assessment of drought-tolerant rice in South Asia. Mitig. Adapt. Strat. Gl. 2017;**22**(6):879-901

[79] Li XM, Chao DY, Wu Y, Huang X, Chen K, Cui LG, et al. Natural alleles of a proteasome α2 subunit gene contribute to thermotolerance and adaptation of African rice. Nature Genetics. 2015;**47**(7):827-833

[80] Sato H, Todaka D, Kudo M, Mizoi J, Kidokoro S, Zhao Y, et al. The A rabidopsis transcriptional regulator DPB 3-1 enhances heat stress tolerance without growth retardation in rice. Plant Biotechnology Journal. 2016;**14**(8):1756-1767

[81] Morita S, Wada H, Matsue Y. Countermeasures for heat damage in rice grain quality under climate change. Plant Prod. Sci. 2016;**19**(1):1-1

[82] Plessis A, Hafemeister C, Wilkins O, Gonzaga ZJ, Meyer RS, Pires I, et al. Multiple abiotic stimuli are integrated in the regulation of rice gene expression under field conditions. eLife. 2015;**4**:e08411

[83] Abdullah E, Idris A, Saparon A. Papr reduction using scs-slm technique in stfbc mimo-ofdm. ARPN J. Eng. Appl. Sci. 2017;**12**(10):3218-3221

[84] Phule AS, Barbadikar KM, Madhav MS, Senguttuvel P, Babu MP, Kumar PA. Genes encoding membrane proteins showed stable expression in rice under aerobic condition: novel set of reference genes for expression studies. *3 Biotech*. 2018;8(9):383.

Section 2

# Hybrid Rice

Chapter 2

# Hybrid Rice Research: Current Status and Prospects

*Diptibala Rout, Debarchana Jena, Vineeta Singh, Manish Kumar, Pandurang Arsode, Prakash Singh, Jawahar Lal Katara, Sanghamitra Samantaray and Ramlakhan Verma*

## Abstract

Heterosis is a solitary means of exploiting hybrid vigor in crop plants. Given its yield advantage and economic importance, several hybrids in rice have been commercialized in more than 40 countries, which has created a huge seed industry worldwide. India has made commendable progress and commercialized 117 three-line *indica* hybrids for different ecology and duration (115–150 days), which accounted for 6.8% of total rice area in the country. Besides, several indigenous CMS lines developed in diversified genetic and cytoplasmic backgrounds are being utilized in hybrid rice breeding. NRRI, which has been pioneering to start with the technology, has developed three popular rice hybrids, viz., Ajay, Rajalaxmi, and CR Dhan 701 for irrigated-shallow lowland ecosystem. Biotechnological intervention has supplemented immensely in excavating desirable genomic regions and their deployment for further genetic enhancement and sustainability in rice hybrids. Besides, hybrid seed production creates additional job opportunity (100–105 more-man days) and comparatively more net income (70% more than production cost) than HYVs. Hence, this technology has great scope for further enhancement in *per se* rice productivity and livelihood of the nation.

**Keywords:** hybrid rice, CMS, genetic gain, heterosis, restorer, breeding value

## 1. Introduction

Heterosis is the superiority of $F_1$ offspring over either parent, a solitary means of harnessing complete hybrid vigor in crop plants. This phenomenon has aided agriculture and captivated geneticists for over centuries for the development of superior cultivar in many crops [1]. Suitable allelic combination and manipulation has made yield advantage in hybrid than HYVs. It covers large acreage for many crops, including rice, and has affected agrarian practices and the seed business across the world. Heterosis had been exploited in several practical ways for centuries before Darwin provided an early scientific explanation in maize. In rice, heterosis was first reported by Jonse [2]. However, owing to its self-pollinating nature (0.3–3.0% out-crossing), heterosis could be realized during middle of second half of the twentieth century after identification and development of the cytoplasmic male sterile (CMS) source. Subsequently, China, under the leadership of Yuan Long Ping, started work on the development of hybrid rice (HR) with a vision to make it possible to be commercial. He identified a natural male sterile mutant plant in rice (*indica*) and pollen abortive genotypes in the wild rice (Oryza rufipogon; Li 1970), which later served as donor of male sterile source (male sterile cytoplasm) for CMS development. In 1973, through recurrent back-cross breeding, several promising *indica* wild abortive CMS, viz., Erjiunan1A, Zhenshan 97A, and V20A CMS-WA, and good restorers, viz., Taiyin1, IR4, and IR1, were developed. Later during 1974, first *indica* rice hybrid, Nanyou 2, was released for cultivation in China. Afterward, relatively more heterotic hybrid rice (HR) breeding approaches like two-line system (1987 AD) and super hybrids (1996 AD) were adopted which supplemented substantially toward Chinese food security and livelihood.

In India, systematic research on hybrid rice was initiated during 1989 when the Indian Council of Agricultural Research (ICAR) launched a special goal-oriented and time-bound project, "Promotion of Research and Development Efforts on Hybrids in Selected Crops," for rice at 12 network centers. Around 4 years (1989–1993) of rigorous research efforts have rewarded substantially, and India became the second country after China to develop and commercialize hybrid rice. The first hybrid variety APRH-1 was released by APRRI, Maruteru, for Andhra Pradesh in 1993–1994. So far, 117 rice hybrids (36 from public organization and 81 from private sector) were developed, suitable for different ecology and duration ranging from 115 to 150 days, covering 3.0 mha, which accounted for ~7.0% of the total rice acreage in India (Varietal Improvement, Progress report) [3].

Hybrid rice technology has substantial yielding ability that is able to enhance farm productivity ~15–25% more than inbred varieties. Given its yield advantage and economic importance, several hybrids in rice have been commercialized in more than 40 countries, which has created a huge seed industry worldwide. Moreover, this venture also has great service opportunity and generates additional employment for the poorer [1]. However, it has some limitations in generation of

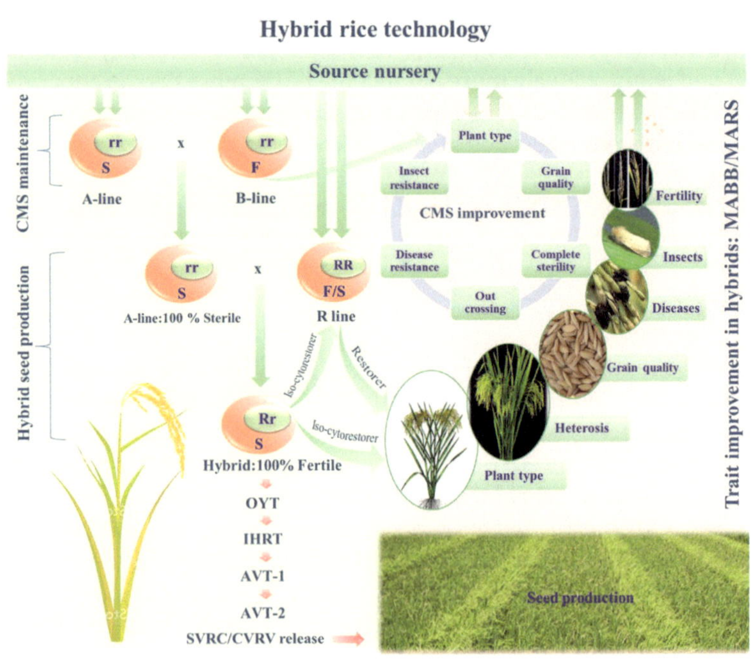


**Figure 1.**
*A schematic representation of hybrid rice technology (seed production, trait improvement, yield evaluation, etc.).*

hybrids, seed production, and marginal heterosis. Success of the hybrid depends on their parental combination, adaptability, and allelic interactions, and hence, faces several problems like unstable male sterility (MS), non-abundancy in cytoplasmic diversity, inherited CMS load, low seed producibility in seed parent, poor grain and eating quality, lack of responsive parents for biotic and abiotic stresses, hybrid sterility, marginal heterosis in *indica* hybrids, etc. This chapter deals with information on: (i) research status of HR, (ii) breeding system and methods involved in hybrid rice development and production, (iii) trait-specific parental line improvement, (iv) molecular dissection of genes and QTLs for parental line improvement, and (v) economic opportunity (**Figure 1**).

## 2. Breeding component and system in hybrid rice development

Rice is a strict self-pollinated crop; commercial exploitation of heterosis requires some parental specificity which could excludes manual emasculation. The invention of naturally occurred male sterility (MS) in rice thus played substantial role in realization of heterosis in rice. Following are the genetic tools as mentioned in various heads are required for development and commercialization of hybrid in rice:

### 2.1 Male sterile system

The male sterility (MS) in plants is the condition where the male reproductive organ, anthers, loses its ability to dehisce and produce viable pollen and thus encourages the allogamous nature of reproduction. This is crucial breeding tools to harness heterosis that exclude additional efforts of emasculation which is cumbersome process. In plants, male sterility is conditioned either by mitochondrial or nucleus genome or in associations. The male sterility in plant was first observed by Joseph Gottlieb Kolreuter in 1763 and later it was reported in >610 plant species. In rice, it was reported by Sampath and Mohanty [4] at ICAR-NRRI (formerly CRRI), Cuttack by studying the differences in male fertility in *indica/japonica* reciprocal crosses. The male sterility in plant is found to be determined by several biological as well as environmental factors. In rice, it is conditioned either by cytoplasmic genes in association with nuclear genes (CMS) or nuclear genes alone (GMS) which cause abnormal development in sporogenous tissue (either sporophytic or gametophytic tissue). The sporophytic male sterility is governed by genetic constitutions of sporogenous tissues like tapetal and meiocytes which creates improper nourishing to developing microspores and cause pollen abortion, whereas in gametophytic male sterility, microspore and pollen development get affected. Sporophytic male sterility is quite useful in hybrid rice breeding as it gets fertile in heterozygous state and encourages complete fertility in resulting hybrids. To date, several types of male sterile system, viz., cytoplasmic male sterile (CMS), environment sensitive male sterile (GMS), viz., thermo-sensitive genetic male sterility (TGMS), photo-sensitive genetic male sterility (PGMS) and reverse photo-sensitive genetic male sterility (rPGMS), etc. have been identified and substantially being utilized in hybrid development (**Table 1**).

### 2.2 Diversity in male sterile system and their mechanism

The CMS is a maternally hereditary trait instigated by improper communication between cytoplasmic and nuclear genome [5]. Gene(s)/genic block(s)-conditioned cytoplasmic male sterility is chimeric construct, which evolved due to rearrangement of the mitochondrial genome (**Figure 2**). In rice, several types of CMS have

| CMS group | Associated ORF | Protein | Cytoplasm source | Representative CMS-line |
|---|---|---|---|---|
| 1. Cytoplasmic male sterile line | | | | |
| a. *BT-CMS and their lineage* | | | | |
| BT-CMS (G) | *B-atp6-orf79* | Membrane protein | Chinsurah Boro II/ Taichong 65 | Liming A, Xu 9201A |
| LD-CMS (G) | UK | UK | Lead Rice (Burmese *indica* variety) × Fujisaka 5 (*japonica* variety) | Fujisaka 5A |
| Dian1-CMS (G) | UK | UK | Yunnan high altitude landrace rice (*indica*) cytoplasm | Yongjing2A, Ning67A |
| HL-CMS (G) | *atp6-orfH79* | Membrane protein | Red-awned wild rice (*Oryza rufipogon*) cytoplasm | Yuetai A, Luohong 3A4 |
| b. *WA-CMS and their lineage* | | | | |
| WA-CMS (S) | *rpl5-WA352* | Membrane protein | Wild abortive rice (*Oryza rufipogon*) cytoplasm | Zhenshan97 A, V20A, IR58025A, CRMS31A, etc. |
| Kalinga-I-CMS (S) | UK | UK | Kalinga-I (*indica*) cytoplasm | CRMS 32A |
| D-CMS (S) | UK | UK | *Indica* rice Dissi D52/37 | D-Shan A, D62A |
| DA-CMS (S) | UK | UK | Dwarf abortive rice (*Oryza rufipogon*) cytoplasm | Xieqingzao A |
| GA-CMS (S) | UK | UK | Gambiaca (*indica*) cytoplasm | Gang 46A |
| ID-CMS (S) | UK | UK | Indonesia paddy rice (*indica*) cytoplasm | II 32A, You1A |
| K-CMS (S) | UK | UK | K52(*japonica*) cytoplasm | K-17A |
| CMS-RT102 (S) | *rpl5-orf352* | Membrane protein | *Oryza rufipogon*, W1125 | RT102A |
| CMS-RT98A (G) | *orf113-atp4-cox3* | Membrane protein | *Oryza rufipogon* Griff, W1109 | RT98A |

| CMS group | Associated ORF | Protein | Cytoplasm source | Representative CMS-line |
|---|---|---|---|---|
| LX-CMS | UK | UK | Luihui rice (*indica*) cytoplasm | Yue 4A |
| Maxie-CMS | UK | UK | MS mutant of Maweizhan (*indica*) with Xieqingzao (*indica*) | Maxie A |
| NX-CMS | UK | UK | Selected from F2 male sterile plants in the progeny of Wanhui 88 (*indica*) × Neihui 92–4 (*indica*) nucleus | Neixiang 2A, Neixiang5A |
| Y-CMS | UK | UK | Yegong (*indica* landrace) cytoplasm | Y Huanong A |
| CW-CMS (G) | *orf307* | Mitochondrial protein | *Oryza rufipogon* Griff. | IR24A, IR64A |
| 2. Environment-sensitive genetic male sterility (EGMS) | | | | |
| PGMS | *pms3* | Noncoding RNA | Nongken 58S, PGMS mutant of *japonica* cultivar Nongken 58 | 7001S, N5088S |
| P/TGMS | *p/tms12–1* | noncoding RNA | Photoperiod and temperature sensitive genic male sterile (P/TGMS) derived from Nongken 58S | Pei'ai 64S |
| TGMS | *tms5, RNase* $Z^{S1}$(loss in function) | Nuclease enzyme | Spontaneous TGMS mutants of Annong S-1 and Zhu 1S | Guangzhan 63S5, Xinan S |
| rPGMS | *csa* OsMST8 | MYB transcript regulator | Carbon starved anther *(csa)* mutant of *japonica* cultivar 9522 | 9522S |

*Note: "S" stands for sporophytic male sterility and "G" stands for gametophytic male sterility.*

**Table 1.**
*Cytoplasmic diversity in rice CMS.*

been identified and characterized, having diversified mechanism in MS expression. Wild abortive (WA-CMS), a sporophytic MS system, is widely utilized in hybrid development. It is found to be caused by a constitutive mitochondrial gene WA352c

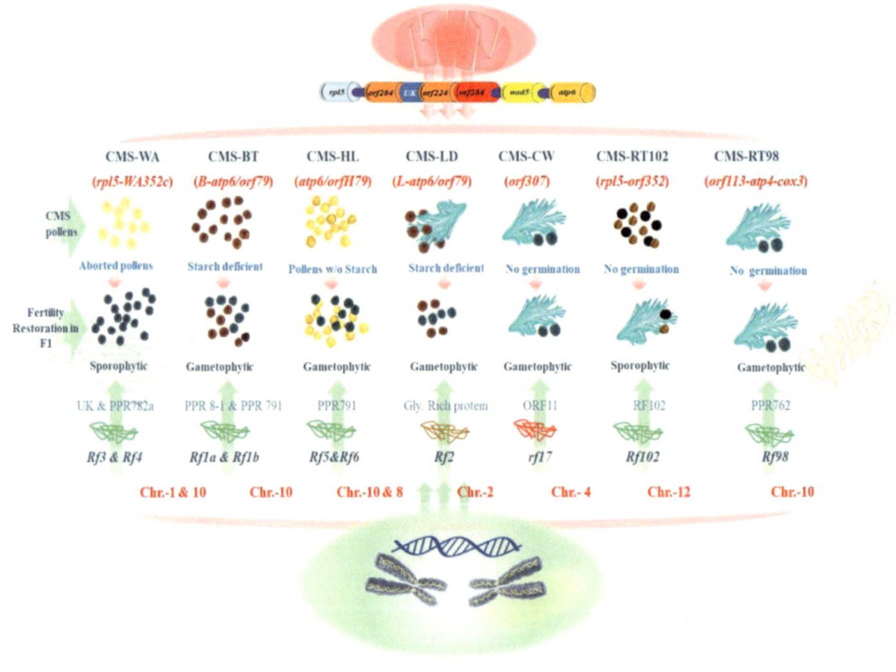


**Figure 2.**
*Schematic presentation of rice CMS types, where WA stands for wild abortive, BT is for boro type, HL for Honglian, LD for lead rice, CW is for Chinese wild rice, RT102A and RT98A, respectively.*

located downstream of *rpl5* (comprised four mitochondrial genomic segments, *orf284*, *orf224*, *orf288*, and *cs4-cs6*) and encodes a putative protein (352-residue) with three transmembrane segments. The WA352c inhibits nuclear-encoded mitochondrial protein *COX11* (essential for the assembly of cytochrome c oxidase, TCA) and triggered premature tapetal programed cell death and pollen abortion [6]. In contrast, BT-CMS is a gametophytic MS reported in the Indian rice variety, Chinsurah Boro-II, in which pollen development get arrested at the tri-nucleate stage. The mitochondrial chimeric (dicistronic) gene *B-atp6-orf79* encodes a trans-membrane protein, cytotoxic peptide*ORF79* [7], which accumulates preferentially in the microspore, was found to be responsible for male sterility. The *orf79* reside downstream to the *atp6* and interact with P61 and mitochondrial complex III and impair the activity of this complex which lead to dysfunctional energy metabolism and elevate oxidative stress and thus causing sterility. However, in HL-CMS, which is also a gametophytic MS system, pollen development gets arrested at di-nucleate stage. A chimeric aberrant transcript of the mitochondrial gene*atp6-orfH79*, located downstream of *atp6*is confirmed as candidate gene of this MS. Transcript of *orfH79* gene preferentially accumulates in mitochondria which interacts with P61 (a subunit of ETC complex III) and impairs mitochondrial function [8] and leads to MS. The MS in CW-CMS is conditioned by mitochondrial *orf307*, which causes anther-specific mitochondrial retrograde regulation for nuclear gene expression. It is a gametophytic MS in which pollen grain appears normal but unable to germinate.

## 2.3 Genetic male sterility (GMS)

The GMS in rice is conditioned generally by recessive nuclear genes and exert showing normal Mendelian inheritance. Owing to difficulties in their maintenance (occurrence of only 50% sterility in $F_1$), GMS could not be part of rice hybrid breeding program. Some GMS lines has shown threshold nature in MS expression where male sterility occurs in specific environmental regime (high temperature and long day length); hence called environment sensitive genetic male sterile (EGMS). The GMS line shows male sterility at elevated temperature, that is, >30°C is called temperature sensitive male sterility (TGMS) whereas male sterility in long day length, that is, >13.5 h is called photoperiod-sensitive genetic male sterility (PGMS). The male sterility in EGMS line is found to be revert into male fertile in favorable temperature (<30°C) and day length (<12.5 h) which provide its unique opportunity to be utilized in hybrid rice breeding program. The rice lines exert MS impression under long photoperiod and elevated temperature are referred as P/TGMS, for example, Pei'ai 64S. The EGMS lines, PGMS-Nongken 58S (NK58S) and TGMS-Annong S-1 and Zhu1S or derivatives are utilized extensively in majority (>95%) of the two-line hybrid program. Among, derivatives of NK58S are exerts either P/TGMS or even TGMS (e.g., Guangzhan 63S), the mechanism underlying to such dramatic changes yet to be revealed. Recently, a novel type of EGMS (*csa*-carban starved anther mutant) in rice called rPGMS (reverse PGMS). These lines expresses MS under short photoperiod (<12.5 h) and revert to normal fertile when exposed to long days (>13.5 h). This is found to be suitable for seed production of two-line hybrids in tropics and subtropics [9].

### 2.4 Transgenic cytoplasmic male sterility

The genetically engineered male sterile line M2BSin rice is developed by transformation of *indica* rice maintainer M2B with partial-lengthHcPDIL5-2a (Hibiscus cannabinus protein disulfide isomerase-like) genetic construct. Male fertility in this CMS is reported to be arrested due to tapetum degeneration which leads pollen abortion. The genetic analysis revealed this MS a maternally inherited inability as of CMS. Besides, by combining cysteine-protease gene (BnCysP1) of Brassica napus with rice anther-specific P12 promoter (promoter region of *Os12bglu38* gene), a transgenic MS system was successfully created which is restored by transgenic rice plants carrying BnCysP1Si silencing system [10]. Zhou and co-workers [11] could develop 11 "transgene clean" TGMS lines by editing most widely utilized TGMS gene *tms5* through CRISPR/Cas9.

### 2.5 Genetics of fertility restorer gene

The rice CMS is found to be restored by nuclear genome, that is, mono or oligo nuclear loci called restorer gene. In rice, a total of 10 *Rf* genes (*Rf1a*, *Rf1b*, *Rf2*, *Rf3*, *Rf4*, *Rf5*, *Rf6* and *Rf17, Rf98 and Rf102)* have been identified, of those seven (*Rf1a, Rf1b, Rf2, Rf4, Rf5*, *Rf17*, and *Rf98*) are characterized. All *Rf* genes are found to be dominant in nature (except *Rf17,* restores fertility in CW-CMS), which can restore male fertility in heterozygous state. Restorer genes are very specific to male sterile genome in the mechanism of fertility restoration. Genes *Rf1a* and *Rf1b* (Chr.-10) encode pentatricopeptide-repeat (PPR)-containing proteins and have functional affinity of fertility restoration in *BT-CMS*; RF1A promotes endonucleolytic cleavage of the *atp6-orf79* mRNA andRF1B promotes degradation of *atp6-orf79* mRNA [7] and revert the male sterility into fertility. Whereas, HL-CMS is restored either by *Rf5 or Rf6* gene, these genes can produce 50% normal pollen grains in F1 plants individually; however, both genes in complementation could restore more than 80% spikelets' fertility in hybrids. The *Rf5* encodes a PPR family protein PPR791 and which bind with GRP162 (glycine rich protein) and *atp6-orfH79* transcripts and makes a RFC (restoration of fertility complex). The RFC cleave the aberrant transcript of *atp6-orfH79*at 1169 nucleotides position [12]. The*Rf6* gene encodes a novel PPR family protein (duplicate PPR motif 3–5)

| S. No. | *Rf genes* | Locality | Marker | CMS system | Restorer line | Causative gene | Encoded product | Reference |
|---|---|---|---|---|---|---|---|---|
| 1 | *Rf1a*, *Rf1b* | Chr-10 | InDel-Rf1a | CMS-BT | BTR, IR24, MTC10R; C 9083 | PPR8–1, PPR791, *Rf1A*, *Rf1B* | PPR | [13] |
| 2 | *Rf2* | Chr-2 | CAPS42–1 | CMS-LD | Kasalath, Minghui 63 | LOC_ Os02g17380.1 | Gly. Rich protein | [14] |
| 3 | *Rf3* | Chr-1 | DRRM-Rf3–10 | CMS-WA | Swarna, PUSA 33 | — | PPR | [15] |
| 4 | *Rf4* | Chr-10 | RM6100 | CMS-WA | IR 24, Pusa 33, CRL 22R | PPR782a | PPR | [15] |
| 5 | *Rf5(t)* | Chr-10 | RM3150 | CMS-HL | Milyang 23 | PPR791 | PPR | [16] |
| 6 | *Rf6* | Chr-10 & 8 | RM5373 | CMS-HL | — | — | — | [16] |
| 7 | *Rf17* | Chr-4 | AT10.5–1, SNP 7–16 | CMS-CW | CWR | PPR2 | RNA interference | [17] |
| 8 | *Rf98* | Chr-10 | UK | CMS-RT98A | RT98C | PPR762 | PPR | [18] |
| 9 | *Rf102* | Chr-12 | UN | CMS-RT102A | RT102C, K102-*Oryza rufipogon. T* | UK | UK | [19] |

**Table 2.**
*Restorer genes in rice plants.*

which in association with hexokinase (*osHXK6*) targets mitochondria and process defective transcript of *atp6-orfH79* at 1238 nucleotide position. Thus, PPR protein family cause editing of aberrant transcript, inhibit their translation, and at the end, fertility restoration. Besides, male fertility in WA-CMS is found to be counteracted by *Rf3* and *Rf4* genes (chrom.-1 and 10, respectively). The genes *Rf3* and *Rf4* encode a pentatricopeptide protein (PPR) where RF4 cleave the abnormal *WA352* transcript and RF3 suppress translation of *WA352* into polypeptide and helps in restoring fertility in WA-CMS. Fertility in LD-CMS is reported to be restored by either *Rf1* or *Rf2*. The *Rf2* gene encodes a glycine-rich protein in mitochondrial; replacement of isoleucine by threonine at amino acid 78 of the *RF2* protein causes functional loss of the *rf2* allele. Moreover, CW-CMS is reported to be restored by a single recessive gene (*Rf17)* which is a retrograde-regulated male sterility (*rms*) gene (**Table 2**) [20].

## 2.6 Breeding system

Commercial hybrid seed production in rice where natural out-crossing (ranged only 0.3–3.0%) is very low, cumbersome, and an expansive task. To be practical and readily adoptable, it requires some specific parental requirements and agro-management practices. Invention of male sterile lines thus provided unique opportunity to start with the technology in rice. Based on mechanism of male sterility, threshold nature in male sterility expression and number of parental lines used, three types of hybrid seed production system namely three-line system (involving three parents, A, B, and R), two-line system (two parents, A and R), and one-line system (apomictic-based) exist. Among them, CGMS-based three-line system is more suitable, hence widely utilized (>90% of world's hybrids developed utilizing this) in hybrid rice varietal development and seed production.

### *2.6.1 CGMS system*

This system involves three parents such as male sterile line (A-line, cytoplasmic male sterile), B-line (maintainer), and R (restorer) lines and two steps in seed production, that is, CMS multiplication and hybrid seed production under strict isolation (spatial or temporal or physical barrier). Male sterile line (A-line), because of their eliminated manual emasculation needs, served as seed parent and facilitates large-scale seed production. A suitable CMS line to be utilized as seed parent should have complete and stable male sterility, substantial seed producibility, wide compatibility, and good combining ability with minimum CMS load. The wealthy panicle and narrow semi-erect leaf configuration in seed parent has additional impact, assures more seed production. In Indian perspective, hybrid seed production is a major dilemma, generally keen to *Rabi* season, hence, CMS lines should have substantial cold tolerance at seedling stage and heat at flowering stage.

The maintainer (B-line), on the other hand, is an isogenic to the CMS line (differs only for fertility/sterility) in their genetic constitution, able to produce functional pollen and maintain the sterility in male sterile line/seed parent. The maintainer line can maintain 100% male sterility in seed parent thus utilized to perpetuate CMS with their inherent male sterile ability.

In contrast, restorer line can restore male fertility in $F_1$s produced on male sterile parent, thus utilized as pollen parent in hybrid seed production. A good restorer should have substantial genetic distance with seed parent which is prerequisite and major determinant of the extent of heterosis in hybrids (more genetic distance more heterosis and *vice-versa*). Restorer is the major contributor of heterosis in three-line hybrids, hence, should have good combining, strong fertility restoration ability (dominant *Rf* gene(s) responsible for fertility restoration in CMS). Besides, restorer line with ideal plant type, acceptable grain quality parameters, substantial source-sink balance, heavy pollen load, and broad spectrum of resistance/tolerance against multiple biotic/abiotic stresses is imperative in maximization of genetic gain in hybrids.

### *2.6.2 EGMS system*

This system is a simple and more efficient hybrid breeding system in rice, involves only two parents, that is, A and R line in seed production, thus, referred as two-line system. This is a threshold of genetic male sterility (EGMS)-based hybrid rice breeding system, where male sterility is conditioned in specific environmental regimes such as long photoperiod (>13.5 h day length) and at elevated temperature (>30°C). In this system, male sterile parents are to be maintained by selfing under favorable conditions (below critical sterility point, i.e., <30°C temperature and at below CSP of photoperiod length, <12.5 h.).

Two-line hybrid seed production system is an easy and effective alternative to CMS and has specific advantages as it requires only one step for seed production. In this system, any good combiner genotype irrespective of their fertility restoration ability can be utilized as a pollen parent. EGMS system is normal and does not exert any ill effect in the growth and development of carrier plant, and thus, exploits comparatively higher extent of heterosis (up to 5–10%) in $F_1$ than the CGMS-based three-line system. The EGMS traits are governed by major genes, thus are easily transferable to any genetic background; besides, no CMS load could be helpful in reducing the potential vulnerability among the hybrids. Because of its eliminating needs for restorer genes in the male parents, this is ideal for developing inter-subspecific (*indica/japonica*) hybrids.

*2.6.3 One-line system (apomictic-based)*

In this system, seeds of rice hybrid once generated need not to be further produced in the hybrid seed production plot. This system is solely based on apomixes phenomenon (embryo developed apart from mixing of sexual gametes/ fertilization) where the embryo developed without fertilization. In this system, hybrid seeds once generated will be maintained through apomixes in their original heterozygous form. The apomictic embryo is formed in the ovule via two fundamentally different pathways, sporophytic or gametophytic, which define the origin of the apomictic embryo [21]. In sporophytic apomixes, the embryo arises directly from the nucellus or the integument of the ovule in a process generally called adventitious embryony. In gametophytic apomixis, two mechanisms are generally recognized, diplospory and apospory. In both of these, an embryo sac is formed and the two mechanisms are distinguished by the origin of the cells that give rise to the apomictic embryo sac. In diplospory, the embryo sac originates from megaspore mother cells either directly by mitosis and/or after interrupted meiosis. In apospory, the embryo sac originates from nucellar cells. In both gametophytic mechanisms, the resulting nuclei forming the embryo sac are of the same ploidy as those found in the female parent because the reduction division cycle of meiosis does not occur. The embryo arises autonomously from one of the cells in the embryo sac.

In a recent adventure, Delphine et al. reported three genes such as SPO11–1, REC8, and OSD1 in the sexual model plant *Arabidopsis thaliana*, which were combiningly mutated to turn meiosis into mitosis and its nourishing tissue from the female gametophyte without contribution of a male genome. This results in the production of clonal male and female gametes, but leads to doubling of ploidy at each generation when self-fertilized. Crossing a *MiMe* plant as male or female with a line whose genome is eliminated following fertilization (lines expressing modified CENH3) leads to the production of clonal offspring [22]. The *MiMe* technology was also implemented in rice to get diploid gametes. Furthermore, a study was conducted by Reda et al. to induce apomixis and fix heterosis in the sterile Egyptian Hybrid1 line using 0.2% colchicines [23]. It was observed that as colchicine is an alkaloid, which during cell division binds to tubulin protein of the spindle fiber and stops microtubules formation, and during meiosis, it prevents chromatids separation and inhibits cytokinesis. So ultimately, colchicines lead to meiosis aberrations, which produce aberrant microspores, pollen sterility, ovule sterility, as well as loss of fertility. Recently, a strategy based on the advanced technique, that is, CRISPR/Cas 9, has been utilized to introduce apomixis into rice (*Oryza sativa*) by mutating the three combined genes OsSPO11-1, OsREC8, OsOSD1, and OsMATL to get a MiMe phenotype [24].

## 3. Progress in hybrid rice research and development

### 3.1 International status

Hybrid technology is one of the greatest innovations in the modern era, contributed greatly in yield enhancement in several important crops. Over the decades of rigorous research, Chinese could develop parental lines, that is, cytoplasmic male-sterile line, maintainer line, and restorer line which assisted in the realization of heterosis exploitation in rice. Subsequently, hybrid seed production system was refined and world's first hybrid rice was released for commercial cultivation during 1974 AD. The first generation wild abortive CMS line, that is, Zhenshan 97A was widely utilized and several elite hybrid rice varieties were commercialized. Besides, several CMS with altered genetic mechanism of male sterility expression were also identified and characterized.

At beginning, low seed producibility with WA-CMS was a concern for its commercialization. However, with the keen interest of agronomist, management practices for hybrid seed production were sustainably rationalized. The Chinese government has supported this venture in pilot mode and established large and effective hybrid rice seed businesses in the late 1970s at all levels. Besides, intensive mechanization of hybrid seed production helped in modification of planting ratio (2R: A as 6–8 rows to 40–80 rows) and reducing the cost of production. Therefore, China could achieve seed yield by 2.7–3.0 t/ha on a large scale in hybrid rice seed production, which is further enhanced to 3400 kg/ha and maximizes their acreage.

Over past three decades, hybrid rice varieties have been substantial for national food security in the China which accounted for approximately 57% of the total 30-million-hectare rice planting area. The Ministry of Agriculture, China, has launched project on super hybrid rice development during 1996 which resulted altogether 73 super hybrids (52 three-line hybrids and 21 two-line hybrids) for commercial cultivation. Super hybrid P64S/E32 released recently has recorded new height of yield potential of17.1 t/ha with some striking characteristics [25].

Beside China, this technology has also been introduced and promoted by more than 40 countries around the world. At beginning, IRRI helped technically and supplied prerequisite parental materials. Later, most of the countries could establish their own hybrid rice breeding program and developed several heterotic hybrids. India was the second country after China that adopted this technology in 1989 and made substantial progress. At present, hybrid rice covers around 3.0 mha in India that has 6.8% of total rice area. Vietnam was the next to adopt this technology in 1992, harnessing yield of 6.3–6.8 t/ha from 0.7 mha, which covers around 10% of their rice area. In Philippines, it was introduced in 1993. Several popular hybrids like Magat, Mestizo, Mestizo 2, Mestizo 3, Bigante, Magilla, SL8H, Rizalina 28, etc. were developed and commercialized. Hybrid seed production in Philippines has been handled by "seed growers" cooperatives that are to produce around 60–70%. In Bangladesh, several rice hybrids were introduced and commercialized from China, India, and Philippines. They are almost self-sufficient in hybrid seed production, producing around 8000 tons to cover about 800,000 ha. In order, Indonesia also has substantial hybrid rice area, developed several good rice hybrids like Hipa7, Hipa 8, Hipa9, Hipa10, Hipa11, Hipa12 SBU, Hipa13, Hipa14 SBU, Hipa Jatim1, Hipa Jatim2 and Hipa Jatim3 were extensively commercialized, having yield superiority of 0.7–1.5 tons/ha over the lowland inbred varieties.

USA has adopted this technology during 2000 and has developed and commercialized several two-line and three-line hybrids. Most of the hybrid rice cultivars in USA employed Clearfield (CL) technology offering selective control of weedy red rice. Rice hybrids, viz., Clearfield XL729, Clearfield XL745, Clearfield XP756 (a late maturing) and Clearfield XP4534 (new plant type) has shown yield advantage ranging from 16 to 39% over inbred cultivars are being commercialized by RiceTec.

### 3.2 National status

In India, systematic hybrid rice research was initiated in 1989.The first hybrid rice was released in Andhra Pradesh during 1993–1994 and India became the second country after China to commercialize hybrid rice. India has made substantial progress and developed total 117 (*indica/indica*) rice hybrids having 15–20% yield superiority with 115–150 days duration for various rice ecosystems. Recently, Savannah Private Limited from India has made another landmark by developing two two-line rice hybrids, viz., SAVA-124and SAVA-134, for commercial cultivation. In addition, more than 100 CMS in diversified genetic and cytoplasmic backgrounds have been developed and utilized. Among, the promising CMS lines CRMS 31A, CRMS 32A, CRMS 8A, PMS10A, PMS 17A, APMS 6A, DR8A, PUSA 5A, PUSA6A, RTN 12A, etc. are substantially being utilized in development of rice hybrids in India and abroad. Notably, medium-duration seedling stage cold-tolerant CMS, CRMS 32A, developed at NRRI under Kalinga-I cytoplasm is more suitable for development of hybrids for *boro* ecosystem. Two popular hybrid rice varieties, namely, Rajalaxmi and KRH 4 were developed using CRMS 32A as one among the parent.

Hybrids released in India having unambiguous specificity like specific to ecosystem, tolerant to several abiotic/biotic stresses and consumer preferences (**Table 3**). These hybrid varieties can be utilized to up scale the hybrid rice cultivation and productivity enhancement *per se* in the respective area.

Hybrids like CRHR 105, CRHR 106, 25P25, 27P31 are suitable for high-temperature regime which has a more deleterious effect on seed development in hybrids. The hybrid varieties, US 382, Indam 200–17, US 312, DRRH3, and JKRH 401 having high N use efficiency are thus found suitable for cultivation in N-deficient soil. Besides, hybrids PNPH 24, RH 1531, and Arize Tej are under mid-early maturity group which can sustain substantially under drought situations. The problems of coastal and shallow lowland ecosystem sharing around 32% of total rice area can be addressed by adopting long-duration hybrids like CRHR 32, Arize Dhani, CRHR 34, CRHR 102, and Sahyadri 5 (**Table 4**).

### 3.3 ICAR-National Rice Research Institute's contribution

The ICAR-National Rice Research Institute, Cuttack has been pioneer to start with the technology in late of seventh decade of last century, quite before the beginning of their project mode program in 1989 by ICAR. In the beginning, ICAR-NRRI has acquired all the prerequisite materials (CMS lines, viz., V 20A, Yar Ai Zhao A, Wu10A, MS 577A, *Pankhari* 203A, V 41A, Er-Jiu nanA, respective maintainers, nine other maintainers, and 13 restorers) from the IRRI (NRRI annual report 1981–1982). Systematic hybrid rice breeding was initiated in an interdisciplinary mode with objectives to develop desirable parental lines, viz., cytoplasmic genetic male sterile (CGMS) lines, maintainers, and restorers for the development of rice hybrids for irrigated and shallow submergence. The farmers

| S. No. | Stress | Promising hybrids |
|---|---|---|
| 1 | Rain-fed upland | DRRH-2, Pant Sankar Dhan-1, Pant Sankar Dhan-3, and KJTRH-4 |
| 2 | Salinity | DRRH-28, Pant Sankar Dhan-3, KRH-2, HRI-148, JRH-8, PHB-71, and Rajalaxmi |
| 3 | Alkalinity | Suruchi, PHB-71, JKRH-2000, CRHR-5, DRRH-2, DRRH-44, and Rajalaxmi |
| 4 | *Boro*/Summer season | Rajalaxmi, CRHR-4, CRHR-32, NPH 924–1, PA 6444, Sahyadri, and KRH 2 |
| 5 | BB resistant | BS 6444G, Arize Prima, Rajalaxmi, Ajay, CR Dhan 701, PRH 10, etc. |

**Table 3.**
*Rice hybrids tolerant to various stresses.*

| | |
|---|---|
| Aerobic condition | PSD 3, PSD 1, Rajalaxmi, Ajay, ADTRH 1, PRH 122, DRRH 44, HRI 126, JKRH 3333, and KRH 2 |
| Early duration | CRHR 105, CRHR 106, 25P25, 27P31 (heat-tolerant), US 382, Indam 200–17, US 312`, DRRH3, JKRH 401high N use efficient; PNPH 24 and RH 1531, Arize Tej-mid-early drought-tolerant; DRRH2, and KJTRH-4 (upland) |
| Long duration | CRHR 32, CRHR 34, CRHR 100, and Sahyadri 5 |
| SRI | TNRH CO-4, KRH 4 |
| Idly making | VNR 2355+ |
| MS grains | CRHR 32, DRRH 3, 27P63, 25P25, and Suruchi |
| Aromatic | PRH 122 (slight aroma), PRH 10 |

**Table 4.**
*Hybrids suitable for specific condition.*

in the rain fed shallow lowland ecosystem would be extremely benefited if the hybrid rice technology can be extended to this ecosystem, which need hybrids of Swarna duration. Keeping in views, ICAR-NRRI has developed three rice hybrids, viz., Ajay, Rajalaxmi, and CR Dhan 701 for this fragile ecosystem. Among them CR Dhan 701 is the country's first long-duration hybrid, substitute for popular variety Swarna. Besides, NRRI has developed several promising CMS lines which have stable male sterility (WA, Kalinga-I and O. perennis, etc. cytoplasmic background), maintainers, and effective restorers. More than 45 CMS lines in diverse genetic and cytoplasmic backgrounds have been developed among Sarasa A, Pusa 33A (WA), Annada A (WA), Kiran A (WA), Deepa A (WA), Manipuri A (WA), Moti A (WA), Krishna A (O. perennis), Krishna A (Kalinga I), Mirai (Kalinga I), Padmini A, PS 92A and Sahabhagidhan A, etc., which are more prominent to be utilized in hybrid development. The medium-duration CMS, CRMS 31A (WA) and CRMS 32A (Kalinga-I) are significantly utilized for hybrid development at NRRI and elsewhere in the country. The CRMS 24A and CRMS 40A, developed under the nucleus background of Moti and Padmini are found suitable for late-duration hybrid breeding. Moreover, short-duration CMS, CRMS 8A, CRMS 51A and CRMS 52A and CRMS 53A having drought tolerance are also being used for development of hybrids for drought prone ecosystem.

The latest release CR Dhan 701 (CRHR32) found suitable for irrigated shallow low land of Bihar, Gujarat and Odisha having MS grain type with an average yield capacity of 7.5 t/ha. This hybrid shows substantial tolerance to low light intensity, thus having great scope in eastern Indian states where low light limits the potential expression of hybrids/varieties during wet season. Moreover, hybrid Rajalaxmi (125–130 days) was developed utilizing native CMS line CRMS 32A, released by SVRC 2006/CVRC 2010 for irrigated-shallow lowland of Odisha and *boro* ecosystem of Odisha and Assam as it has seedling stage cold tolerance. Ajay is a medium-duration, long slender grain-type hybrid, released for cultivation in irrigated-shallow lowland of Odisha. As these hybrids are adaptable for eastern Indian climatic condition with assured remuneration, 12 private seed agencies over five states have commercialized them.

To make this technology more sustainable and amenable to farmers, trait development strategy among the parental lines becomes mandatory. The parents of ICAR-NRRI bred hybrids Ajay, Rajalaxmi and CR Dhan 701 has been improved for bacterial blight, the most devastating disease of rice [26]. The submergence and salinity are the major abiotic stresses occur frequently in rain-fed shallow lowland area and causes substantial yield loss in rice. Hence, to cope up with the problems, and make hybrid rice more sustainable during these adversity, ICAR-NRRI has successfully stacked submergence and salinity-tolerant QTLs in the seed parents CRMS 31A and CRMS 32A. To enhance the seed producibility in seed parents, introgression of stigma exsertion trait from O. longistaminata into CRMS 31A and CRMS 32A, are under progress. To excavate the genetic region responding heterosis in rice, transcriptomic analysis of hybrids Rajalaxmi and Ajay are completed and interpreted. Availability of restorers for WA-CMS lines is very stumpy in nature, only 15% of total rice genotypes having the ability to restore complete fertility in WA-CMS-based hybrid rice [15]. Hence, good combiner genotypes having partial fertility restorers Mahalaxmi and Gayatri were improved by introgressing fertility restorer gene(s) *Rf3* and *Rf4* through MABB approach. Further, to make clear cut identity and ensure pure seed of parents/hybrids to the stack-holder, 12 signature markers that unambiguously distinguish 32 rice hybrids were developed, which can be utilized for DNA fingerprinting and genetic purity testing of hybrids.

## 4. Potential application of OMICS approaches in hybrid rice breeding

Recent advancement in molecular biology has offered tremendous opportunities to the breeder and breeding *per se* in enhancement in their efficacy and speed up the varietal development process. It has diverse applications like mapping, tagging, amplification-based cloning, gene pyramiding, marker-assisted selection (MAS/MARS), fingerprinting applications, including varietal identification, ensuring seed purity, phylogeny and evolution studies, diversity analysis, and elimination of germplasm duplication. The progress in research related to application of DNA marker technology in hybrid rice improvement may be valuable in following way.

### 4.1 DNA fingerprinting and genetic purity testing

Varietal identity of hybrids and parents is imperative to assure the ownership (IPR issue) and pure seeds to the stakeholders. The genetic purity testing of hybrid seed is done by conducting Grow-Out-Test (GOT) which is time taking (needs one full growing season), tedious and very expensive. Molecular markers in this context found to be a suitable alternative, provide an unbiased means of identifying crop varieties. Among available DNA-based markers, sequence-tagged microsatellites (STMSs), which are co-dominant in nature, are widely used for speedy genetic purity assessment of the hybrids and parental lines [27, 28]. Besides, ICAR-NRRI has developed another set of nine signature markers which can distinguish parents CRMS 31A, CRMS 32A; and hybrids Ajay, Rajalaxmi and CR Dhan 701, unambiguously.

### 4.2 Trait improvement in parental lines and hybrids

Hybrid rice has been one of the innovations that led the quantum jump in rice productivity last century. However, the challenge of meeting the increasing demand for rice and making hybrid more sustainable under impeding climatic changes, trait development in parental lines for ideal plant type with substantial yield, grain quality, and resistance/tolerance to multiple biotic and abiotic stresses is necessary. In this context, conventional breeding is more cumbersome, time taking and less précised. The advancement in molecular breeding techniques makes it convenient to improve the parents and hybrids for desirable traits with great precision. Marker-assisted selection/MABB has provided strong utensils for indirect selection/trace the trait of interest at any plant growth stage. The bacterial blight and blast are the two-major destructive diseases affecting rice plant at different growth stages and caused substantial yield loss. Resistant genes for BB diseases have been deployed successfully in popular hybrids like Rajalaxmi, Ajay [26], BS 6444G, PRH 10 [29], Shanyou 63, Guangzhan63-4S; seed parent of CR Dhan 701; restorers Minghui 63 and Mianhui 725 [5, 26], Zhonghui 8006 and Zhonghui 218, etc. The popular CMS line Rongfeng A, Pusa 6A female parent of popular basmati hybrids PRH 10, RGD-7S, and RGD-8S [30] were successfully stacked with blast and BB resistant gene(s). Besides, CRMS 31A and CRMS 32A were deployed with submergence and salinity tolerance QTLs (NRRI newsletter 2015). Grain and eating quality in hybrids are concerns which are addressed by stacking QTLs/genes for quality traits in parents. Zhenshan 97A seed parent of several hybrids in China has been stacked with QTLs of AC, GC and GT [31]. Efforts were made toward quality improvement of both the parental lines of popular indica hybrids, viz., Xieyou57, using marker-assisted selection for *Wx* locus [32]. Yield-enhancing QTLs, *yld1.1* and *yld2.1*, from *O. rufipogon* to restorer "Ce64" [33] are successfully stacked. Hybrid sterility in inter-subspecific (*indica/japonica*) hybrids is reported to be effectively addressed by utilizing genome editing tool "CRISPR/Cas9" [34].

### 4.3 Screening of Rf genes in parents

Limited availability of fertility restorer system in rice makes three-line system very selective and less heterotic. Rice genotypes have fertility restorer ability can only be utilized as pollen parent in three-line hybrid breeding. Identification of genetically compatible, well combining restorers is tedious process, involve laborious test cross generation and evaluation steps. However, prior information on fertility restorer genes in the pollen parent excludes test cross steps thus make it convenient for saving time of hybrid development. Plenty of co-segregating molecular markers (tightly linked or functional markers) for fertility restorer gene(s) having functional specificity to diverse CMS systems are available (**Table 2**). The genic/functional markers, RM6100 and DRRM Rf3–10 of restorer gene(s) *Rf4* and *Rf3*, respectively, are widely utilized for screening the fertility restoration efficacy of unknown pollen parents for WA and lineage CMS well in advance [15].

### 4.4 Screening of parental lines for wide compatibility genes

Hybrid sterility is common nuisance menacing breeder to exploiting heterosis in inter-subspecific (5–10% more heterosis) hybrids. Generally, *indica* × *japonica* hybrids are sterile due to lack of wide compatibility (*WC*) between parents. It is reported that hybrid sterility in inter-subspecific crosses is mainly affected by the genes at *Sb*, *Sc*, *Sd*, and *Se* [35] loci causes male sterility in $F_1$and the gene at *S5* locus cause female sterility in $F_1$. Presence of these genic regions in at least one parent ensures complete fertility in resulting hybrids. These gene(s) can be assessed in advance by utilizing co-segregating markers (S5-InDel, functional marker to $S5^n$) [36] and G02–14827 (genic marker) PSM8, PSM12, and PSM180 (linked SSR); IND19 and ID5 (indel markers) to *Sb*, *Sc*, *Sd*, and *Se*, loci). Thus, it helps breeder in selection of *WC*-positive parent in more predictable way which circumvents laborious test-cross and their evaluations steps.

### 4.5 Prediction of heterosis

Genetic distance and level of genetic gain/breeding value in parents are major determinants of extent of heterosis in the resulting hybrid. Molecular markers help in assessing the genetic diversity among parents and breeding values in progenies (through genomic selection, high-density SNP genotyping) with great convenient. There are abundant STMS and SNP markers available which can be utilized for assessment of genetic diversity/genetic distance between parents and genomic selection in progenies easily [37]. Hence, this is helpful in the selection of diverse parents with maximum breeding values in turn higher heterosis or genetic gain in hybrids.

### 4.6 Determination of heterotic group and heterosis pattern

The extent of genetic variation and selection strategies are keys to the success of heterosis breeding. Accurate assessment and assignment of parental lines into heterotic groups "group of genotypes (related or unrelated) having similar combing ability and heterosis response when crossed with the genotypes of other diverse group" are fundamental prerequisites. Usually it is evaluated by combining ability analysis of parents and hybrids in multi-environment trials. However, advances in molecular marker technology have made it possible to combine information on parental pedigree and field trials with molecular marker data to detect and establish heterotic groups. Several heterotic groups have been developed and utilized for three-line and two-line hybrid development in rice [38].

### 4.7 Excavating QTLs/gene(s) responses heterosis

Omics techniques reported to have great potential in excavation of QTLs/gene(s) responses heterosis in rice. By utilizing genomics tools, many QTLs/genes for several important traits has been mapped, validated, and deployed in trait development in rice. The transcriptomics, an emerging technique helps in genome-scale comparisons of the transcripts of different individuals within the same species/population. It helps in understanding the level of variation for gene expression, as measured by transcript abundance that exists within plant species and between hybrids and their parents. This is useful for identification of transcript and gene *per se* involves in heterotic expression. Moreover, epigenetics, a posttranslational biochemical regulation of gene is found to be playing substantial role in trait expression. Individuals of the same species can have epigenetic variation in addition to genome and transcriptome content variation. A potential role for epigenetic regulation in heterosis has been proposed. It is possible for epigenetic variation to affect heterosis by creating stable epialleles that would behave similarly to the genomic or transcriptomic differences. Alternatively, hybrids may exhibit unique epigenomic states that lead to heterosis.

## 5. Major challenges and potential research opportunities

### 5.1 Major challenges

Despite of being remunerative and varietal abundancy, HR technology could not make substantial dent in the rice farming system outside China. The following are the inherited void led poor acceptability and acreage expansion of hybrids:

#### *5.1.1 Lack of cytoplasmic diversity in countries outside China*

Outside China, WA-CMS or their lineages are commonly utilized as seed parent in more than 90%rice hybrids. Several alternative MS cytoplasmic sources such as BT-CMS, HL-CMS, and CW-CMS are identified in China, but the hybrid breeding program of other countries relied only on WA-CMS which has several inherited abnormalities. These narrowed genetics of sterile cytoplasm limits the extent heterosis exploitation and make hybrids vulnerable to many biotic and abiotic stresses.

#### *5.1.2 Marginal heterosis in intra-subspecific hybrids*

Two-lines and inter-subspecific (*indica/japonica*) hybrids are comparatively more heterotic (5–10%) than three-line *indica* hybrids. But owing to several inevitable difficulties in seed production of two-line hybrids and poor grain and eating quality in inter-subspecific hybrids, both could not be exploited in the countries like India who has vast climatic and food affection diversity. We are utilizing three-line *indica* hybrids which are comparatively less heterotic hybrid breeding systems giving low yields. Hence, focused and intensive research is proposed to make above said hitches be addressed in future.

#### *5.1.3 Poor grain and eating quality*

In hybrids, consumable parts are $F_2$ grains, segregating for various quality traits hence very poor in quality limits its acceptability among stakeholders. Therefore, make hybrids more sustainable and popular, quality traits in hybrids needs to be addressed urgently in the country like India where people have vast category of food fondness. Hence, a strong breeding strategy for quality concern in hybrids is needs to be devised and implemented.

#### *5.1.4 Subtle information on QTLs/gene(s) responding heterosis*

Although heterosis, or hybrid vigor, is widely exploited in agriculture, but despite extensive investigation, complete description of its molecular underpinnings has remained elusive. It appears that there is not a single, simple explanation for heterosis. Instead, it is likely that heterosis arises in crosses between genetically distinct individuals because of a diversity of mechanisms. Hence, mining factors responding heterosis in rice will have a substantial role in development and exploiting heterosis in most precise way.

#### *5.1.5 Inter-subspecific hybrid sterility*

Hybrid sterility is key nuisance in inter-subspecific hybrids, limiting development and commercialization of more heterotic *indica/japonica* hybrid in rice. The sterility in hybrids (*inter-subspecific*) generally occurs due to non-functional pollens as well as sterility in female reproductive organs. It is reported that mutant of *S-i* alleles at *Sb, Sc, Sd,* and *Se* loci produce sterile pollens; and mutants of *S5*locus causes sterility in female gamete. Hence, trait development for wide compatibility in either parent has great opportunity in addressing the hybrid sterility in rice.

### 5.2 Potential research opportunity

#### *5.2.1 Exploitation of inter-specific heterosis*

Inter-subspecific (*indica/japonica*) hybrids as discussed in earlier section are more heterotic than intra-subspecific hybrids. However, owing to hybrid sterility and poor grain quality, this genetic pool remains untapped. Grain quality of inter-subspecific hybrids proposed to be improved by utilizing parental combinations having good combining ability but similar in quality parameters, might reduce the concern of segregation for quality traits. Hybrid sterility problem in inter-subspecific hybrids can be addressed by stacking *indica* allele (*S-i*) at *Sb, Sc, Sd,* and *Se* loci and the neutral allele (*S-n*) at *S5*locusin to*japonica* genetic background [35] or by silencing the *S-i* and *S5* mutant loci through genome editing tools [34].

### 5.2.2 Utilization of Iso-cytoplasmic restorers

In three-line hybrid system, cytoplasm of CMS exerts various unwanted effect (called CMS penalty) and reduces the complete heterosis expression (up to 5–10%) in CGMS hybrids. Iso-cytoplasmic restorer is fertile transgressive segregant of CGMS hybrid, having same cytoplasm as of CMS. In combination with iso-cyto-CMS, it can normalize the fatal cyto-nuclear conflicts, hence enhances the heterosis to substantial extent. In rice, several iso-cytoplasmic restorers has been developed and utilized in hybrid rice research [39].

### 5.2.3 Out-crossing enhancement in seed parent

Low seed producibility (1.5–2.5 t/ha) in the CMS remains a concern, restricts seed abundancy, and area expansion in India. Trait development in seed parent for out-crossing traits like stigma exertion, complete panicle exertion is important and needs to be addressed strategically. Recently, a CMS line, IR-79156A possessing more than 50% out-crossing, developed by IRRI showed seed producibility of 3.5 t/ha.

### 5.2.4 Ideotype hybrid breeding

To maximize genetic gain in rice, breeding of ideal plant type was started long back in Japan and subsequently adopted by China. Through morphological improvement and adopting inter-subspecific (*indica/japonica*) hybrid strategies, substantial progress in ideotype hybrid breeding "super hybrid" have been achieved. China, indeed has made considerable progress and released more than 100 high yielding super hybrids [25]. Hence, inclusion of inter-subspecific quality type inbreds "super rice" in hybrid development will have substantial impact in attaining quantum genetic gains in hybrids.

## 6. Economic importance

Inspite of being more cumbersome and high input intense practice, hybrid rice seed production is a profitable venture. It creates additional job opportunity (requires 100–105 more-man days) and provides more net income (around 1050 USD/ha net income, 70% more than the unit production cost) as compared to seed production of HYV (192.0 USD/ha, only 18% more than production cost) (**Table 5**). The market price of hybrid seed is 3.5–4.25 USD per kg. The farmers producing the hybrid seed get only 1.15–1.30 USD per kg. In case of low production (<5 quintal/acre) farmers get minimum 635.0 USD as compensation from seed production agencies.

| Item | | Quantity/number (per hectare) | Cost/income (USD) | |
|---|---|---|---|---|
| | | | Hybrid seed | HYV |
| Seed cost | Male | 5 kg @ 0.71 USD/kg | 4 | 28 |
| | Female | 15 kg @ 5.65 USD/kg | 42 | Nil |
| Labor cost | | 250/145 @ 2.83 USD/labor/day | 707 | 410 |
| FYM and fertilizer cost | | N:P:K (100:50:50) (based on market price) | 76 | 76 |
| Irrigation | | 18–20 Irrigation (weekly) @ 21.20USD/ha/irrigation | 425 | 425 |
| Gibberellic acid | | | 28 | Nil |
| Others | | | 212 | 141 |
| Total cost | | | 1494 | 1080 |
| Average production | | | 2.0 t/ha | 4.5 |
| Gross income | | Price @ 1.27USD/kg and 0.28USD/kg[a] | 2544 | 1272 |
| Net income | | | 1050 | 192 |

[a]*Price of seed is the price given to the farmer.*
*Source: Verma et al., [40].*

**Table 5.**
*Cost analysis of hybrid rice seed.*

## 7. Way forward

Hybrid technology has been substantial in enhancement of rice productivity *per se* production in temperate countries, however, owing to low photo-intensity during growing period in tropics, its impact remains meager. Under changing climatic and agriculture scenario, rice hybrid is likely to face stiff competition to sustain in future. Despite having great potential to enhance production and productivity, it has not been adopted on large scale as was expected. This is due to several constraints like lack of acceptability of hybrids in some regions such as Southern India due to region-specific grain quality requirement. Moderate (15–20%) yield advantage in hybrids is not economically very attractive and there is a need to increase the magnitude of heterosis further. Lower market price offered for the hybrid rice produce by millers/trader sis acting as a deterrent for many farmers to take up hybrid rice cultivation. Higher seed cost is another restrain for large-scale adoption and hence there is a need to enhance the seed yield in hybrid rice seed production plots. Efforts for creating awareness and for technology transfer were inadequate in initial stages. Involvement of public sector seed corporations in large-scale seed production has been less than expected. Hybrid rice for aerobic/upland, *boro* season and long-duration hybrids for shallow lowland conditions are to be developed. Most of the constraints mentioned above are being addressed with right earnestness through the ongoing research projects and transfer of technology efforts.

## 8. Conclusion

Since inception, this technology has a substantial impact in enhancing the productivity and production in crop plant and livelihood of the farming community.

In rice, it is adopted worldwide over 40 countries; however, it could not make a substantial dent in outside of China. This chapter has represented the holistic status of hybrid technology in rice along with future research and developmental road map to make this venture more substantial and sustainable for benefiting all stakeholder involves. This chapter identifies the ambiguities held responsible for slow adoption of this technology and probable strategies to get rid of those. Therefore, this chapter will be helpful for researchers and students in planning of future hybrid rice breeding strategies.

## Author details

Diptibala Rout[1], Debarchana Jena[1], Vineeta Singh[2], Manish Kumar[3], Pandurang Arsode[3], Prakash Singh[4], Jawahar Lal Katara[1], Sanghamitra Samantaray[1] and Ramlakhan Verma[5]*

1 Crop Improvement Division, National Rice Research Institute, Odisha, India

2 Acharya Narendra Dev University of Agriculture and Technology, U.P., India

3 Department of Genetics and Plant Breeding, I.Ag.Sc., Banaras Hindu University, U.P., India

4 Department of Plant Breeding and Genetics, Veer Kunwar Singh College of Agriculture (Bihar Agricultural University-BAU, Sabour), Bihar, India

5 Hybrid Rice Breeding, Crop Improvement Division, National Rice Research Institute, Odisha, India

*Address all correspondence to: ram.pantvarsity@gmail.com

## References

[1] Verma RL, Katara JL, Sah RP, Sarkar S, Azaharudin TP, Patra BC, et al. CRMS 54A and CRMS 55A: Cytoplasmic male sterile lines with enhanced seed producibility. 2018, NRRI Newsletter. January-March 2018;**39**:1-14

[2] Jonse JW. Hybrid vigour in rice. Journal of American Society of Agronomy. 1926;**18**:423-428

[3] IIRR, Indian Institute of Rice Research, Technical Bulletin, 2019. IIRR: Hyderabad

[4] Sampath S, Mohanty HK. Cytology of semi-sterile rice hybrid. Current Science. 1954;**23**:182-183

[5] Chen S, Lin XH, Xu CG, Zhang Q. Improvement of bacterial blight resistance of "Minghui 63," an elite restorer line of hybrid rice, by molecular marker-assisted selection. Crop Science. 2014;**40**:239-244

[6] Tang H, Zheng X, Li C, Xie X, Chen Y, Chen L, et al. Multi-step formation, evolution, and functionalization of new cytoplasmic male sterility genes in the plant mitochondrial genomes. Cell Research. 2017;**27**:130-146

[7] Wang ZH, Zou YJ, Li XY, Zhang QY, Chen LT, Wu H. Cytoplasmic male sterility of rice with Boro II cytoplasm is caused by a cytotoxic peptide and is restored by two related PPR motif genes via distinct modes of mRNA silencing. Plant Cell. 2006;**18**:676-687

[8] Wang K, Gao F, Ji Y, Liu Y, Dan Z. ORFH79 impairs mitochondrial function via interaction with a subunit of electron transport chain complex III in Honglian cytoplasmic male sterile rice. New Phytologist. 2013;**198**:408-418

[9] Zhang H, Xu C, He Y, Zong J, Yang X. Mutation in CSA creates a new photoperiod-sensitive genic male sterile line applicable for hybrid rice seed production. Proceeding National Academy of Science, USA. 2013;**110**:76-81

[10] Rao GS, Deveshwar P, Sharma M, Kapoor S, Rao KV. Evolvement of transgenic male-sterility and fertility-restoration system in rice for production of hybrid varieties. Plant Molecular Biology. 2018;**96**:35-51

[11] Zhou H, He M, Li1 Z, Chen L, Huang Z, Zheng S, et al. Development of commercial thermo-sensitive genic male sterile rice accelerates hybrid rice breeding using the CRISPR/Cas9-mediated TMS5 editing system. Scientific Reports. 2016;**6**:373-395

[12] Hu J, Wang K, Huang W, Liu G, Gao Y. The rice pentatricopeptide repeat protein *RF5* restores fertility in Honglian cytoplasmic male-sterile lines via a complex with the glycine rich protein GRP162. Plant Cell. 2012;**24**:109-122

[13] Tao C, Ming L, Zya Z, Zhen Z, Ling Z, Qing Y, et al. Development and application of a functional marker associated with fertility restoring gene for BT-type cytoplasmic male sterility (CMS) in *japonica* rice. Chinese Journal of Rice Science. 2013;**27**(3):259-264

[14] Itabashi E, Iwata N, Fujii S, Kazama T, Toriyama K. The fertility restorer gene, *Rf2* for lead rice-type cytoplasmic male sterility of rice encodes a mitochondrial glycine-rich protein. The Plant Journal. 2011;**65**:359-367

[15] Katara JL, Verma RL, Nayak D, Umakanta N, Ray S, Subudhi H, et al. Frequency and fertility restoration efficiency of *Rf3* and *Rf4* genes in Indian rice. Plant Breeding. 2017;**136**:74-82

[16] Liu XQ, Xu EX, Li SQ, Hu J, Huang JY, Yang DC, et al. Inheritance

and molecular mapping of two fertility-restoring loci for Honglian gametophytic cytoplasmic male sterility in rice (Oryza sativa L.). Molecular Genetics and Genomics. 2004;**271**:586-594

[17] Fujii S, Toriyama K. Molecular mapping of the fertility restorer gene for ms-CW-type cytoplasmic male sterility of rice. Theoretical and Applied Genetics. 2005;**111**:696-701

[18] Igarashi K, Kazama T, Toriyama K. A gene encoding pentatricopeptide repeat protein partially restores fertility in RT98-type cytoplasmic male-sterile rice. Plant Cell Physiology. 2016;**57**(10):2187-2193

[19] Okazaki M, Kazama T, Murata H, Motomura K, Toriyama K. Whole mitochondrial genome sequencing and transcriptional analysis to uncover an RT102-type cytoplasmic male sterility-associated candidate gene derived from *Oryza rufipogon*. Plant Cell Physiology. 2013;**54**(9):1560-1568

[20] Toriyama K, Kazama T. Development of cytoplasmic male sterile IR24 and IR64 using CW-CMS/*Rf17* system. Rice. 2016;**9**:22-26

[21] Koltunow AM, Bicknell RA, Chaudhury AM. Apomixis: Molecular strategies for the generation of genetically identical seeds without fertilization. Plant Physiology. 1995;**108**(4):1345

[22] Mieulet D, Jolivet S, Rivard M, Cromer L, Vernet A, Mayonove P, et al. Turning rice meiosis into mitosis. Cell Research. 2016;**26**(11):1242-1254

[23] Gaafar RM, El Shanshoury AR, El Hisseiwy AA, AbdAlhak MA, Omar AF, Abd El Wahab MM, et al. Induction of apomixis and fixation of heterosis in Egyptian rice Hybrid1 line using colchicine mutagenesis. Annals of Agricultural Science. 2017;**62**(1):51-60

[24] Xie E, Li Y, Tang D, Lv Y, Shen Y, Cheng Z. A strategy for generating rice apomixis by gene editing. Journal of Integrative Plant Biology. 2019;**61**(8):911-916

[25] Yuan LP. Progress in super-hybrid rice breeding. The Crop Journal. 2017;**5**:100-102

[26] Dash AK, Rao RN, Rao GJN, Verma RL, Katara JL, Mukherjee AK, et al. Phenotypic and marker-assisted genetic enhancement of parental lines of rajalaxmi, an elite rice hybrid. Frontiers in Plant Science. 2016;**7**:1005

[27] Behera L, Patra BC, Sahoo RK, Nanda A, Patnaik A, Singh ON. Assessment of genetic diversity in medicinal rice using microsatellite markers. Australian Journal of Crop Science. 2012;**6**(9):1369-1376

[28] Verma RL, Singh S, Singh P, Kumar V, Singh SP, Singh S, et al. Genetic purity assessment of *indica* rice hybrids through DNA fingerprinting and grow out test. Journal of Environmental Biology. 2017;**38**(6):1321-1331

[29] Basavaraj SH, Singh VK, Singh A, Singh AK. Marker-assisted improvement of bacterial blight resistance in parental lines of Pusa RH10, a superfine grain aromatic rice hybrid. Molecular Breeding. 2010;**26**(2):293-305

[30] Liu W, Wang FM, Jin S. Improvement of rice blast resistance in TGMS line by pyramiding of Pi1 and Pi2 through molecular marker-assisted selection. Acta Agronomica Sinica. 2008;**34**:1128-1136

[31] Zhou PH, Tan YF, He YQ, Zhang Q. Simultaneous improvement for four quality traits of Zhenshan 97, an elite parent of hybrid rice, by molecular marker-assisted selection. Theoretical and Applied Genetics. 2003;**106**:326-331

[32] Ni D, Zhang S, Chen S, Xu Y, Li L, Li H, et al. Improving cooking and

eating quality of Xieyou57, an elite *indica* hybrid rice, by marker-assisted selection of the Wx locus. Euphytica. 2011;**179**(2):355-362

[33] Duan MJ, Sun ZZ, Shu LP. Genetic analysis of an elite super-hybrid rice parent using high-density SNP markers. Rice. 2013;**6**(1):21-36

[34] Shen R, Wang L, Liu X, Wu J, Jin W, Zhao X, et al. Genomic structural variation-mediated allelic suppression causes hybrid male sterility in rice. Nature Communications. 2017;**8**:1310-1320

[35] Guo J, Xu X, Li W, Zhu W, Zhu H, Liu Z, et al. Overcoming inter-subspecific hybrid sterility in rice by developing *indica*-compatible *japonica* lines. Scientific Reports. 2016;**6**:268-278

[36] Priyadarshi R, Hari Prasad SA, Singh AK, Khandekar D, Ulaganathan K, Shenoy V. Molecular stacking of wide compatibility gene, S5 n and elongated uppermost internode (*eui*) gene into IR 58025B, an elite maintainer line of rice. Journal of Plant Biochemistry and Biotechnology. 2017;**26**(4):425-435

[37] Soni SK, Yadav VK, Bhadana VP, Yadav MC, Sundaram RM. Prediction of heterosis using hypervariable SSR markers in tropical *japonica* x *indica* rice hybrids (*Oryza sativa* L). International Journal of Current Microbiology and Applied Sciences. 2017;**6**(10):1419-1427

[38] Lu ZM, Xu B. On significance of heterotic group theory inhybrid rice breeding. Rice Science. 2010;**17**:94-98

[39] Kumar A, Bhowmick PK, Singh VJ, Malik M, Gupta AK, Seth R, et al. Marker-assisted identification of restorer gene(s) in iso-cytoplasmic restorer lines of WA cytoplasm in rice and assessment of their fertility restoration potential across environments. Physiology and Molecular Biology of Plants. 2017;**23**(4):891-909

[40] Verma et al. NRRI Technology Bulletin-114. 2015. p. 16

Chapter 3

# Hybrid Rice in Africa: Progress, Prospects, and Challenges

*Samuel Oppong Abebrese and Alex Yeboah*


## Abstract

Hybrid rice varieties exploit the phenomenon of heterosis to out-yield their inbred counterpart to increase productivity per unit area. Unlike inbred rice varieties whose seeds could be used for replanting season after season, farmers will have to purchase seeds of hybrids every season to obtain the expected yield and other quality attributes. While the hybrid system is a disadvantage to farmers in terms of mandatory seed purchase, it serves as a motivation and opportunity for private seed companies to recoup their investment and therefore encourages their involvement in seed production research and development. Hybrid rice technology originated from China and is well commercialized in Asia and the Americas. Africa is among the few places where hybrid rice is still not fully commercialized. Besides Egypt, many African countries have just begun to exploit the benefits of hybrid rice technology. A number of introduced hybrids have exhibited a 15–20% yield advantage over the available top inbred varieties. Most African countries also have a conducive environment for viable hybrid rice seed production. Besides other pertinent challenges, some which can be addressed technically, there should be a recommended production package that will give farmers the best of yield to enable hybrid rice contribute to attaining rice self-sufficiency in Africa.


**Keywords:** rice, hybrid rice, Africa, increasing rice productivity, hybrid rice adoption

## 1. Introduction

*Although over 90% globally is produced and consumed in Asia, rice is the fastest growing food source in Africa.* In Sub-Saharan Africa (SSA), consumption is estimated to be increasing at a rate of 6% per annum, the highest in the world [1]. The rate of increase in consumption of rice in Africa has not been matched by corresponding increases in production and the demand-supply gap keeps widening. It is estimated that Africa imports 36% of its annual rice requirement (18 million metric tons) at a cost of US$7 billion per annum (FAO, 2019). The continuous dependence on imports places a heavy demand on scarce foreign currency reserves of the countries in the region, which are among the poorest in the world. In order to address this situation, national and international institutions as well as bilateral and multilateral donors formed a consultative group on rice known as the Coalition for African Rice Development (CARD). The goal of CARD was to double rice production in Africa by 2018. Nevertheless, due to outdated production systems, biotic and abiotic constraints as well as low investment in production technologies, only about 60% of the consumer demand is still met through local production [2]. As a response to the 2008 rice crisis and other challenges such as population growth, rapid urbanization, climate change, and natural resource degradation, national, regional, and international institutions have increased investment toward boosting rice production on the continent. Consequently, many countries have national strategic plans and favorable policies for increasing rice production in place. The current situation is encouraging and requires strengthening existing partnerships to address the challenges across the whole rice value chain to enhance sustainable food security across the continent [2].

Promoting the adoption of productivity enhancing technologies including hybrid rice technology is key to the continent's rice self-sufficiency agenda. Hybrid rice has the penitential of increasing rice productivity and encouraging private sector involvement in seed production research and development in Africa. Since there is currently no proper plant variety protection (PVP) system in place in most of Sub-Saharan countries; the hybrid system could serve as a form of biological intellectual property through the control of the hybrid parents. This chapter examines the procedures of hybrid rice development, current state of development and commercialization in Africa as well as its prospects and challenges.

## 2. Rice breeding methods

Rice breeding is an effective mechanism for delivering the benefits of science and technology to millions of resource-poor farmers. Rice is considered one of the crops that has achieved remarkable success through breeding. Notable success includes the contribution to the green revolution with semi-dwarf varieties that averted a looming hunger in Asia in the 1960s [3, 4]. The methods of breeding rice can be categorized into conventional selection, in-vitro, molecular and transgenic. The available conventional selection breeding methods include mass selection, pure line selection, pedigree method, bulk method, backcross method, recurrent selection, and single seed descent [5, 6]. Mass and pure line selections are mostly used for purifying heterogeneous varieties and are rarely used in present day breeding programs. The pedigree method is the most widely used method to develop rice varieties. More than 85% of the released rice varieties were developed through pedigree selection [4]. Backcrossing is commonly used to incorporate one or a few genes into an adapted or elite variety [4–6]. Although hybrid breeding is primarily applicable to outcrossing species such as maize, it has been successfully added to the rice breeding portfolio. The aim of hybrid rice breeding is to raise the yield ceiling of rice beyond what is currently achieved by the semi dwarf varieties [7, 8]. Mutagenesis to some extent has also been employed to develop some valuable rice varieties [4, 9]. The in-vitro methods include tissue culture techniques such as anther culture to develop doubled haploids, somaclonal variation to identify useful variants and embryo rescue to assist in wide hybridization such as the one that led to the development of the NERICAs. In the NERICA development, embryo rescue was used to obtain viable progeny between *Oryza sativa* and *Oryza glaberrima* crosses [10]. Molecular breeding methods mostly involve the use of molecular markers in marker assisted selection to increase the efficiency and precision of conventional breeding [6, 11]. Genetic engineering (transgenic technique) allows addition of alien genes from any living organism to the rice gene pool to impart a useful function. This technique allows breeders to accomplish objectives which cannot be achieved through conventional plant breeding [12]. Quite recently, genome editing technique has been added to the rice breeding methods. By genome editing, specific modification can be made at targeted locations of the rice genome. Unlike genetic engineering, this method does not involve the introduction of foreign genes into the rice genome [13].

### 2.1 Rice varietal types

The above breeding methods lead to the development of one of the three main rice varietal types i.e. inbreds (pure lines), hybrids or GM (transgenic) rice [6]. Inbreds are the most commonly used rice varietal type. Since offspring or succeeding generations produced by these varietal types are of the same genetic makeup, seeds harvested from an inbred variety can be used for succeeding planting without losing their varietal identity provided cross pollination with other varieties is avoided. Hybrids are products of crossing two genetically diverse inbred lines. As a result, seeds harvested from the hybrid plants are not recommended for replanting because some vigor is lost resulting in lower yield and genetic segregation [8]. Farmers are recommended to buy new hybrid seeds for each planting season from accredited sources. The increased profits resulting from increased yields of hybrids versus pure line varieties offset the cost of the hybrid seed. Transgenic (GM) rice results from the use of genetic engineering. Though the resulting varieties breed true, they are separated from the conventional (nontransgenic) inbreds due to the involvement of a transgene. Though several transgenic rice lines have been developed, they are yet to find their way into commercial cultivation. Notable transgenic (GM) rice is the Golden rice; a variety engineered to produce beta carotene to help combat vitamin A deficiency [12].

## 3. Hybrid rice technology

Hybrid rice is the commercial rice crop grown from $F_1$ seeds of a cross between two genetically dissimilar parents. This could only be possible through the use of a male sterility system due to the strictly self-pollination nature of the rice plant [4, 8]. Hybrid varieties exploit the phenomenon of hybrid vigor (heterosis) to increase the yield potential of rice beyond the level of modern inbred rice varieties. A yield advantage of 15–30% over conventional inbred varieties grown under similar conditions has been reported [14]. The hybrid rice technology concept dates back to 1964 in China. However, it was only after 1976, when a wild abortive pollen plant was identified in Southern China, did the idea begin to materialize [4, 8]. Since the expression of heterosis is confined to the first generation only, farmers have to buy fresh seeds every season to raise commercial crop. Since the hybrids yield 15 to 30% more than pure line varieties, farmers prefer hybrid seeds if the price is economically beneficial and seeds are readily available. Hybrids can offer biological intellectual property protection which attracts and encourages private-sector involvement in seed production research and development [15].

### 3.1 Genetic basis of heterosis

For a long time, two theories; the dominant and overdominance theories were put forward to explain the phenomenon of heterosis [14]. According to the dominant theory, hybrid vigor is due to the action and interaction of favorable dominant alleles. The overdominance theory on the other hand suggests that heterozygous loci are superior to homozygous loci. Thus, two alleles complement each other and there is overexpression of genes in the heterozygous state [16]. Recently, epistasis or interactions among loci has been recognized as a major contributor to heterosis [17]. Estimates based on mating designs of the relative magnitude of additive, dominance and epistatic components of variance indicate that the magnitude of epistatic variance is small compared to additive and dominance components. This is because statistical designs cannot predict epistasis. Using molecular markers,

Yu *et al.* [18] provided evidence of the importance of epistasis as a possible genetic basis of heterosis in rice. Since none of these theories exclusively explains heterosis, suggestions are to ascribe the phenomenon to a combination of the three models.

## 4. Hybrid rice breeding methods

### 4.1 The CMS (three-line) method

Hybrids rice is currently produced either by the three-line (CMS) or the two-line (EGMS) method [19]. The three-line method employs three lines; the CMS (A line), maintainer (B line) and restorer (R line). An A-line is genetically identical to the B-line except that the B-line has a normal (N) cytoplasm while the A-line has a male sterile (S) cytoplasm. An R-line has dominant fertility restorer genes in its nucleus which restore male fertility to the $F_1$ hybrid. The B-line is crossed to the A-line to maintain/produce seeds of the A-line whiles the R-line is crossed to the A-line to produce hybrid seeds.

#### *4.1.1 Organization of hybrid rice breeding program using the CMS system*

Due to the relatively complex nature of the three-line (CMS) breeding system, breeding materials are grouped into separate nurseries or stages for efficient handling and development of parental lines. Experimental hybrids normally pass through the required stages and are finally tested on farmers' fields before their release. The nurseries include the source nursery, the testcross nursery, backcross nursery, re-testcross nursery and combining ability nursery [19]. The source nursery contains elite breeding and CMS lines with the potential of becoming parents of commercial hybrids. In the testcross nursery, $F_1$s from the CMS and tester lines in the source nursery are evaluated for their pollen and spikelet fertility to identify prospective maintainers (B-lines) and restorers (R-lines). Re-testcross nursery is for confirming and purifying prospective R-lines and backcross conversion of prospective B lines into CMS lines. Combining ability nursery evaluates general and specific combining abilities of selected CMS and R lines. This nursery helps identify higher yielding hybrids and is very crucial in the hybrid rice development process [17, 19]. Various modifications of the IRRI's system described above are made to improve breeding efficiency. Elite tester inbreds are selected from R-lines to cross onto new A-lines and select testers from A and B-lines to identify new R-lines. Once commercial R and A-lines are identified, they are used as testers instead of crossing a bunch of new R-lines by new A-lines. For instance, in China, IRRI's elites IR24, IR26 were directly used as restorer lines. These restorers were also used as restorer gene resources to develop new restorer lines [17]. Conversion of identified maintainer into a CMS line. Often lines introduced to new areas are not adapted to the target environment. They may be susceptible to particular local pests and diseases or may lack the desirable grain quality attributes. This necessitates the conversion of the available CMS lines into locally adapted and desirable ones. Introduced CMS lines maintainer identified from a cross of local lines onto a CMS source should be backcrossed to develop adapted CMS lines [19]. Backcrossing is continued to the 5th -6th generation ($BC_5$-$BC_6$) where the nuclear content of the original CMS source is replaced almost completely with that of its corresponding maintainer. Development of new CMS lines is normally difficult due to the limited chances of identifying stable sterile lines from local germplasm [19]. Developing new CMS lines using African cultivars has proven difficult due to sterility instability in the $BC_1$ and $BC_2$ generations [15].

### 4.2 The two-line method

In this system, male sterility is conditioned by the interaction of nuclear genes with environmental factors such as photoperiod, temperature or both. Such lines are referred to as environment-genetic male sterility (EGMS). The ones conditions by lines temperature are technically known as thermosensitive-genetic male sterility (TGMS), ones by photoperiod are photo-genetic male sterility (PGMS) and by both temperature and photoperiod are referred to as photo-thermo-genetic male sterility [17, 20]. Organization and seed production are simpler in this system than the CMS system. The EGMS lines are multiplied by sowing these lines in such a way that the sensitive period coincides with photoperiod/temperature that is conducive for inducing fertility. Hybrid seed production is taken up by sowing these lines in such a way that the sensitive stage coincides with the photoperiod or temperature conducive for inducing complete male sterility [20]. Magnitude of heterosis in two-line hybrids is 5 to 10% higher than in three-line hybrids. The major constraints to developing and using TMGS lines in the tropics are limited availability of stable TGMS germplasm. Since the PGMS, TGMS and PTGMS are controlled by recessive gene (s), when these lines are crossed with a fertile line, the hybrids are fully fertile, irrespective of the day length and temperature conditions prevailing during the growth season. Although attractive and potential as a tool for exploiting heterosis, the EGMS system has some advantages and disadvantages. Compared to the two-line system, the use of the three-line system is expensive and labor intensive but much more reliable. Since the two-line system does not require a maintainer line, any line can be used as pollen donor. This increases the chances of identifying higher yielding hybrids in the two-line system than the three-line system. However, any sudden changes in the environmental conditions during the hybrid production season affect the sterility of the temperature sensitive line and the requirement of additional land in different day length areas limits the ability to reliably produce hybrid rice using the two-line system [20].

## 5. Identification of high-yielding hybrids

After a large number of prospective parental lines (CMS, restorers or EGMS) have been identified, there is the need to select the most promising ones by their ability to give superior hybrids. This is normally achieved using the line by tester design [17]. Although it is still not well understood, the positive correlation between genetic distance of hybrid parents and the resulting $F_1$ heterosis is accepted phenomenon. Heterosis levels in rice are reported to increase in the direction of *Japonica* × *japonica* < *indica* × *indica* < japonica × *javanica* < *indica* × *javanica* < *indica* × *japonica* [14]. Utilization of intersubspecific heterosis has been regarded as a promising strategy for increasing rice productivity. Large efforts have been invested in the last decades for breeding *indica–japonica* hybrids. However, such efforts have been hindered by hybrid sterility that frequently occurs in intersubspecific crosses. Discovery of Wide Compatibility Varieties (WCVs) brought hope for breaking the sterility barrier between *indica* and *japonica* subspecies and provided a possibility for exploiting the strong heterosis between them. The WCVs could produce fertile $F_1$ hybrids when crossed with *Indica* or *Japonica* lines. The key approach was to introduce wide compatible genes into the restorer or CMS lines for developing widely compatible restorer or CMS lines which will permit the production of fertile $F_1$ hybrids from either subspecies. Through marker-assisted selection, Guo *et al.* [21] successfully pyramided the *indica* allele (*S*-i) at four loci (*Sb*, *Sc*, *Sd* and *Se*) and the neutral allele (*S*-n) at *S5* locus in *japonica* genetic background to develop

*Indica*-Compatible *Japonica* Lines (ICJLs). These lines have a great promise of overcoming the intersubspecific hybrid sterility and exploiting the high heterosis between them. There is also a super hybrid rice breeding project in China trying to exploit intersubspecific heterosis in combination with ideal plant type [17].

## 6. Enhancing outcrossing in hybrid rice seed production

Various natural and artificial methods are employed to increase outcrossing rate in hybrid rice seed production. Among the artificial ones are flag leaf clipping, gibberellin application and supplementary pollination [14]. Besides, there are several traits that naturally contribute to hybrid rice seed production efficiency. These include days to heading, pollen load, pollen longevity, and morphological traits of floret such as size of stigma and style, stigma exertion, stigmatic receptivity and spikelet opening angle. Stigma exertion rate is emphasized as a key factor for efficient hybrid rice seed production [14]. The extent to which the stigma is exerted in the female parent (male sterile line) increases the chances of outcrossing thereby increasing hybrid seed set. Stigma exertion is a genetic trait and not all male sterile lines possess high expression of it. It is therefore possible to enhance the trait through specific breeding efforts using appropriate donors. Both qualitative and quantitative modes of inheritance have been reported for stigma exertion. Besides several quantitative trait loci (QTL) controlling stigma exertion have been mapped to different rice chromosomes [22] with molecular markers and can easily be introgressed into other parents.

## 7. Attempts at development and commercializing hybrid rice in Africa

Egypt is the only African country that has successfully developed local rice hybrids in commercial scale. Hybrid rice research in Egypt started around 1982 and intensified since 1995 with the launching of a mission-oriented project through a cooperation between Rice Research Program of Egypt and International Rice Research Institute (IRRI). Since then, hybrid rice varieties have been developed and released to help Egyptian farmers improve productivity and increase production [23, 24]. In conjunction with Africa Rice Center (AfricaRice), several African countries started introducing and evaluating hybrid rice mostly from Asia. AfricaRice started a hybrid rice breeding program to developed hybrid varieties for the sub-Saharan Africa region starting from the year 2000 [15]. For a start, the AfricaRice program tried to build on hybrids developed by the Green Super Rice project jointly coordinated by the Chinese Academy of Agricultural Sciences (CAAS), AfricaRice and IRRI. Hybrids from the project were evaluated for yield and general adaptability in the rainfed and irrigated ecologies of eight African countries (Liberia, Mali, Mozambique, Nigeria, Rwanda, Senegal, Tanzania and Uganda) [15]. Although promising hybrids out-yielded the best inbred checks, they were susceptible to the major pest and diseases in the region [15, 25]. This necessitated the development of AfricaRice's in-house hybrids program with the following objectives: (i) develop new parental lines from local varieties; (ii) determine adaptability of some CMS lines in Africa; and (iii) establish a hybrid rice seed production system in some African countries [15]. About 50 high-yielding hybrid rice lines were developed and evaluated in several African countries by the AfricaRice program [25]. Notable among them is the aromatic hybrid rice variety (AR051H) released in 2017 by the Senegalese Institute of Agricultural Research under the name ISRIZ 09. AfricaRice and CORAF through donor support projects have been showcasing and promoting the adoption of hybrid rice technology in some Sub-Saharan African countries. Initial studies on local hybrid rice seed production in some West African countries indicate no technical hindrance since a seed yield of about 1–2 t/ha could be obtained in many of these countries [15].

Aside the AfricaRice's program, the African Agricultural Technology Foundation (AATF) is working with partners to develop new indigenous rice hybrids that can increase rice yields and improve productivity for farmers in the Eastern and Southern Africa. Unlike AfricaRice's program which uses the CMS (three-line) system, the AATF project dubbed "breeding by design" uses the thermosensitive (two-line) system [26]. The partnership intends to develop hybrid rice germplasm that is adapted to African conditions using the 2-line hybrid rice system. Under the partnership, an information technology tool with interpolated weather surfaces to predict temperature regimes and manage 2-line hybrid rice production risk is being established. The project as well intends to train a network of researchers and seed production specialists interested in 2-line hybrid rice. The broad objective of the AATF partnership is to develop and expand 2-line hybrid rice technology in selected African countries and ensure that through private companies and public institutions in Africa, this technology reaches farmers and increases their rice yields and income. As support for hybrid rice on the African continent increases, the AATF and its partners including IRRI have formed the Alliance for Hybrid Rice in Africa (AHyRA). The key objectives for AHyRA initiative are to:

1. Promote collaboration and business integration between stakeholders of rice value chains to develop, produce and market hybrid rice seeds.

2. Create a consolidated advocacy for an enabling environment for the sustainable use of hybrid rice in Africa.

3. Conduct joint adaptability testing of available hybrid rice varieties

4. Generate a robust database for parental lines and hybrid rice for the use of stakeholders, and private seed companies, particularly in Sub-Saharan Africa.

5. Strengthening the capacity of partners on the hybrid rice technologies.

Other National Agricultural Research systems (NARs) in collaboration with multi-national and local private seed companies also initiate their own in-country hybrid rice development and commercialization endeavor. For instance, NARS in Ghana in collaboration with international and local private seed companies (WIENCO, ADVANTA and SEED CO.) have conducted studies into farmer preferred traits and potential for adoption of hybrid rice [27] in Ghana, evaluated introduced rice hybrids for yield, rection to disease and grain quality attributes [28]. A study has also been conducted to identify CMS maintainers and restorers for local hybrid rice development [29]. Although these efforts have led to the release of two hybrid rice varieties with 15–20% yield advantage over the best inbred check, promotion and commercialization is still low. Plans for in-country hybrid rice seed production has also not been materialized and seeds of these hybrid are still imported. Mali in collaboration with IRRI initiated its own hybrid rice development program in early 2011 [15]. A summary of countries involved in hybrid rice research and development and their status and collaborators is presented in **Table 1**. Examples of some released hybrid rice varieties on the African continent is presented in **Table 2**.

| Country | Status | Collaborating institutions/ projects |
|---|---|---|
| Egypt | Grow locally developed hybrid rice varieties | Rice Research and Training Centre, IRRI, FAO |
| Ghana | Identification of farmer preferred traits, evaluation of introduced hybrids, identification of CMS maintainers and restorers, | Advanta, Seed Co, Wienco-Ghana |
| Mali | Evaluation of introduced hybrids, identification of CMS maintainers and restores | IRRI/AfricaRice hybrid rice projects |
| Mozambique | Evaluation of introduced hybrids | AfricaRice hybrid rice breeding program |
| Nigeria | Evaluation of introduced hybrids | AfricaRice hybrid rice breeding program |
| Senegal | Evaluation of introduced hybrids | AfricaRice hybrid rice breeding program |
| Tanzania | Evaluation of introduced hybrids | AfricaRice/AATF hybrid rice breeding programs |
| Uganda | Evaluation of introduced hybrids | AfricaRice hybrid rice breeding program |
| Kenya | Evaluation of introduced/developed hybrids | AATF hybrid rice project |

*IRRI, International Rice Research Institute; AATF, African Agricultural Technology Transfer; AfricaRice, Africa Rice Center; FAO, Food and Agricultural Organization.*

**Table 1.**
*African countries involved in hybrid rice research and development and their status.*

| Hybrid | Released country | Institution |
|---|---|---|
| AR051H (ISRIZ 09) | Senegal | Senegalese Institute of Agricultural Research |
| PAC 801 | Ghana | CSIR-Savanna Agricultural Research Institute |
| Arize 6444 Gold | Ghana | CSIR-Crops Research Institute |
| EHR1 (SK2034) | Egypt | Rice Research and Training Centre |
| EHRI2 (SK2046) | Egypt | Rice Research and Training Centre |
| EHR3 (SK2151H) | Egypt | Rice Research and Training Centre |

**Table 2.**
*Examples of released hybrid rice varieties on African continent.*

## 8. Prospects of hybrid rice in Africa

Hybrid rice has the potential to help increase rice production and productivity as well as reducing rice imports in Africa. The main tenet of hybrid rice technology is to employ the phenomenon of heterosis to out yield the available inbred semi-dwarf varieties. A yield advantage range of 15–20% over available inbred checks reported in other parts of the world has also been recorded in Africa [15, 24, 25, 28]. Scaling up and promoting adopting of these hybrids, will contribute to food security in general on the African continent. It will also enable majority of the populace afford to buy their staple food at reasonable prices thereby helping to maintain political stability. Improving rice productivity on the African continent has relied heavily on increasing land area which is unsustainable. Hybrid rice technology has the ability to help save land required for rice production and put into other productive uses. The technology also could encourage private sector involvement in the rice seed research and development. One challenge for multi-nationals and other local private seed companies engaged on the African continent has been the unavailability of a functional plant variety protection (intellectual property) systems to protect their new varieties. Since the hybrid system provides some sort of biological intellectual property through the control hybrid parents, seed companies could operate effectively to recoup their investment. This in turn will assure farmers access to quality rice seed which normally is a challenge to improve rice productivity on the African continent. Hybrid rice cultivation requires fresh $F_1$ seed for every cropping season. This will require the development of a functional hybrid seed production, processing and marketing infrastructure in the form of seed enterprises (public, private or NGOs). This has the potential to create additional rural employment opportunities for the rural folk mostly youth and women. Introduced hybrids were found to exhibit substantial yield advantage under both rainfed and irrigated lowland ecologies [25]. These two ecologies form a large portion of the available area for rice cultivation. Thus, hybrids varieties could be promoted widely among rice production regions of Africa. There seems to be an emerging interest by donor and international agencies working toward achieving rice self-sufficiency on the African continent to position hybrid rice as a technology that can contribute to a food secured Africa. Promotion of locally adapted high yielding varieties developed by AficaRice currently enjoys some donor support through technologies for African agricultural transformation (TAAT) rice compact and the West and Central African Council for Agricultural Research and Development (CORAF). This involves working toward the right policies and infrastructural requirements to make this technology thrive and such a concerted effort is bound to chalk some success.

## 9. Challenges of hybrid rice adoption

Like other continents, Africa will have to overcome the major constraints to large scale adoption of hybrid rice. These include high cost of seeds, poor grain quality issues, pest and diseases susceptibility, human capacity for hybrid rice development and the difficulty of identifying higher heterosis (> 25%) at the field level [30]. Most African rice farmers have built a tradition of using self -saved seeds. This tradition, maybe, was to adapt to the unavailability of adequate reliable seed companies to ensure timely supply of quality rice seeds. Most rice farmers still use traditional varieties. For the few that use improved varieties, it is through the informal system. The business of rice seed production is at the budding stage and will have to be developed before hybrid varieties could have a place. Most African governments still ensures that their farmers get the benefits of improved seeds through the subsidy. The major challenge for hybrid rice includes whether African rice farmers will be willing to buy hybrid seeds for every cropping season and to do so at a higher price. The hybrid rice technology itself could also act as a catalyst to creating a sustainable rice seed business by providing varieties with high yield advantage which can attract farmers to patronize improved rice varieties. That means the realizable yield advantage of hybrids on the field should be high enough. Although initial studies suggest that farmers anticipate above 30% yield advantage over existing best inbred to guarantee adopting a variety that requires seasonal seed purchase [27], there is a technical limitation to the realizable field heterosis by hybrid varieties.

Sub-Saharan Africa is among the few places globally where rice yields are still low with average yield of about 2.2 t/ha. This is largely due to the use of less productive cultivation technologies and inputs particularly fertilize [31]. Although hybrid varieties perform relatively better under stress conditions, the full potential could be realized under optimal inputs. For African rice farmers to realize the yield benefit of hybrids, there will be the need for hybrid seeds be accompanied with a well-researched and recommended agronomic package that will give farmers the best yield.

Hybrid rice is new to the African continent and knowledge on the technology is limited. There will be the need to train a critical mass of scientists and technicians on hybrid rice research and seed production to spearhead the development, demonstration and adoption of the technology. Aside that, farmers should be educated not to save and replant hybrid seeds.

The African continent is fragmented in terms of essential traits. There will be the need to develop different hybrid combinations for different rice production regions depending on their important traits. Hybrid rice breeding uses concepts, skills, and procedures which are strikingly different from those used for conventional inbred rice development. Unlike conventional inbred rice development which accumulates productive genes that perform well in a homozygous state year after year, hybrid rice breeding exploits hybrid vigor which provides additional genes that add to the yield obtained by the productive genes in the homozygous varieties. Most introduced rice hybrids tend to have problem of susceptibility to local pest and diseases such as blast, rice yellow mottled virus (RYMV) disease and African rice gall midge (AfRGM) [25]. Aside these stresses, poor grain quality was one of the major constraints for large scale adoption of hybrid rice in Asia. Varieties with long slender aromatic grains that cook soft are mostly preferred by consumers within the West African sub-region. Breeders will have to battle with these traits which are mostly quantitatively inherited and difficult to combine in in developing products.

## 10. Conclusion

There is an emerging interest to position hybrid rice as a technology that can contribute to attaining rice self-sufficiency in Africa. Some African countries have responded by testing promising introduced and in-house hybrids developed by the pan-African rice research institute (AfricaRice). Some of these hybrids have shown favorable yield advantages over the best available inbred lines, but there are reports of susceptibility to some local pest and diseases. Initial studies on seed production potential indicates that most countries have conducive environment for local hybrid rice seed production. Nonetheless, there is the need to train a critical mass of scientists and technicians to champion the hybrid rice agenda and educate farmers not to save and re-plant hybrid rice seeds. There are a number of traits that hybrid breeders should also consider in developing products for the various market segments. For farmers to appreciate the need for buying seeds every season and at a higher price, hybrid varieties with higher field level heterosis should be developed and demonstrated. Also, promotion of hybrid rice seeds should go along with a well-researched recommended crop management package that will give farmers the best of yield. The hybrid rice agenda should form part of the overall strategy to develop a sustainable integrated rice seed sector to benefit African rice farmers.

## Acknowledgements

This chapter is part of the review of a PhD thesis research funded by Alliance for a Green Revolution in Africa (AGRA) through the West Africa Centre for Crop Improvement (WACCI), University of Ghana. The advice and suggestions of my supervisors are greatly acknowledged.

## Author details

Samuel Oppong Abebrese* and Alex Yeboah
CSIR-Savanna Agricultural Research Institute, Tamale, Ghana

*Address all correspondence to: sam555oppa@yahoo.com

## References


[1] Soullier G, Demont M, Arouna A, Lançon F, Mendez P. The state of rice value chain upgrading in West Africa. Global Food Security. 2020;**25**:100365. DOI: 10.1016/j.gfs.2020.100365

[2] Zenna N, Senthilkumar K, Sie M. Rice production in Africa. In: Chauhan BKJ, Mahajan G, editors. Rice Production Worldwide. Cham: Springer; 2017

[3] Khush GS. What it will take to feed 5.0 Rice consumers in 2030. Plant Molecular Biology. 2005;**59**:1-6

[4] Guimaraes EP. Rice breeding. In: Cereals. New York, NY: Springer; 2009. pp. 99-126

[5] Shastry SVS. Rice breeding in retrospect. Current Science (00113891). 2006;**12**:9

[6] Breseghello F, Coelho ASG. Traditional and modern plant breeding methods with examples in rice (*Oryza sativa* L.). Journal of Agricultural and Food Chemistry. 2013;**61**:8277-8286. DOI: 10.1021/jf305531j

[7] Singh S, Singh P, Singh DK, Singh AK. Hybrid rice development: Two and three line system. BIOLOGIX. 2013. p. 178

[8] Cheng SH, Cao LY, Zhuang JY, Chen SG, Zhan XD, Fan YY, et al. Super hybrid rice breeding in China: Achievements and prospects. Journal of Integrative Plant Biology. 2007;**49**:805-810

[9] Viana VE, Pegoraro C, Busanello C, De Oliveira AC. Mutagenesis in rice: The basis for breeding a new super plant. Frontiers in Plant Science. 2019;**10**:1-28. DOI: 10.3389/fpls.2019.01326

[10] Arouna A, Lokossou JC, Wopereis MCS, Bruce-Oliver S, Roy-Macauley H. Contribution of improved rice varieties to poverty reduction and food security in sub-Saharan Africa. Global Food Security. 2017;**14**:54-60

[11] Collard BCY, Cruz CMV, Mcnally KL, Virk PS, Mackill DJ. Rice molecular breeding laboratories in the genomics era: Current status and future considerations. International Journal of Plant Genomics. 2008;**2008**:524847. DOI: 10.1155/2008/524847

[12] Sah SK, Kaur A, Kaur G, Cheema GS. Rice research: Open access genetic transformation of rice: Problems, progress and prospects. Rice Research. Open Access; 2014;**3**:1-10. DOI: 10.4172/2375-4338.1000132

[13] Miglani GS. Genome editing in crop improvement: Present scenario and future prospects. Journal of Crop Improvement. 2017;**31**:453-559. DOI: 10.1080/15427528.2017.1333192

[14] Virmani S. Heterosis and Hybrid Rice Breeding. Vol. 22. Berlin: Springer Science & Business Media; 1994. ISBN: 9783662061688

[15] El-namaky RA, Demont M. Hybrid rice in Africa: Challenges and prospects. In: Wopereis MCS, Johnson DE, Ahmadi N, Tollens E, Jalloh A, editors. Realizing Africa's Rice Promise. UK: CAB international; 2013. pp. 173-178

[16] Duvick DN, Coors JG, Pandey S. Heterosis: Feeding people and protecting natural resources. In: The Genetics and Exploitation of Heterosis in Crops. Madison, WI: American Society of Agronomy. 1999. pp. 19-29

[17] Cao L, Zhan X. Chinese experiences in breeding three-line, two-line and super hybrid rice. Rice—Germplasm, Genetics and Improvement. 2014;2: 279-308. DOI: 10.5772/56821

[18] Yu SB, Li JX, Xu CG, Tan YF, Gao YJ, Zhang Q, et al. Importance of epistasis as the genetic basis of heterosis in an elite rice hybrid. Proceedings of the Proceedings of the National Academy of Sciences. 1997;**94**(17):9226-9231

[19] IRRI (International Rice Research Institute). Hybrid Rice Breeding Manual. Los Banos, Laguna, Philippines: International Rice Research Institute; 1997. p. 26

[20] Virmani SS, Sun ZX, Mou TM, Ali AJ, Mao CX. Two-Line Hybrid Rice Breeding Two-Line Hybrid Rice Breeding Manual. Los Ba nos, Philippines: IRRI; ISBN: 9712201856

[21] Guo J, Xu X, Li W, Zhu W, Zhu H, Liu Z, et al. Overcoming inter-subspecific hybrid sterility in rice by developing indica-compatible japonica lines. Scientific Reports. 2016;**6**:1-9. DOI: 10.1038/srep26878

[22] Lou J, Yue GH, Yang WQ, Mei HW, Luo LJ, Lu HJ, et al. Mapping QTLs influencing stigma exertion in rice. Bulgarian Journal of Agricultural. Science. 2014;**20**:1450-1456

[23] El-Mowafi H, Bastawisi AO, Abdelkhalik AF, El-namaky RA. Hybrid rice technology in Egypt. In: Xie F, Hardy B, editors. Accelerating Hybrid Rice Development. Philippines: Int. Rice Res. Inst, Los Baños; 2009. pp. 593-608

[24] El-Mowafi H, Bastawisi AO, Attia KA, Abdelkhalik AF, Abdallah RM, Reda AM, et al. EHR3 (Egyptian hybrid Rice 3): A new high yielding hybrid variety ehr3 (Egyptian hybrid rice 3). Egyptian Journal of Plant Breeding. 2019;**23**:11-23

[25] El-namaky R, Coulibaly MMB, Alhassan M, Traore K, Nwilene F, Dieng I, et al. Putting plant genetic diversity and variability at work for breeding: Hybrid rice suitability in West Africa. Diversity. 2017;**9**(3):1-12. DOI: 10.3390/d9030027

[26] AATF. Hybrid Rice Breeding by Design Project. Nairobi, Kenya: AATF; 2012. p. 6

[27] Abebrese SO, Martey E, Dartey PKA, Akromah R, Gracen VE, Offei SK, et al. Farmer preferred traits and potential for adoption of hybrid rice in Ghana. Sustainable Agriculture Research. 2019;**8**:38. DOI: 10.5539/sar.v8n3p38

[28] Abebrese SO, Yeboah A, Dogbe W, Kofi P, Dartey A, Akromah R, et al. Evaluation of yield , reaction to diseases, and grain physical attributes of some introduced Rice hybrids in Ghana. International Journal of Agronomy. 2019;**2019**:3926765

[29] Abebrese SO, Kofi P, Dartey A, Akromah R, Gracen VE, Offei SK, et al. Identification of CMS maintainers and restorers for hybrid rice development in Ghana. Journal of Crop Improvement. 2018;**00**:1-15. DOI: 10.1080/15427528.2018.1488158

[30] Spielman DJ, Kolady DE, Ward PS. The prospects for hybrid rice in India. Food Security. 2013:651-665. DOI: 10.1007/s12571-013-0291-7

[31] Kanfany G, El-namaky R, Ndiaye K, Traore K, Ortiz R. Assessment of rice inbred lines and hybrids under low fertilizer levels in Senegal. Sustainability. 2014:1153-1162. DOI: 10.3390/su6031153

Section 3

# Rice Metabolites and Biology

Chapter 4

# Phenolic Compounds and Potential Health Benefits of Pigmented Rice

*Prinya Wongsa*

## Abstract

Rice (*Oryza sativa* L.) is one of the most important staple plant foods for global population especially in Asian countries. Pigmented rice including red rice, black, and purple contains a range of bioactive compounds including phenolics acids and flavonoids. Anthocyanins and proanthocyanidins are recognized as a major functional component in pigmented rice. Recently, pigmented rice varieties have received increasing attention from consumers due to its high nutritional values and bioactive compounds, providing its potential health benefits including antioxidant, anti-inflammatory, anticancer, and antidiabetic. Therefore, the objective of this chapter is to provide an up-to-date coverage of a systematic and advanced isolation, extraction and analytical methods, and potential health benefit studies related to antioxidant, anti-inflammatory, antidiabetic cardiovascular disease risk inhibition potential and anti-neurodegenerative potential of pigmented rice.

**Keywords:** black rice, red rice, phenolics, anthocyanins, potential health benefits

## 1. Introduction

Rice (*Oryza sativa* L.) is a cereal food and consumed, providing energy and nutrients for more than half of the world's population, particularly in Asian countries [1]. An increasing in rice consumption in Africa and Latin America has been observed in the past decade, mainly due to urbanization and changes in eating habits. In addition, European, US and Australian citizens are eating more rice, possibly due to an increased interest in global traveling and Asian cuisines. The world production of rice (paddy) was 782.00 million metric tons in 2018 [2]. The most common rice consumed by human is white rice (85%) and the rest is pigmented rice.

Rice paddy consists of hull and caryopsis. The hull or the outer layer constitutes about 20% of the total paddy rice contain minerals and cellulose. The hulling process also separated the hull from the grain or kernel. After removing of husk, whole-grain rice is the unpolished version of the grains consisting of the germ (2-3%), bran (6-7%), and endosperm (about 90%), and is also called brown rice [3]. Consequently, the process of milling is carried out to obtained white rice, also referred to as milled or polished rice, by removing 8-10% of external layers (mainly bran) from brown rice [4]. The endosperm of rice contains water 12%, carbohydrate 75–80%, and protein 7% [5]. Rice carbohydrate is primarily a starch which is made up of linear amylose and branched amylopectin [6–8]. According to the amylose content in rice, rice cultivars can be classified into five groups including waxy (1-2%), very low (2-9%), low (10-20%), intermediate (20-25%) and high (25-33%) [9].

White rice is a common source of starches and phtochemicals including phenolic compounds, sterols, γ-oryzanol, tocotrienols and tocopherols which locate particularly in the outer layer such as pericarp and aleurone of rice grains [10]. In addition, pigmented or colored rice including black, red and dark purple rice is well known by the rice pericarp or grain having a black, red brown or dark purple color in its covering layers. The pigments, which are located in the aleuronic layer of rice grain, have been reported to consist of a mixture of anthocyanin compounds [6, 10–13]. The pigmented rice has been cultivated and consumed in Asia including China, Japan, Korea, and South East Asian countries for a long time and is widely known as "forbidden rice." Royal families and kings used to consume these special varieties of rice in order to have their health improvement and to enhance their wealth [14]. In addition, many varieties of dark pigmented rice including Japonica black rice, Chinese black rice, Thai black rice and Indonesian black rice have been widely studied. Similar to white rice, pigmented rice is found in short-grained and long-grained, waxy (glutinous) and non-waxy (non-glutinous), husked and early maturing stages. The black and red rice have been used as a functional food and their extracts are also used as food colorants in breads, ice cream and liquor [8, 15].

Interestingly, an increased incidence in non-communicate including heart disease, Type II diabetes, obesity, and cancers in both developing and developed countries has been observed. Recently, pigmented rice varieties have gained an increase in attention from consumers due to its high nutritional value and bioactive compounds, providing its potential health benefits including antioxidant, anti-inflammatory, anticancer, and antidiabetic [8, 11, 16, 17]. In addition, many epidemiological and interventional studies have reported that the consumption of fruits and vegetables, cereal, whole grains, wine, cocoa, coffee and tea can reduce the risks of non-communicable disease including obesity, neurodegenerative diseases, Type II diabetes, and cardiovascular diseases [18, 19]. For the most part, the biological functional effects in acute and chronic diseases of plants have been traced to their phenolic compounds. Various studies have shown that phenolic compounds act as antioxidants [20]. Antioxidants control and reduce the oxidative damage in foods by delaying or inhibiting oxidation caused by reactive oxygen species (ROS), extending the shelf-life and quality of food products [21]. In addition, beta carotene, ascorbic acid, and phenolic compounds play important roles in delaying aging, reducing inflammation, and preventing certain cancers in human. The health benefits of whole grains are mainly contributed by one of its major constituents of phenolic compounds including phenolic acids, anthocyanins, and proanthocyanidins, which are the most common secondary metabolites in cereal grains, exist in pigmented grains and are considered to be the most effective antioxidants in nature [12]. Therefore, the objective of this chapter is to provide an up-to-date coverage of pigmented rice in regard to bioactive constituents, extraction and analytical methods, and their potential health benefits. Special attention is paid to bioactivities of pigmented rice and its main anthocyanins.

## 2. Phytochemicals in cereal grains

Cereal grains are composed of nutrients and phytochemicals. Phytochemicals are bioactive, non-nutritional plant compounds, which typically occur in small quantities in plants foods, for example, fruits and vegetables, whole grains and cereals. The most important groups of dietary phytochemicals can be divided into general categories as phenolics, alkaloids, nitrogen-containing compounds, organosulfur compounds, phytosterols, and carotenoids [18]. In addition, bioactive compounds including phenolic compounds, phytosterols, tocols, dietary fibers (mainly beta-glucan), lignans, alkylresorcinols, phytic acid, γ-oryzanols, avenanthramides, inositols and betaine are obtained from whole grain cereals [19, 22]. Some bioactive compounds are quite specific to certain cereals, for example, γ-oryzanol in rice, avenanthramide and saponins in oats, beta glucans in oats and barley and alkylresorcinol in rye, although these are also present in other cereals like wheat but relatively in fewer amounts.

## 3. Phenolic compounds in pigmented rice

Phenolics compounds or polyphenols contain one or more aromatic rings with one or more hydroxyl groups and generally are categorized as phenolic acids, flavonoids, stilbenes, coumarins and tannins [18]. Phenolic compounds are the secondary metabolites plants, providing essential function in the reproduction and growth of the plant, acting as defense mechanisms against pathogens, parasites and environments (light, chilling, pollution, etc.), and contributing to the color of plant [23, 24]. In addition to their roles in plants, phenolic compounds in human diet provide health benefits associated with reduced risk of chronic diseases [22].

### 3.1 Phenolic acids

Phenolic acids in rice grain present in two forms including soluble and insoluble forms. The soluble form, including free and conjugated forms, the former of which can be extracted by solvent, such as aqueous methanol, ethanol, and acetone, and the latter can be hydrolyzed from soluble phenolics by alkali. However, the insoluble form, also called bound phenolics, which esterify to the cell walls. Among these three phenolic fractions, the most abundant is insoluble bound phenolic acids, followed by soluble conjugated phenolic acids and the least is soluble free phenolic acids [25]. In addition, phenolic acids mainly exist in rice bran with trace amounts in endosperm in different rice genotypes. Phenolic acids are widely distributed into two sub-classes including hydrobenzoic and hydroxycinnamic acids and derivatives. **Figure 1** shows the structure of some common phenolic acids detected in rice. Hydroxybenzoic acid derivates include *p*-hydroxybenzoic, protocatechuic acid, vannilic, syringic and gallic acids. In addition, hydroxycinnamic acid derivates like *p*-coumaric, caffeic, ferulic and sinapic acids. Phenolic acids in many plants are

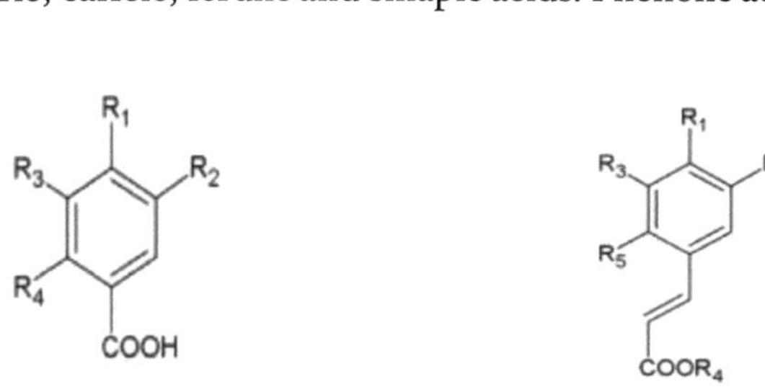


| Hydrobenzoic acids | R1 | R2 | R3 | R4 | Hydroxycinnamic acids | R1 | R2 | R3 | R4 | R5 |
|---|---|---|---|---|---|---|---|---|---|---|
| *p*-Hydroxybenzoic acid | -OH | -H | -H | -H | *p*-Coumaric acid | -OH | -H | -H | -H | -H |
| Vanillic acid | -OH | -H | -$OCH_3$ | -H | Caffeic acid | -OH | -OH | -H | -H | -H |
| Gallic acid | -OH | -OH | -OH | -H | Sinapic acid | -OH | -$OCH_3$ | -$OCH_3$ | -H | -H |
| Syringic acid | -OH | -$OCH_3$ | -$OCH_3$ | -H | Ferulic acid | -OH | -$OCH_3$ | -H | -H | -H |

**Figure 1.**
*The chemical structures of the main phenolic acids.*

polymerized into larger molecules such as the proanthocyanins. Moreover, phenolic acids may arise in food plants as glycosides or esters with other natural compounds such as sterols, alcohols, glucosides and hydroxy fatty acids.

### 3.2 Flavonoids

Like as phenolic acids, flavonoids are secondary metabolites of plants with polyphenolic structure. Flavonoids consist of a 15-carbon skeleton organized by a three-carbon chain (C6–C3–C6 structure) and they are the most diverse compounds in the plant kingdom. Flavonoids can be classified in to several sub-classes including flavanols, flavones, flavones, isoflavones and anthocyanins. The most common flavonoids of rice belong to a wide variety of sub-families such as flavonols, flavones, flavanols, flavanons, and anthocyanins.

#### *3.2.1 Anthocyanins and proanthocyanidins*

Anthocyanins and proanthocyanidins are known as color pigments found in several varieties of rice as bioactive compounds. These colorful pigment bioactive compounds are located in the aleurone layer of rice grain [26]. Pigmented rice is diverse in the color, mainly due to the grain's high anthocyanin content. Several pigmented rice including black, brown, dark brown, dark purple and red-grain rice have been reported have been reported, which its color is depend upon the kinds of pigment colors [17].

##### *3.2.1.1 Anthocyanins*

Anthocyanins is a group of reddish to purple water-soluble flavonoids existing in pigmented rice and other cereal grains [27, 28]. The anthocyanidins or aglycons, the basic structure of anthocyanins, consist of an aromatic C6 (A ring) that bonded to a heterocyclic C3 (C ring) that contains oxygen, which is bonded by a carbon-carbon bond to a third aromatic C6 (B ring). When the anthocyanidins are bonded to a sugar moiety in the glycosidic linkage, they are known as anthocyanins [26]. In plants, they are found in mono, di, or tri of O-glycosides and acylglycosides of anthocyanidins [17]. Individual differences in anthocyanidins are related to the number of hydroxyl groups; the nature, number and position of sugars linked to the molecule; and the presence of aliphatic or aromatic acids attached to the sugar molecule. Anthocyanins are derived from the most common six anthocyanidins (aglycones) including cyanidin, delphinidin, petunidin, peonidin, malvidin and pelargonidin. Several anthocyanins have been isolated and identified from pigmented rice including cyaniding 3-glucoside, cyaniding 3-galactoside, cyaniding 3-rutinoside, cyaniding 3, 5-diglucoside, malvidin 3-galactoside, peonidin 3-glucoside, and pelargonidin 3, 5-diglucoside [6, 13, 28] and the basic chemical structures of the main anthocyanidins are shown in **Figure 2**. Cyanidin-3-O-glucoside has been identified in black rice as the significantly higher than others [6, 13].

##### *3.2.1.2 Proanthocyanidins*

Proanthocyanidins are a group of polymeric phenolic compounds consisting mainly of flavan-3-ol units such as afzelechin, epiafzelechin, catechin, epicatechin, gallocatechin, and epigallocatechin (**Figure 3**) [26]. More complex proanthocyanidins, having the same polymeric building block, form the group of tannins. Proanthocyanidins can be A-type or B type structure with flavan-3-ol units doubly linked by C4-C8 and C2-O7 or C4-C6 and C2-O7 for the former, and linked mainly

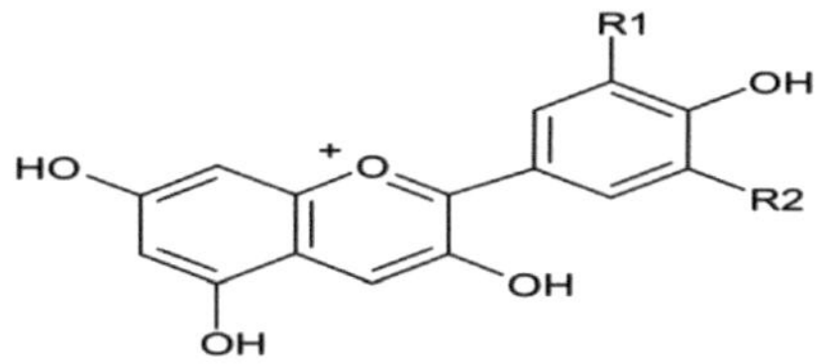


| Anthocyanidins | R1 | R2 |
|---|---|---|
| Cyanidin | -OH | -H |
| Delphinydin | -OH | -OH |
| Pelargonidin | -H | -H |
| Malvinidin | -$OCH_3$ | -$OCH_3$ |
| Peonidin | -$OCH_3$ | -H |
| Petunidin | -OH | -$OCH_3$ |

**Figure 2.**
*The chemical structures of the main anthocyanidins.*

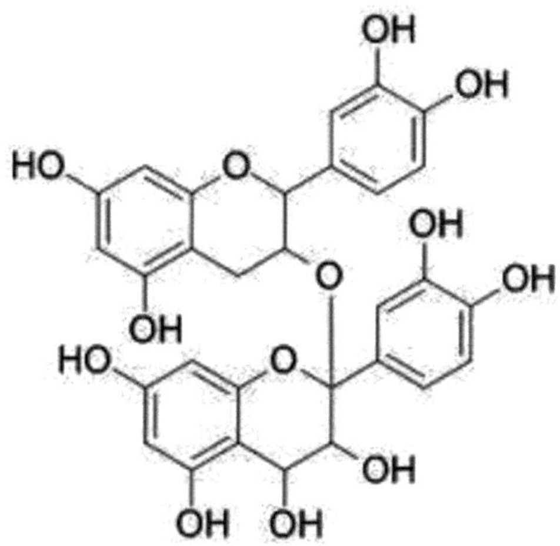


**Figure 3.**
*The chemical structures of the main proanthocyanidins.*

through C4-C8 or C4-C6 for the latter B-type proanthocyanidin is very common in nature. In red rice, the block unit of proanthocyanidin consists of catechin and epicatechin. Proanthocyanidins are synthesized in plants by using anthocyanidins as key intermediates. These pigmented compounds are also responsible for red and purple color in rice.

## 4. Extraction, identification, and quantification

The determination of phenolic compounds is a necessary prerequisite not only to define the nutritional qualities of whole grain rice, but mostly to investigate on the health benefits associated to the consumption of these food plants [29]. Therefore, the most recent techniques for the extraction of the target compounds from rice along with the analytical approaches adopted for the separation, identification and quantification of phenolic acids, flavonoids, anthocyanins, and proanthocyanidins must be fully studied. Extraction is a process used for separating bioactive compounds from solutions using specific solvents by applying standard procedures. In addition, extraction of bioactive compounds can be obtained by using either conventional or non-conventional methods [30, 31].

### 4.1 Conventional solvent extraction bioactive compounds

Conventional extraction is being used at a small-scale level to extract bioactive components from several plant materials. This technique is usually based on the extraction efficiency of different solvents, which are being used for this purpose. The manual solvent extraction at ambient temperature is the most commonly used method in extracting bioactive compounds from grains. The solvents included acidified methanol with 1.0 N HCl (85:15, v/v), acidified methanol with 1 M phosphoric acid (95:5 v/v), acidified methanol with trifluoro acetic acid (99.8:0.2, v/v), acidified methanol with glacial acetic acid (95:5, v/v), and acetone/water (80:20, v/v). The extraction ratio was a material to solvent at 1:10 (w/v) [30]. In addition, in cold conditions, methanol (85%) and HCl (1 mol/L) was found to be an appropriate extraction solvent for anthocyanins, along with 85% methanol or 70:29.5:0.5, v/v acetone:water:acetic acid for free proanthocyanidins [17].

According to Shao et al., [12] soluble-free, soluble-conjugated and insoluble-bound phenolics of white, red and black rice were extracted by using 80% methanol. The soluble phenolics mixture was extracted and concentrated to obtain soluble phenolics. In order to get soluble-free phenolics, the concentrated soluble phenolics were further extracted by ethyl acetate three times, and then dried by a rotary evaporator, and dissolved in 5 mL of 50% methanol. To get soluble-conjugated phenolics, the concentrated soluble phenolics were hydrolyzed using 4 M NaOH for 2 h followed by adjusting pH to 1.5–2.0, extraction with ethyl acetate, drying using a vacuum evaporator, and then dissolving in 5 mL of 50% methanol. After the extraction of soluble phenolics, the residues were used to extract insoluble-bound phenolics. Similarly, the soluble-conjugated phenolics could be prepared from the concentrated soluble phenolics extracts by using 4 M NaOH and ethyl acetate.

In addition, our group [13] also used solid phase extraction (SPE) techniques to purify and prepare soluble-free (unbound fraction) and soluble-conjugated (polyphenol-rich bound fraction) phenolic compounds of pigmented rice. The crude extracts of colored rice were purified by applied to $C_{18}$ solid phase extraction unit. The solid phase cartridge was pre-washed in 0.2% (v/v) formic acid in acetonitrile and then pre-equilibrated in 0.2% (v/v) formic acid in water. The unbound materials including free sugars, organic acids and vitamin C were collected. The SPE unit was then washed with a unit volume of 0.2% (v/v) aqueous formic acid and then with 2 volumes of ultra-pure water. The polyphenol-rich bound fraction was eluted with a unit volume of 80% (v/v) acetonitrile in water.

### 4.2 Non-conventional extraction techniques

The longer extraction time, costly and high purity solvent, evaporation of the huge amount of solvent, low extraction selectivity, and thermal decomposition of thermolabile compounds are major challenges of conventional extraction. These limitations of conventional extraction methods can be improved by introducing the promising techniques or non-conventional extraction techniques, for example, ultrasound-assisted extraction (UAE) and microwave-assisted extraction (MAE) [31].

#### *4.2.1 Microwave-assisted extraction (MAE)*

**Microwave-assisted extraction** has been implemented as an alternative technique for extracting anthocyanins from pigmented rice because of its ability to reduce both consumption time and solvent volume. For the MAE method, a combination of 70°C, 300 W, with 10 min in MAE was the most effective in extracting anthocyanins from blue wheat and purple corn compared with 50°C, 1200 W, and 20 min for black rice [30]. Moreover, this technique reduced the losses of the biochemical compounds being extracted.

#### *4.2.2 Ultrasonic-assisted extraction*

Ultrasound-assisted extraction (UAE) has been used in applications of food-processing technology to extract bioactive compounds from plant materials. Ultrasound at levels greater than 20 kHz is used to disrupt plant cell walls. It helps to improve the solvent's ability to penetrate the cells and obtain a higher extraction yield. The UAE operates at a low operating temperature through processing and maintain a high extract quality for compounds. Recently, Setyaningsiha et al. (2019) reported the optimization of the UAE conditions for individual phenolic compounds extraction from rice grains using 80% methanol in water for 25 min at 45°C with amplitude 47%, cycle 0.4 s – 1, pH 4.25 and sample-to solvent ratio of 1:5 [32]. The developed method presented the acceptable value for linearity and precision (RSD). Therefore, the proposed UAE method is an effective technique for the determination of individual phenolic compounds including caffeic, *p*-coumaric, syringic, chlorogenic, isovanillic, isoferulic ferulic, *p*-hydroxybenzoic, sinapic, *p*-hydroxybenzaldehyde, protocatechuic, vanillic acids, protocatechuic aldehyde and quercetin in rice samples. However, the UAE has two main negative properties mainly related to experimental repeatability and reproducibility [31].

### 4.3 Identification

After the extraction of bioactive compounds, the separation, identification and quantitation are necessary to sudied. In the past few decades, there are a huge number of published reports on HPLC analysis of extracted bioactive compounds from rice grains describe as the most widely used analytical method. Recently, Prabhakaran et al. (2019) reported the analyzed method of selected phenolic compounds in rice grains and its by-products using liquid chromatography-electrospray ionization-tandem mass spectrometry (LC-ESI-MS/MS) [33].

In addition, our group developed an identification and quantification techniques for phenolic acids and anthocyanins in pigmented rice by using UPLC-ESI-QqQ-MS/MS analysis [13]. The analysis was performed using a UPLC coupled with a mass spectrometer. The separation was carried out by UPLC HSS T3 column 1.8 μm, 2.1 × 100 mm. Column temperature was maintained at 35°C. The mobile phase consisted of 0.1% formic acid (solvent A) and 0.1% formic acid in acetonitrile (solvent B) and the flow rate was set at 0.4 mL/min. The injection volume was 2.0 μL. A stepwise gradients B (%) including an initial isocratic at 2.0% for 1 min, then linear gradient to 98% in 5 min, and by return to the initial condition of 2% B in 7 min. Therefore, the total operation time was 12 min. The solvents and extracts were previously filtered through a 0.45 μm filter membrane. Mass spectral data were obtained in positive or negative mode with a mass range between *m/z* 0 to *m/z* 500. The Multiple Reaction Monitoring (MRM) transitions and compound parameters for the target phenolic compounds were developed. Identification was confirmed by comparing *m/z* values, retention times and fragmentation patterns with those of references standards. In addition, the concentration of phenolic compounds was quantified using external standard method. Our study showed that eight target phenolic compounds were detected and identified in both the unbound and polyphenol-rich bound fractions of pigmented rice [13]. The identification of compounds was carried out by applying one quantification transition (quantifier ion) and/or one or two confirmation transitions (qualifier ions) to assess the detection and quantification specific to each compound (**Table 1**). Positive ionization mode was selected for caffeic, ferulic acids, (+)-catechin and anthocyanins while negative ionization mode was applied for *p*-coumaric acid and quercetin, due to the chemical structures of the analyses and their ionization behavior observed in ion mode.

### 4.4 Quantitation

Pigmented rice is diverse in color, mainly due to the grain's high anthocyanins content. The contents of phenolic compounds and anthocyanins are summarized in **Tables 2** and **3**, respectively, as examples of the phytochemicals that were isolated and analyzed from pigmented rice including black, red, and purple rice varieties. A range of phenolic compounds including vanillic, *p*-coumaric, protocatechuic, caffeic and ferulic acids has been detected in pigmented rice. In pigmented rice, the major phenolic acid was ferulic acid [6, 13, 25, 37]. Moreover, major flavonoids present in pigmented rice are quercetin and catechin. In addition, cyanidin-3-O-glucoside is the most predominant anthocyanins in pigmented rice. A mean of cyanidin-3-O-glucoside content in black rice was about higher than that of red rice [6, 13, 25, 37]. However, malvidin-3-O-glucoside was not observed in pigmented rice.

Our UPLC-ESI-QqQ-MS/MS analysis for phenolic acids and anthocyanins in pigmented rice showed that the retention times for the target phenolic compounds ranged between 2.9 and 4.6 min An example of a UPLC-ESI-QqQ-MS/MS chromatogram for cyanidin-3-O-glucoside for the polyphenol-rich bound fraction of black rice cv. *Hom nil* is shown in **Figure 4**. Hydroxycinnamic acids including caffeic, *p*-coumaric and ferulic acids were characterized by the loss of the carboxylic acid group (-COO; −44 *m/z*) [38]. The loss of the carboxylic acid group was observed for *p*-courmaric acid giving *m/z* at 119 [M-H-44]−, as a characteristic ion. For acids with a methoxy moiety, such as ferulic acid, the -COO loss was found by a loss of the methyl group (-CH3; −15 *m/z*, -COO; − 44 *m/z* and -CH3; −59 *m/z*). The peak with *m/z* 195 was contributed to the ferulic acid's precursor ion while the peak of product ion at *m/z* 136 was characterized to -COO and -CH3 moiety. In addition, the characterization of anthocyanin was mostly based on the loss of glucose (162 *m/z*). A molecular ion at *m/z* 449 [M + H] + corresponding with glycoside derivative of cyanidin and a major fragmentation occurring at *m/z* 287 [M + H-162] + corresponding with a cyanidin aglycone was observed in the polyphenol-rich bound fraction of black rice cv. *Hom nil* (**Figure 5(a)**). The MS/MS spectrum of pelargonidin-3-glucoside in

| Qualified compound | Retention time (min) | m/z Precursor ion | m/z Quantifier | m/z Qualifier |
|---|---|---|---|---|
| (+)-Catechin | 3.04 ± 0.01 | 291.10 | 130.01 | 123.03 |
| Caffeic acid | 3.29 ± 0.01 | 181.02 | 163.03 | 145.01 |
| *p*-Coumaric acid | 3.72 ± 0.01 | 162.95 | 118.96 | 92.95 |
| Ferulic acid | 3.87 ± 0.01 | 195.02 | 134.00 | 145.02 |
| Quercetin | 4.59 ± 0.01 | 300.90 | 178.05 | 150.92 |
| Cyanidin-3-glucoside | 2.87 ± 0.01 | 449.10 | 287.05 | 137.00 |
| Pelargonidin-3-glucoside | 3.03 ± 0.01 | 433.10 | 271.05 | 121.00 |
| Peonidin-3-glucoside | 3.10 ± 0.01 | 463.10 | 301.08 | 286.03 |

*Reference: [13].*

**Table 1.**
*Retention time and fragment of phenolic compounds.*

| Phenolic acids | Pigmented rice varieties | Extracting solvent | Analysis method | Content (mg/g) | Reference |
|---|---|---|---|---|---|
| Vanillic acid | Purple Rice | Methanol | HPLC | 0.19 ± 0.06 | [34] |
| | | 95% Ethanol | HPLC | 0.77 ± 0.02 | |
| | | Deionized water | HPLC | 1.15 ± 0.08 | |
| | Purple rice | 80% Methanol | Reversed-phase HPLC | 0.34 ± 0.02 | [35] |
| | Red rice | 80% Methanol | Reversed-phase HPLC | nd | |
| p-Coumaric acid | Black rice cv. Hom nil | Acidified methanol | UPLC-ESI-QqQ-MS/MS | 0.05 ± 0.01 | [13] |
| | Black rice cv. Rice berry | Acidified methanol | UPLC-ESI-QqQ-MS/MS | 0.04 ± 0.00 | |
| | Purple rice cv. Khoa kum | Acidified methanol | UPLC-ESI-QqQ-MS/MS | 0.02 ± 0.00 | |
| | Red rice cv. Hom deang | Acidified methanol | UPLC-ESI-QqQ-MS/MS | 0.06 ± 0.00 | |
| | Purple rice | 80% Methanol | Reversed-phase HPLC | 0.31 ± 0.02 | [34] |
| | Red rice | 80% Methanol | Reversed-phase HPLC | 0.01 ± 0.20 | |
| Protocatechuic acid | Purple Rice | Methanol | HPLC | 0.24 ± 0.05 | [34] |
| | | 95% Ethanol | HPLC | 1.02 ± 0.08 | |
| | | Deionized water | HPLC | 1.26 ± 0.01 | |
| Protocatechuic acid | Purple rice | 80% Methanol | Reversed-phase HPLC | 1.60 ± 0.02 | [35] |
| | Red rice | 80% Methanol | Reversed-phase HPLC | 0.31 ± 0.02 | |
| Caffeic acid | Black rice cv. Hom nil | Acidified methanol | UPLC-ESI-QqQ-MS/MS | nd | [13] |
| | Black rice cv. Rice berry | Acidified methanol | UPLC-ESI-QqQ-MS/MS | nd | |
| | Purple rice cv. Khoa kum | Acidified methanol | UPLC-ESI-QqQ-MS/MS | nd | |
| | Red rice cv. Hom deang | Acidified methanol | UPLC-ESI-QqQ-MS/MS | 0.02 ± 0.00 | |

(Continued)

| Phenolic acids | Pigmented rice varieties | Extracting solvent | Analysis method | Content (mg/g) | Reference |
|---|---|---|---|---|---|
| | Purple rice | 80% Methanol | Reversed-phase HPLC | 0.98 ± 0.32 | [35] |
| | Red rice | 80% Methanol | Reversed-phase HPLC | nd | |
| Ferulic acid | Black rice cv. Hom nil | Acidified methanol | UPLC-ESI-QqQ-MS/MS | 0.25 ± 0.01 | [13] |
| | Black rice cv. Rice berry | Acidified methanol | UPLC-ESI-QqQ-MS/MS | 0.40 ± 0.03 | |
| | Purple rice cv. Khoa kum | Acidified methanol | UPLC-ESI-QqQ-MS/MS | 0.24 ± 0.02 | |
| | Red rice cv. Hom deang | Acidified methanol | UPLC-ESI-QqQ-MS/MS | 0.57 ± 0.02 | |
| | Purple rice | 80% Methanol | Reversed-phase HPLC | 1.03 ± 0.02 | [12] |
| | Red rice | 80% Methanol | Reversed-phase HPLC | 0.34 ± 0.05 | |
| Catechin | Purple Rice | Methanol | HPLC | 0.09 ± 0.01 | [34] |
| | | 95% Ethanol | HPLC | 1.73 ± 0.05 | |
| | | Deionized water | HPLC | 0.37 ± 0.09 | |
| | Black rice cv. Hom nil | Acidified methanol | UPLC-ESI-QqQ-MS/MS | 0.12 ± 0.01 | [13] |
| | Black rice cv. Rice berry | Acidified methanol | UPLC-ESI-QqQ-MS/MS | 0.10 ± 0.01 | |
| | Purple rice cv. Khoa kum | Acidified methanol | UPLC-ESI-QqQ-MS/MS | 0.01 ± 0.00 | |
| | Red rice cv. Hom deang | Acidified methanol | UPLC-ESI-QqQ-MS/MS | 0.91 ± 0.08 | |
| Quercetin | Purple rice | 80% Methanol | Reversed-phase HPLC | 0.29 ± 0.22 | [35] |
| | Red rice | 80% Methanol | | n.d | |
| | Black rice cv. Hom nil | Acidified methanol | UPLC-ESI-QqQ-MS/MS | 0.07 ± 0.03 | [13] |
| | Black rice cv. Rice berry | Acidified methanol | UPLC-ESI-QqQ-MS/MS | 0.08 ± 0.00 | |
| | Purple rice cv. Khoa kum | Acidified methanol | UPLC-ESI-QqQ-MS/MS | 0.10 ± 0.01 | |

*n.d: not detectable.*

**Table 2.**
*Phenolic compound contents in pigmented rice.*

| Anthocyanins | Pigmented rice varieties | Extracting solvent | Analysis method HPLC | Content (mg/g) | Reference |
|---|---|---|---|---|---|
| Cyanidin 3-glucoside | Black rice cv. Niaw Dam Pleuak Khao | Acidified methanol | HPLC | 137.41 ± 16.66 | [6] |
| | Black rice cv. Niaw Dam Pleuak Dam | Acidified methanol | HPLC | 19.39 ± 0.09 | |
| | Purple Rice | Methanol | HPLC | 21.09 ± 1.09 | [34] |
| | | Ethanol | HPLC | 8.40 ± 0.00 | |
| | | Deionized water | HPLC | n.d | |
| | Thai black rice | Acidified methanol | HPLC | 142 ± 1.5 | [36] |
| | Thai Jasmine red rice | Acidified methanol | UPLC-ESI-QqQ-MS/MS | n.d | |
| | Black rice cv. Hom nil | Acidified methanol | UPLC-ESI-QqQ-MS/MS | 132.96 ± 393 | [13] |
| | Black rice cv. Rice berry | Acidified methanol | UPLC-ESI-QqQ-MS/MS | 47.80 ± 0.44 | |
| | Purple rice cv. Khoa kum | Acidified methanol | UPLC-ESI-QqQ-MS/MS | 11.75 ± 0.32 | |
| | Red rice cv. Hom deang | Acidified methanol | UPLC-ESI-QqQ-MS/MS | 0.06 ± 0.00 | |
| Pelargonidin 3-glucoside | Black rice cv. Hom nil | Acidified methanol | UPLC-ESI-QqQ-MS/MS | 0.23 ± 0.01 | [13] |
| | Black rice cv. Rice berry | | UPLC-ESI-QqQ-MS/MS | 0.07 ± 0.00 | |
| | Purple rice cv. Khoa kum | | UPLC-ESI-QqQ-MS/MS | 0.03 ± 0.00 | |
| | Red rice cv. Hom deang | | | n.d | |
| Malvidin 3-glucoside | Purple Rice | Methanol | HPLC | n.d | [34] |
| | | Ethanol | HPLC | n.d | |
| | | Deionized water | HPLC | n.d | |

(Continued)

| Anthocyanins | Pigmented rice varieties | Extracting solvent | Analysis method HPLC | Content (mg/g) | Reference |
|---|---|---|---|---|---|
| Peonidin-3-glucoside | Black rice cv. Niaw Dam Pleuak Khao | Acidified methanol | HPLC | 11.07 ± 0.97 | [6] |
| | Black rice cv. Niaw Dam Pleuak Dam | Acidified methanol | HPLC | 12.75 ± 0.51 | |
| | Black rice cv. Hom nil | Acidified methanol | UPLC-ESI-QqQ-MS/MS | 19.56 ± 0.39 | [13] |
| | Black rice cv. Rice berry | Acidified methanol | UPLC-ESI-QqQ-MS/MS | 6.94 ± 0.26 | |
| | Purple rice cv. Khoa kum | Acidified methanol | UPLC-ESI-QqQ-MS/MS | 5.29 ± 0.10 | |
| | Red rice cv. Hom deang | Acidified methanol | UPLC-ESI-QqQ-MS/MS | 0.01 ± 0.0 | |
| | Thai black rice | Acidified methanol | HPLC | 98 ± 0.5 | [36] |
| | | | | n.d | |
| | Thai Jasmine red rice | Acidified methanol | HPLC | n.d | |

*n.d: not detectable.*

**Table 3.**
*Anthocyanins contents in pigmented rice.*

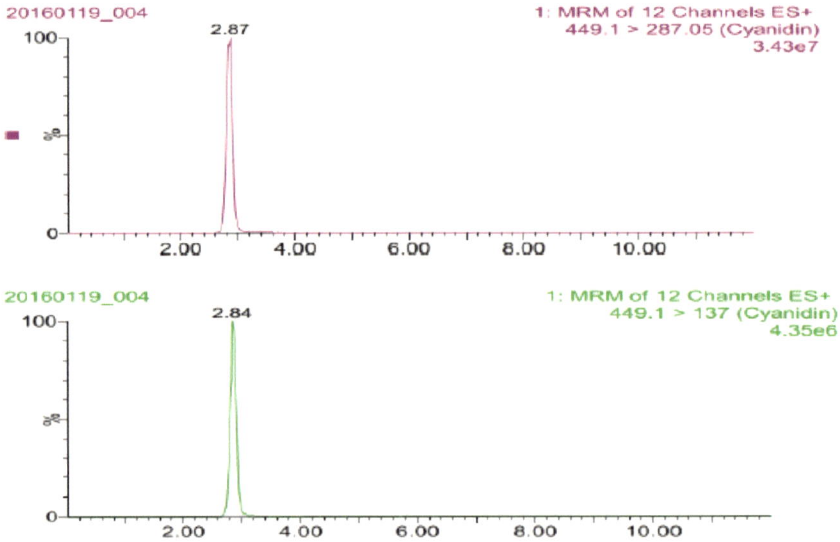


**Figure 4.**
*Chromatogram of cyanidin 3-O-glucoside [13].*

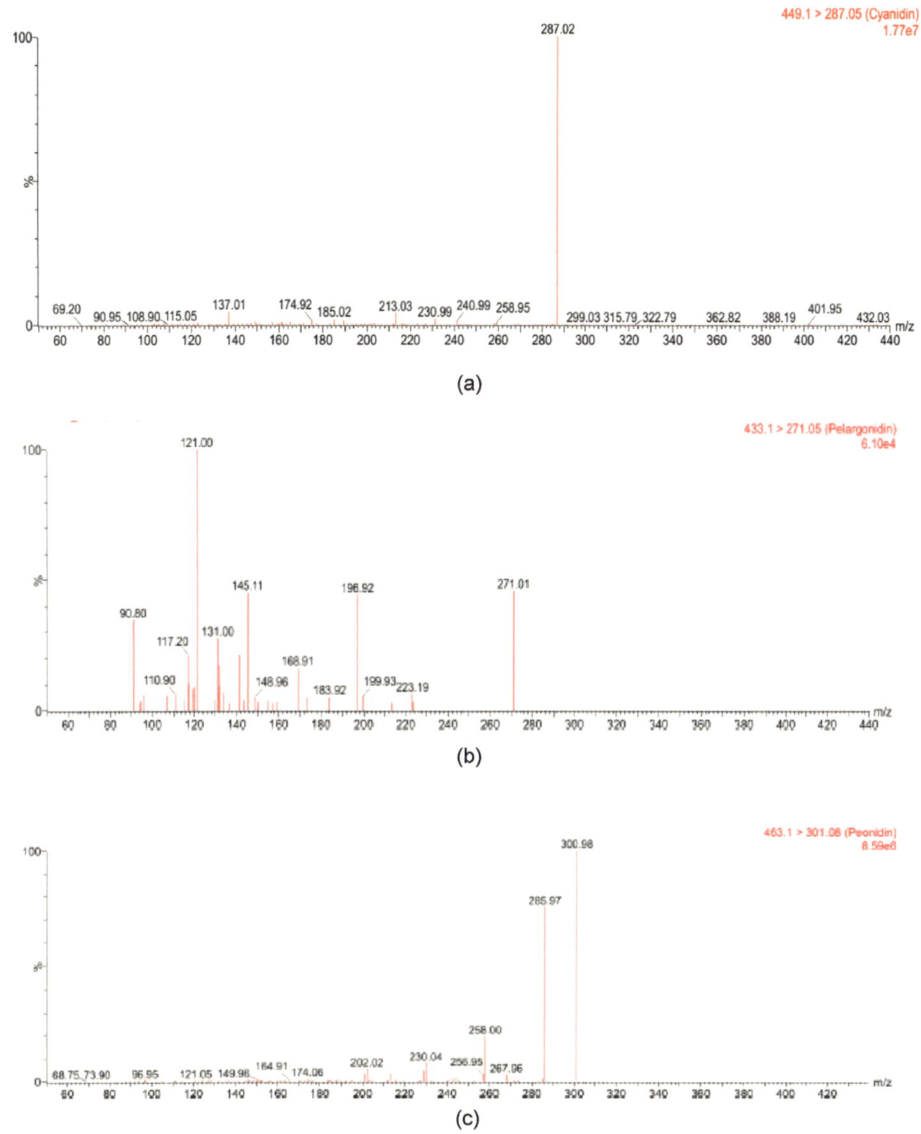


**Figure 5.**
*(a) MS/MS spectrum of cyanidin 3-O-glucoside. (b) MS/MS spectrum of pelargonidin-3-O-glucoside. (c) MS/MS spectrum of peonidin 3-O-glucoside [13].*

the positive ionization mode (**Figure 5(b)**) of the polyphenol rich bound fraction of black rice cv. *Hom nil* showed protonated molecular ions at *m/z* 433 [M + H]+. The product ion of aglycone at *m/z* 271 [M + H-162] + corresponds to the loss of a pelargonidin moiety. The positive ion mass spectrum of the polyphenol-rich bound fraction of black rice cv. *Hom nil* (**Figure 5(c)**) showed its molecular ion at *m/z* 463 [M + H] + and a major fragment at *m/z* 301 [M + H-162]+. These data indicated the presence of a peonidin aglycone and a hexose moiety.

Proanthocyanidins are high molecular weight of flavan-3-ol polymers that consist of catechin, epicatechin, gallocatechin, and epigallocatechin units that can also be found in rice germ and bran, particularly in pigmented rice. Proanthocyanidins were only detected in red and black rice, and their contents were dependent on the genotypes which differed from 15.57 mg/kg to 1417.12 mg/kg, respectively. In red rice, the majority of proanthocyanidins are oligomers of 5–8 mers (40%), whereas the polymers (DP > 10) accounted for 29% [12]. Moreover, total proanthocyanidins of the traditional Sri Lankan red-grained rice varieties ranged from 11.95 to 24.70 mg/g in bran samples and from 1.07 to 2.27 mg/g in brown rice samples [39].

## 5. Health benefit potentials of pigmented rice

Phenolic compounds have many biological activities, for instance, antioxidant [40, 41] and anti-inflammatory properties [42]. Several epidemiological studies suggested that high dietary consumption of polyphenols is associated with decreased risk of a range of diseases including antidiabetic [43, 44], cardiovascular disease (CVD) [45], and neurodegenerative diseases [46].

### 5.1 Antioxidant activity

Phenolic acids and flavonoids function as reducing agents, free radical scavengers, and quenchers of singlet oxygen formation. The antioxidant property of the phenolic ring depends on the number and location of the hydroxy groups [47]. Pigmented rice varieties showed high antioxidant capacity along with the highest flavonoid and polyphenol content [42]. Flavonoids and phenolic acids components play important roles in the control of oxidative stress, which is considered to be substantial in an initiation and development of many current conditions and diseases including inflammation, hyperglycemia, hypertension, heart diseases, neurodegenerative diseases, cancer and in the aging process. However, antioxidants are substances that significantly delay or prevent the oxidation of an oxidisable substrate when present in low concentrations compared to the substrate.

### 5.2 Antidiabetic activity

Hyperglycaemia, or high blood sugar, is one of the most common disorders. Antidiabetic activity of phenolic acids and flavonoids has been reported by several studies. It has been reported that these anthocyanins have various beneficial effects, including inhibitory effects against carbohydrate hydrolysing enzymes, including in α-amylase and α-glucosidase. Our study on α-amylose inhibitory action of cyanidin-3-glucoside, ranged from 0.1 mg/100 g DM in the red rice sample to 138.8 mg/100 g DM in the black rice sample, showed that the IC50 value for potential inhibition against α-amylase activity of crude extract of colored rice samples ranged from 2.5 μg/mL to 14.4 μg/mL. The potential inhibition against α-amylase activity in black rice samples were higher than those in red and purple rice samples, respectively [13]. In addition, the percentage inhibition on α-glucosidase and α-amylase of fragrant black rice fractions had higher potential than acarbose, a synthetic antidiabetic drug [48]. However, variation in the structure of anthocyanins makes it difficult to determine their effects on Type II diabetes. Understanding the absorption and metabolism of anthocyanins is important for understanding their role in the improvement of this disease. Previous literature suggested that anthocyanins may lower blood glucose by improving insulin resistance, protecting β cells, increasing secretion of insulin and reducing digestion of sugars in the small intestine. The mechanisms of action are primarily related to their antioxidant properties, but enzymatic inhibition and other pathways may also be relevant [49].

### 5.3 Anti-inflammatory potential

Overproduction of free radicals and/or oxidants can cause oxidative stress and oxidative damage to biological macromolecules including lipid, protein and nucleic acids, which are associated with chronic diseases such as cancer, cardiovascular disease and neurological disease [42]. In addition, free radicals including superoxide anion, (NO) and peroxynitrite play important roles in the inflammatory process. According to the study carried out by Chalermpong et al. (2012) on the anti-inflammatory activities of bioactive rich extracts from Thai glutinous purple rice bran [50], the extracts of five purple rice bran sample showed a strong anti-inflammatory activity through inhibitory effect on nitric/oxide (NO) production in combined LPS-IFN-γ-activated RAW 264.7 murine macrophage cells. Moreover, the results also indicated that γ-oryzanol rich extracts from Thai glutinous purple rice bran were acting as a lipophilic radical scavenger better than a hydrophilic radical scavenger. Moreover, these extracts exert a potent anti-inflammatory activity.

### 5.4 Cardiovascular disease risk inhibition potential

Cardiovascular disease (CVD) is a chronic disease related to a range of genetic and environmental factors including smoking, high saturated fat diets and physical inactivity. Moreover, coronary heart disease and stroke are a major cause of mortality in developed countries Dyslipidemia is an abnormal high level of dietary lipids such as plasma cholesterol, triglycerides, or a low level of high-density lipoproteins (HDL). In addition, dyslipidemia often increases the risk factor for atherosclerosis, which further increases the risk for developing coronary heart diseases that are the leading cause of death among the aged population [51]. Several studies indicated a correlation between the intake of flavonols, flavones and flavanols and a reduced risk of coronary artery disease and anthocyanin and flavanone intake and reduced CVD related mortality. Lo et al., (2016) demonstrated the effects of the newly bred Korean non-germinated and germinated pigmented rice cultivars on ovariectomized (OVX) rats in comparison with the nonpigmented normal brown rice [45]. All rat groups were fed with 20% rice per total diet and showed that the supplementation of germinated rice for some groups, particularly pigmented rice cultivars, resulted in better a lipid profile compared to the groups that consumed non-germinated rice cultivars. In addition, germination increased the quantities of the bioactive compounds that are responsible for the hypolipidemic activities of these rice grains. Showed a low total cholesterol levels, high levels of high-density lipoproteins-cholesterol, high fecal lipid output, low hepatic lipid values, and low hepatic adipocyte accumulation. There were also an increase in the rate of lipolysis and decrease in lipogenesis based on the lipid-regulating enzyme activity profiles obtained for the groups that fed on germinated rice. Also, results revealed that pigmented rice cultivars had superior effects in improving the lipid metabolism relative to the non-pigmented normal brown rice variety. The application of germinated rice, blackish-purple cultivars enhanced potential for the prevention and occurrence of dyslipidemias.

### 5.5 Anti-neurodegenerative potential

Neurodegenerative disorders including Parkinson's and Alzheimer's diseases have been increasing in our aging societies. Flavonoids may act to protect the brain in a number of ways, including by protection of vulnerable neurons, the enhancement of existing neuronal function or by stimulating neuronal regeneration. Vargas et al. (2018) reported the of studies using neuron-like cells, such as the SH-SY5Y cell line, allowed the evaluation of the antioxidant activity of compounds in human cells with biochemical characteristics similar to neurons, indicated the extract of bioactive compounds in red and black rice brans extracts was able to prevent $H_2O_2$-induced oxidative damage in SH-SY5Y cells [46].

## 6. Conclusion

Pigmented rice especially black rick contains a wide range of biologically active compounds for example cyanidin O-glucoside that can be used in a various in functional foods and in nutraceuticals applications. The advancement of isolation and quail-quantification techniques provides more opportunities for *in-vitro* and *in vivo* studies on potential health benefits, given by pigmented rice and its by-products. In addition, most of the studies on the biological properties of black or red rice bioactive compounds have been conducted through an *in vitro* approach; however, more preclinical or in animal studies should be considered. Further investigations will be needed to study evidence on the efficacy of bioactive compounds including phenolic acids and anthocyanins in pigmented rice in terms of antagonistic, additive/non-interactive, or synergistic effect on potential health benefits.

## Acknowledgements

The author would like to thank Mae Fah Luang University for a financial support of this work.

## Conflict of interest

The author declares no conflict of interest.

## Author details

Prinya Wongsa
Food Science and Technology Program, School of Agro-Industry, Mae Fah Luang University, Chiang Rai, Thailand

*Address all correspondence to: prinya@mfu.ac.th

## References

[1] FAO. World food situation. 2020. Available at http://www.fao.org/worldfoodsituation/csdb/en/accessed on 5 May 2020.

[2] FAOSTAT. Crop production. 2020. Available at http://www.fao.org/faostat/en/#data/QC/ accessed on 5 May 2020.

[3] Juliano, BO. Rice chemistry and quality. 2003. Philippines, Cap: Island Publishing House; p 25-49.

[4] Burlando, B. Cornara, L. Therapeutic properties of rice constituents and derivatives (*Oryza sativa* L.): A review update. Trends in Food Science & Technology. 2014;**40**(1): 82-98.

[5] Juliano BO. Criteria and test for rice grain qualities; in Rice Chemistry and Technology, 2nd Edition. 1985. American Association of Cereal Chemists; p 7-57.

[6] Sompong, R, Siebenhandl-Ehn, S, Linsberger-Martin, G, Berghofer, E. Physicochemical and antioxidative properties of red and black rice varieties from Thailand, China and Sri Lanka. Food Chemistry. 2011;**124**(1): 132-140.

[7] Wongsa P, Landberg R. Rattanapanone, N. Chemical compositions and metabolites profiling of rice varieties from Chiang Rai province Thailand. Chiang Mai Journal of Science. 2018; **45**(7): 2703-2714.

[8] Ito VC, Lacerda LG. Black rice (*Oryza sativa* L.): A review of its historical aspects, chemical composition, nutritional and functional properties, and applications and processing technologies. Food Chemistry. 2019; **301**: 125304

[9] IRRI. Knowledge bank 2020. Available at http://www.knowledgebank.irri.org accessed on 1 August 2020.

[10] Huang YP, Lai HM. Bioactive compounds and antioxidative activity of colored rice bran. Journal of Food Drug Analysis. 2016; **24**: 564-574.

[11] Deng G-F, Xu X-R, Zhang, Y, Li, D, Gan, R-Y, Li, H-B. Phenolic compounds and bioactivities of pigmented rice. Critical Reviews in Food Science and Nutrition. 2013; **53**: 296-306.

[12] Shao Y, Hu Z, Yu Y, Mou R, Zhu Z, Beta T. Phenolic acids, anthocyanins, proanthocyanidins, antioxidant activity, minerals and their correlations in non-pigmented, red, and black rice. Food Chemistry. 2018; **239**: 733-741.

[13] Wongsa, P., Chaiwarith, J., Voranitikul, J., Chaiwongkhajorn, J., Rattanapanone, N. & Lanberg, R. Identification of phenolic compounds in colored rice and their inhibitory potential against α-amylase. Chiang Mai Journal of Science. 2019; **46**(4), 672-682.

[14] Kushwaha UKS. Black rice. Research history and development. 2016; Springer International Publishing, p 21-31.

[15] Pratiwi R, Purwestri YA. Black rice as a functional food in Indonesia. Functional Foods in Health and Disease. 2017; **7**(3): 182-194.

[16] Prasad, BJ Sharavanan, PS, Sivaraj R. Health benefits of black rice – A review. Grain & Oil Science and Technology. 2019; **2**: 109-113.

[17] Verma DK, Srivastav PP. Bioactive compounds of rice (*Oryza sativa* L.): Review on paradigm and its potential benefit in human health. Trends in Food Science & Technology. 2020; **97**: 355-365.

[18] Liu RH. Health-promoting components of fruits and vegetables in

the diet. Advance Nutrition. 2013; **4**: 384S–392S.

[19] Dykes L, Rooney LW. Phenolic compounds in cereal grains and their health benefits. Cereal Foods world. 2013; 105-110. doi: 10.1.1094/CFW-52-3-0.105.

[20] Cory H, Passarelli S, Szeto J, Tamez M, Mattei J. The role of polyphenols in human health and food systems: A mini-rreview. Frontiers in Nutrition. 2018; **5**(87): doi: 10.3389/fnut.2018.00087.

[21] Altemimi A, Lakhssassi N, Baharlouei A, Watson DG, Lightfoot DA. Phytochemicals: Extraction, isolation, and identification of bioactive compounds from plant extracts. Plants. 2017; **6**:42; doi:10.3390/plants6040042

[22] Gani A, Wani SM, Masoodi FA, Hameed G. Whole-grain cereal: Bioactive compounds and their health benefits: A Review. Journal of Food Process and Technology. 2012; **3**:146, doi:10.4172/2157-7110.1000146.

[23] Ghasemzadeh A, Ghasemzadeh N. Flavonoids and phenolic acids: Role and biochemical activity in plants and human. Journal of Medicinal Plants Research. 2011; **5**(31): 6697-6703.

[24] Kabera JN, Semana E, Mussa AR, He X. Plant secondary metabolites: Biosynthesis, classification, function and pharmacological properties. Journal of Pharmacy and Pharmacology. 2014; 2: 377-392.

[25] Shao Y, Xu F, Sun F, Bao J, Beta T. Identification and quantification of phenolic acids and anthocyanins as antioxidants in bran, embryo and endosperm of white, red and black rice kernels (*Oryza sativa* L.) 2014. Journal of Cereal Science; **59**:211-218.

[26] Finocchiaro, F., Ferrari, F. & Gianinetti, A. A study of biodiversity of flavonoid content in the rice caryopsis evidencing simultaneous accumulation of anthocyanins and proanthocyanidins in a black-grained genotype. Journal of Cereal Science. 2010; **51**(1): 28-34.

[27] Mazza G, Gao L. Blue and purple grains. In: Abdel-Aal E-SM, Wood PJ, editors. Specialty grains for food and feed. 2005. St. Paul, MN: American Association of Cereal Chemists, Inc. p. 313-50.

[28] Abdel-Aal E-SM, Young JC, Rabalski I. Anthocyanin composition in black, blue, purple and red cereal grains. Journal of Agricultural and Food Chemistry. 2006; 54: 4696-704.

[29] Ciulu M, de la Luz Cádiz-Gurrea M, Segura-Carretero, A. Extraction and analysis of phenolic compounds in rice: A review. Molecules.2018; **23**: 2890.

[30] Abdel-Aal E-S.M, Akhtar H, Rabalski I, Bryan. M. Accelerated, microwave-assisted, and conventional solvent extraction methods affect anthocyanin composition from colored grains. Journal of Food Science. 2014; **79**(2). C138-C146.

[31] Selvamuthukumaran M, Shi. J. Recent advances in extraction of antioxidants from plant by-products processing industries. Food Quality and Safety.2017; **1**: 61-81.

[32] Setyaningsih W, Saputro IE, Carrera, CA, Palma M. Optimisation of an ultrasound-assisted extraction method for the simultaneous determination of phenolics in rice grains. Food Chemistry. 2019; **288**: 221-227.

[33] Prabhakaran M, Chung, I-M, Son, N-Y, Chi H-Y, Kim, S-Y, Yang Y-J, Kwon C, An, Y-J, Ahmad A, Kim S-H. Analysis of selected phenolic compounds in organic, pesticide-free, conventional rice (*Oryza*

*sativa* L.) using LC-ESI MS/MS. Molecules. 2019; **24**: 67; doi:10.3390/molecules24010067.

[34] Punvittayagul C, Sringarm K, Chaiyasut C, Wongpoomchai R. 2014. Mutagenicity and antimutagenicity of hydrophilic and lipophilic extracts of Thai Northern purple rice. Asian Pacific Journal of Cancer Prevention.2014; **15** (21): 9517-9522.

[35] Owolabi IO, Saibandith B, Wichienchot S, Yupanqui CT. Nutritional compositions, polyphenolic profiles and antioxidant properties of pigmented rice varieties and adlay seeds enhanced by soaking and germination conditions. Functional Foods in Health and Disease. 2018;**8**:(12): 561-578.

[36] Melini V, Acquistucci R. 2017. Health-promoting compounds in pigmented Thai and wild rice. Foods.2017; **6**(9): 1-13.

[37] Hao J, Zhu H, Zhnag Z, Yang S, Li H. Identification of anthocyanins in black rice (*Oryza sativa* L.) by UPLC/Q-TOF-MS and their in vitro and in vivo antioxidant activities. Journal of Cereal Science, 2015; **64**: 92-99.

[38] Lambert M, Meudee E, Verbaere A, Mazerrolles G, Wirth J, Masson G, Cheynier V, Sommerer N. A High-throughput UHPLC-QqQ-MS method for polyphenol profiling in Rosé wines. Molecules, 2015; 20: 7890-7914. DOI 10.3390/molecules20057890.

[39] Gunaratne A, Wu K, Li D, Bentota A, Corke H, Cai Y-Z. Antioxidant activity and nutritional quality of traditional red-grained rice varieties containing proanthocyanidins. Food Chemistry. 2013;**138**: 1153-1161.

[40] Hao J, Zhu H, Zhang Z, Yang S, Li H. Identification of anthocyanins in black rice (*Oryza sativa* L.) by UPLC/QTOF-MS and their *in vitro* and *in vivo* antioxidant activities. Journal of Cereal Science. 2015; **64**: 92-99.

[41] Houa F, Zhanga, R, Zhanga M, Sua D, Wei Z, Denga Y, Zhanga Y, Chia J, Tanga X. Hepatoprotective and antioxidant activity of anthocyanins in black rice bran on carbon tetrachloride-induced liver injury in mice. Journal of Functional Foods. 2013: **5**: 1705-1711.

[42] Petroni K, Landoni M, Tomay F, Calvenzani V, Simonelli C, Cormegna M. Proximate composition, polyphenol content and anti-inflammatory properties of white and pigmented Italian rice varieties. Universal Journal of Agricultural Research. 2017; **5**(5): 312-321.

[43] Sansenya S, Nanok, K. α-glucosidase, α-amylase inhibitory potential and antioxidant activity of agrant black rice (Thai coloured rice). Flavour and Fragrance Journal. 2020;**35**(4):376-383.

[44] Boue SM, Daigle KW, Chen M-H, Cao H, Heiman ML. Antidiabetic potential of purple and red rice (*Oryza sativa* L.) bran extracts. Journal of Agricultural and Food Chemistry. 2016; **64:** 5345-535.

[45] Lo LMP, Kang MY, Yi SJ, Chung SI. Dietary supplementation of germinated pigmented rice (*Oryza sativa* L.) lowers dyslipidemia risk in ovariectomized Sprague–Dawley rats. Food & Nutrition Research.2016; **60**:1: 30092, DOI: 10.3402/fnr.v60.30092

[46] Vargas CG, da Silva Junior JD, Rabelo TK, Moreira JCF, Gelain DP, Rodrigues E, Paul Augusti R, de Oliveira Riosa A, Flôres SH. Bioactive compounds and protective effect of red and black rice brans extracts in human neuron-like cells (SH-SY5Y). 2018. Food Research International; **113**: 57-64.

[47] Sroka, Z, Cisowski W. Hydrogen peroxide scavenging, antioxidant

and anti-radical activity of some phenolic acids. Food and Chemical Toxicology.2003; **41**: 753-758.

[48] Sansenya S, Nanok K. α-glucosidase, α-amylase inhibitory potential and antioxidant activity of fragrant black rice (Thai coloured rice). Flavour and Fragrance Journa.2020;**35**(4):376-386.

[49] Sancho RAS, Pastore GMP. Evaluation of the effects of anthocyanins in type 2 diabetes. 2012; F0od Research International **46**: 378-386.

[50] Saenjum C, Chaiyasut C, Chansakaow S, Suttaj M, Sirithunyalug B. Antioxidant and anti-inflammatory activities of gamma-oryzanol rich extracts from Thai purple rice bran. Journal of Medicinal Plants Research.2012; **6**(6):1070-1077.

[51] Banjerdpongchai R, Wudtiwai B, Sringarm K. Cytotoxic and apoptotic-inducing effects of purple rice extracts and chemotherapeutic drugs on human cancer cell lines. Asian Pacific Journal of Cancer Prevention. 2013;**14** (11): 6541-6548.

Chapter 5

# Computational Analysis of Rice Transcriptomic and Genomic Datasets in Search for SNPs Involved in Flavonoid Biosynthesis

*Rabiatul-Adawiah Zainal-Abidin and Zeti-Azura Mohamed-Hussein*


## Abstract

This chapter describes the computational approach used in analyzing rice transcriptomics and genomics data to identify and annotate potential single nucleotide polymorphism (SNPs) as potential biomarker in the production of flavonoid. SNPs play a role in the accumulation of nutritional components (e.g. antioxidants), and flavonoid is one of them. However, the number of identified SNPs associated with flavonoid nutritional trait is still limited. We develop a knowledge-based bioinformatic workflow to search for specific SNPs and integration analysis on the SNPs and their co-expressed genes to investigate their influence on the gain/loss of functional genes that are involved in the production of flavonoids. Raw files obtained from the functional genomics studies can be analyzed in details to obtain a useful biological insight. Different tools, algorithms and databases are available to analyze the ontology, metabolic and pathway at the molecular level in order to observe the effects of gene and protein expression. The usage of different tools, algorithms and databases allows the integration, interpretation and the inference of analysis to provide better understanding of the biological meaning of the resutls. This chapter illustrates how to select and bring together several software to develop a specific bioinformatic workflow that processes and analyses omics data. The implementation of this bioinformatic workflow revealed the identification of potential flavonoid biosynthetic genes that can be used as guided-gene to screen the single nucleotide polymorphisms (SNPs) in the flavonoid biosynthetic genes from genome and transcriptomics data.


**Keywords:** SNPs, comparative genomics, transcriptome, nutritional traits, colored rice, integrative analysis, bioinformatics, flavonoid, genomics, single nucleotide polymorphism, transcriptomics, bioinformatics workflow

## 1. Introduction

In recent years, high-throughput omics technologies (i.e. genomics, proteomics, transcriptomics and/or metabolomics) provide unprecedented opportunities to discover potential genes, proteins, metabolites, pathways and molecular markers for various applications. The availability of omics data produced from high-throughput omics technologies has facilitated the molecular and genetic improvement of rice varieties with higher yield, quality, nutrient dan resistance to the biotic and abiotic stresses. However, past 15 years have seen the significant increase of omics data in volume and types. This event has challenged the researchers to extract and decipher invaluable information encoded in the data. This challenge can be addressed with the use of bioinformatics in analyzing, integrating and interpreting these massive omics data. There are various bioinformatic tools and algorithms that can be used by the researchers to process, interpret and integrate data in a more efficient and reproducible way. However, lack of connectivity between various tools and algorithms complicates the process of extracting and deciphering the data. Hence there is the need to find procedures to connect these tools to develop workflows that can connect and bring together different techniques that are able to exploit data at many levels.

Here we describe a computational workflow composed of different bioinformatic tools that exploits data from large-scale gene expression experiments and contextualize them at many biological levels. To illustrate the relevance of our workflow, we applied it to data from rice varieties datasets in search for potential SNPs that are associated with flavonoid biosynthetic genes. The workflow started with identification of known flavonoid biosynthetic genes from published articles and database search using several genome and pathway databases such as. KEGG, PlantReactome, RiceCyc) and similarity search analysis.

The potential flavonoid biosynthetic genes were used as a guide-gene to screen for single nucleotide polymorphisms (SNPs) in the flavonoid biosynthetic genes from the genomics and transcriptomics data. Integration of SNP and co-expressed genes was performed via network analysis. The transcriptomics data was used to construct the gene co-expression network followed by the mapping of SNPs onto it. A pathway-network analysis was performed to interpret the biological information related to the flavonoid pathway-network. All information generated from these computational analyses are stored in *MyNutRice*Base (http://www.mynutricebase.org) for knowledge sharing and future use in the functional genomics study and the development of molecular markers.

## 2. Overview of flavonoids in omics and rice breeding improvement

Several types of rice breeding traits such as disease resistance, drought tolerance, salinity tolerance, grain quality, high nutritional content are currently pursue by the breeders and farmers in their effort to improve the traits of rice varieties. However, nowadays breeding for improved rice varieties with high nutritional content has attracted interest among breeders, geneticists and nutritionists. High protein [1], carotenoid [2], micronutrients [3] and antioxidant content [4] are among the preferred nutritional contents improved in rice varieties. Rice breeding for high nutritional content has been carried out extensively by bio-fortification [5, 6] and/or genetic engineering [7]. The improvement of nutritional content in rice is essential due to their benefits for human health. Lack of specific nutrition can lead to several diseases and malnutrition.

Flavonoids are secondary metabolites commonly produced in flowers, fruits, vegetables and pigmented rice. Flavonoids are known as a potent antioxidant beneficial for human health. Consumption of foods with high antioxidant contents may lower the risk of cardiovascular disease, type II diabetes and colon cancer [8]. Additionally, flavonoid has shown to improve plant resistance against abiotic and biotic stress [9].

The flavonoid biosynthetic pathways are found in several crops such as maize, tomato and rice [10]. Genes encoding enzymes involved in the flavonoid biosynthetic pathways are categorized into general phenylpropanoid, early biosynthetic genes (EBGs), late biosynthetic genes (LBGs) and transcription factors [11]. General phenylpropanoid contains three major genes such as phenylalanine ammonia-lyase (PAL), cinnamic acid 4-hydroxylase (C4H) and 4-coumarate CoA ligase (4CL). The EBGs include chalcone synthase (CHS), chalcone isomerase (CHI), flavanone 3-hydroxylase (F3H) and flavanone 3′-hydroxylase (F3'H) [11]. Genes in the general phenylpropanoid category and EBGs are upstream genes that initiate the flavonoid biosynthetic pathways and responsible in the production of secondary metabolites [11]. The LBGs such as dihydroflavonol reductase (DFR), leucoanthocyanidin reductase (LAR), UDP-glucose flavonoid 3-O-glucosyl transferase (UGT) and leucoanthocyanidin oxidase (LDOX) lead to the production of anthocyanin [11]. Genes categorized in EBGs and LBGs are also recognized as structural genes [11].

Four transcription factors involved in the flavonoid biosynthetic pathways are *Kala4* (Os04g0557500), *Rc* (Os07g0211500), *R2R3-MYB* (Os06g0205100) and *WD40* (Os02g0682500). *Kala4* encodes a basic-helix–loop–helix (bHLH) and it activates late biosynthetic genes to produce black pigmentation or anthocyanin in the seed pericarps [12]. *Rc* gene encodes for bHLH that regulates the *Rd* (Os01g0633500) expression to produce red pigmentation [13]. However, *Rc* expression without the participation of *Rd* has resulted to brown pigmentation [13]. Besides the existence of *R2R3-MYB* and *WD40* in the flavonoid biosynthesis, they also responsible to regulate the pigmentation in purple leaves [14, 15].

Anthocyanin and proanthocyanidins are two major flavonoid compounds [11]. Pigmented rice (black and red) is enriched with antioxidant due to the presence of anthocyanin and proanthocyanidin [16]. Understanding the genetic basis of pigmented rice varieties is essential to develop rice varieties with high antioxidant (flavonoid, anthocyanin, proanthocyanidin) content. A study by [17] has dissected the regulation of flavonoid biosynthesis in edible rice tissue using a metabolic engineering approach. Bioinformatics approach was used to identify key flavonoid structural genes in the rice genome by species comparison against maize and sorghum as sequence homologs [17]. A total of six genes encoding enzymes (CHS, CHI, F3H, F3'H, DFR and ANS) in the flavonoid biosynthetic pathway were selected for tBLASTN analysis against Nipponbare rice genome sequence. At least 66% amino acid identities were found in the rice genome sequences. The expression patterns of six flavonoid genes were analyzed to investigate the accumulation of flavonoids level in rice seedlings.

Previous study showed the correlation between sequence polymorphism and metabolite profiling affects the flavonoid accumumation between *indica* and *japonica* rice sub-species [18]. Different accumulation of flavonoids in these two rice sub-species might due to the variation in flavonoid biosynthetic genes [19, 20]. A quantitative trait loci analysis was performed to develop molecular markers associated with antioxidant content in rice [21]. The potential molecular markers associated with antioxidant content can be applied in the marker-assisted breeding towards the improvement of high-level antioxidant content in rice variety.

Understanding gene-based-SNP underlying flavonoid biosynthesis process is crucial in developing rice cultivars with higher flavonoid contents. Integration of single nucleotide polymorphisms (SNPs) and co-expressed genes can be used to identify causal SNPs and to prioritize the functional SNPs involved with flavonoid biosynthetic genes.

However, the study on molecular and genetic improvement of antioxidant content in rice via integration of multi-omics data is still limited. Current data is still inadequate to be applied in rice breeding selection. Furthermore, this constraint limits detailed insight into the underlying mechanism and regulation of antioxidant content on the system level. The association of SNPs in the causative biosynthetic genes might be useful to uncover key alleles that influence the accumulation of flavonoid. Linking the SNPs with their co-expressed genes that involved in the flavonoid biosynthesis process could be a promising approach to prioritize the functional SNPs and causal genes to be used in the experimental validation towards the molecular and genetic improvement of rice variety enriched with flavonoid content.

## 3. Data analysis workflow

### 3.1 Data mining of potential genes and transcription factor related to flavonoid biosynthesis

Pubmed (https://pubmed.ncbi.nlm.nih.gov/), Web of Science (https://www.webofknowledge.com/) and Scopus (https://www.scopus.com/) were used systematically to search for publications related to flavonoid biosynthetic genes in *Oryza sativa* using the following keywords and their combinations such as "flavonoid", "rice", "SNP", "anthocyanin", "proanthocyanidin" and "pigmented". The keywords such as "flavonoid", "C4H", "4CL", "CHS", "F3H", "F3'H", "UGT", "DFR", "CHI", "PAL", "LAR", "ANS", "ANR", "LDOX", anthocyanin" and "proanthocyanidin" were used in the genome (i.e. RAPDB) and pathway (i.e. KEGG, RiceCyc, PlantReactome) databases to search for gene identifier, gene sequences and gene descriptions.

All genes that are related to the flavonoid genes were identified using similarity search analysis from OmicsBox (https://www.biobam.com/omicsbox/). BLASTX [22] analysis was performed with the e-value cut off of less than 1e-10 and sequence identity more than 75%. eggNOG [23] analysis was carried out to characterize whether the sequences are paralogous or orthologous.

Bibliomic, database and similarity search are the first steps in identifying candidate genes related to the biosynthesis pathways. These genes that can be used in downstream analysis1 such as evolutionary study [24] and identification of SNPs in the biosynthetic genes [25]. From these steps, a total of 95 flavonoid related genes (FRGs) and two transcription factors were successfully identified. Structural genes (i.e. F3H, CHI, CHS,DFR, LDOX) are found to be more conserved than transcription factors. These flavonoid related genes were used to screen the SNPs that reside in the sequences. **Figure 1** shows the three steps analysis workflow to identify the potential flavonoid related genes.

### 3.2 Mining of SNPs from genome and transcriptome data

Single nucleotide polymorphism (SNP) has been widely used as a genetic marker tool in crop improvement. Previous studies used SNPs to investigate the evolutionary relationship [26], to facilitate cultivar identification [27] and to associate SNPs with agronomic traits [28]. SNPs located in the intergenic and genic region [29]. The genic SNPs can be classified into the coding region, untranslated region (UTR) and intron. SNPs in the coding region are divided into synonymous and non-synonymous. The non-synonymous SNPs has two effects such as deleterious and tolerance that might represent causal genetic variation which could lead to the phenotypic consequences [30].

SNPs in the coding regions are usually associated with functional SNP and they can influence the phenotypic expression of agronomic traits such as

*DOI: http://dx.doi.org/10.5772/intechopen.94876*

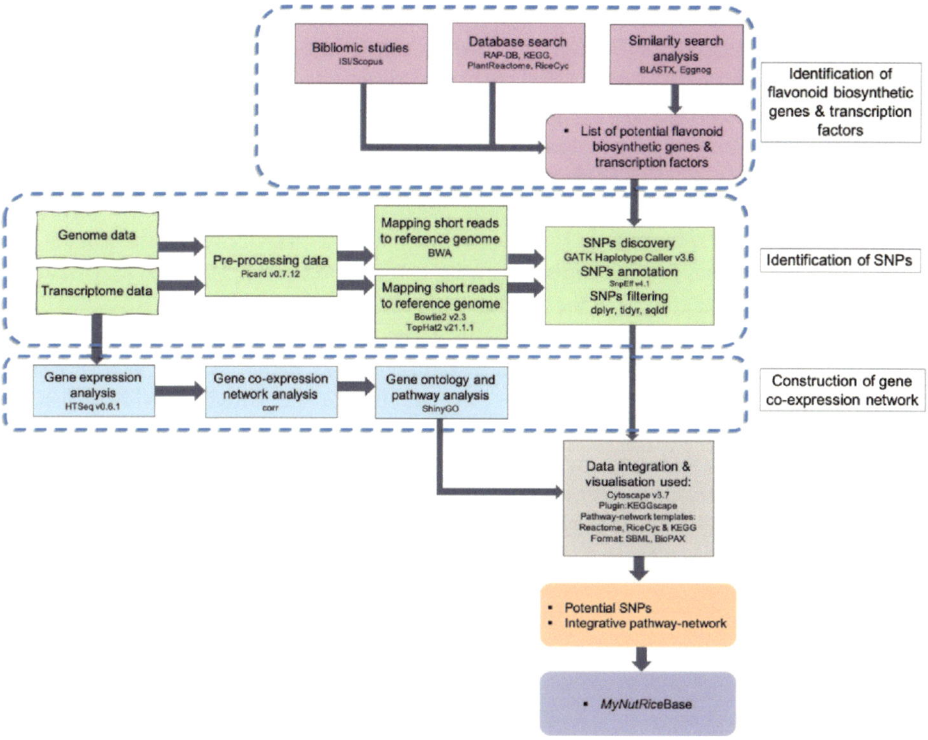


**Figure 1.**
*Bioinformatic analysis workflow on the genome and transcriptomic datasets of six Malaysian rice varieties.*

resistance to disease, resistance to drought and high nutritional content [31]. The non-synonymous SNPs might affect the protein function through amino acid substitution [32] whilst synonymous SNP has the ability to affect gene function by regulating mRNA splicing [33], stability [34] and protein translation [35]. However, synonymous SNP is not a preferred polymorphism to be further validated. Therefore, the utilization of non-synonymous SNPs is important to prioritize their involvement in the biosynthetic pathway and to determine the effect of non-synonymous SNPs to phenotypic expression.

SNPs in the intron region are able to shift the gene splicing or regulate the transcript level by changing the binding sites of miRNA [36]. Lower number of SNPs in untranslated region (UTR) has demonstrated that mutation in this region can change local mRNA structure and will affect the translation process [36]. SNPs in the UTR region are conserved and consist of binding sites for proteins or antisense RNAs that able to modulate transport, RNA stability, cellular localization, expression level and translation [37].

Functional effect of SNPs in the flavonoid related genes can affect the expression and regulation of structural genes and transcription factor during the flavonoid biosynthesis process. Previous study has demonstrated that genetic variation caused by SNP can influenced the variation and accumulation of metabolites in the biosynthetic pathway [19, 20]. By selecting SNPs with specific functional effect such as non-synonymous and deleterious, it can be effectively used as a molecular marker in the development of new and improved rice variety with agronomic traits of interest.

However, it is a challenge to depend only on SNPs to reveal the interaction and relationship that occur in the multiple levels of biological mechanisms [38], especially in the qualitative and complex traits (i.e. abiotic stress, quality, yield).

Therefore, numerous studies have been performed to identify and evaluate the functional effects of SNPs through multi-omics data integration. For instance, the integration of whole genome and transcriptome data offers opportunities to highlight the expressed SNPs [39] and to better understand the biological processes and mechanisms underlying the pigmented rice varieties. Previous study has performed integration of SNPs with gene co-expression network to identify the causal genes and causal SNPs that might be responsible in the mechanism of blast disease resistance [40], salt tolerance [41] and amylose content [42]. Throughout these integration processes, several bioinformatics tools have been applied to identify the putative SNPs and to annotate the SNPs into their functional effects.

The identification of SNPs in the flavonoid related genes started with mining of SNPs from the genome and transcriptome datasets of *O. sativa indica* cv. Bali, PH9, MRM16, MRQ100, MR297 and MRQ76 [43, 44]. Bali, PH9, MRM16 and MRQ100 are pigmented rice varieties that contain high antioxidant contents. MR297 and MRQ76 are white rice varieties with medium tolerance to diseases. **Figure 1** shows the bioinformatic workflow to identify SNPs from the genome and transcriptome datasets. The first step in SNPs identification is to map the sequence reads onto selected reference genome sequences; in this case it was Nipponbare. It was chosen because of it was well-assembled and annotated. The genome and transcriptome of reads mapping were individually aligned using different mapping tools such as BWA [45] for genome reads mapping and TopHat2 [46] for transcriptome reads mapping.

Then PICARD version 0.7.12 [47] was used to add and replace the read groups, marking the duplicate reads and fixing mate information on the mapped reads in order to obtain the high-quality SNPs during SNPs calling. This process is known as post-processing and it is a standard step used to identify potential SNPs from the genome and transcriptome datasets. Once this process is completed, GATK version 3.6 [47] was used in the SNPs calling process and this is a crucial step in SNPs discovery as it helps in obtaining high-quality SNPs as well as reducing false-positive SNPs. In our case, the parameters used are as follow:

i. mapping quality (MQ) more than and equal to 30 (MQ > = 30);

ii. variant quality more than and equal to 50 (VQ > =50);

iii. number of supporting reads for every base (DP) more than and equal to 10 (DP > = 10)

All SNPs obtained from the filtering process were annotated using several annotation tools such as SnpEff version 4.1 [48], Variant Effect Predictor (VEP) [49], CooVar [50] and Annovar [51] for their intergenic, genic, coding region, UTR, intron, synonymous and non-synonymous. Most of these tools can be locally installed to ease the users who perform large-number of SNPs annotation. Those SNPs that were annotated as genic were then filtered using R packages (i.e. dplyr, tidyr, sqldf) for the identification of SNP position, chromosome, allele, gene identifier and SNPs effect. This information was then used in matching the SNPs in genome and transcriptome datasets as well as screening all SNPs that reside in the flavonoid related genes.

Comparative SNPs analysis of six rice varieties were performed to investigate the uniqueness, differences and similarities in the potential SNPs. Currently several rice SNP databases are available to providing the information on SNPs that were mined from various rice varieties and sub-specie such as Rice SNP-Seek [52],

Rice VarMap [53], IC4R [54], and Ensembl Plants Variation database [55]. Comparative SNPs analysis on SNP databases can provide the SNPs information and usability in various rice varieties and sub-species.

SNP occurs in transcription factor often lead to altering or loss of function of key pathway enzymes that are required to regulating the production of anthocyanin [56]. Hence, it could affect the expression levels of flavonoid. Previous research has investigated that mutations were accumulated during the domestication process, which suggests the presence of agronomically valuable genes in landraces as well as in wild relative [15]. Hence, SNPs in transcription factor, such as *Kala*4 and *Rc,* can be prioritized for further integration with genes co-expression network.

### 3.3 Gene co-expression network analysis from transcriptome data

A gene co-expression network analysis was performed to correlate gene and phenotypic expressions, to infer the function of unknown genes and to identify the key regulatory networks in biosynthetic processes [57, 58]. The principle used in the gene co-expression network shows that genes that cluster in the same network represent similar biological process [58]. To date, the increasing number of genes co-expression network databases such as ATTED, STRING, RED facilitate the exploitation of co-expressed genes from different conditions in various crops [59]. In the co-expressed gene databases, the gene identifier or gene name can be used as a query to search for the co-expressed genes of interest. Parameter, such as maximum number of interactors = 0.1 and confidence score cut-off = 0.40 are used to select the significant co-expressed genes. Most of the co-expressed databases obtained the information from microarray gene expression and transcriptome datasets from public databases such as NCBI Gene Expression Omnibus (NCBI GEO) [60] and Sequence Read Archive (SRA) [61].

Gene co-expression network analysis was performed using the expression values of genes in fragments per kilobase of transcript per million mapped reads (FPKM) as an input data. Pearson correlation coefficient (PCC) value was used to measure the co-expression correlation between paired genes in the network. The PCC score represents the confidence level describing the association of the two genes whether they are functionally associated. In this case, the FPKM value with more than and equal to 0.1 (FPKM > = 0.1) was used to perform a gene co-expression network analysis. Pearson Correlation Coefficient (PCC) in corr R package [62] was used to measure the correlation between paired genes in the network. The cut-off value of PCC more than and equal to 0.7 (PCC > = 0.7) was used for selection of co-expressed genes. The flavonoid related genes were selected as guided-genes to identify their interactors. **Figure 1** shows the analysis workflow to perform a gene co-expression network analysis.

Cytoscape v3.7 was used to construct and visualize the gene co-expression network. The Network Analyzer plugin was used to calculate the degree connectivity of each nodes and edges. ShinyGO version 0.6.1 [63] was used in gene ontology and pathway enrichment analysis to elucidate the biological processes and molecular function in clusters of network.

## 4. Data integration

### 4.1 Integration of SNPs with co-expressed genes

There are several publications on the bioinformatics approaches on the functional effect of SNPs where by assessing their effect on the functional site of protein structure [64, 65] and integrate with biological pathway [66]. In this study, SNP, especially with non-synonymous effect and deleterious impact is better appreciated to study its impact through the pathway-network analysis coupled with a different omics data type.

SNPs can be integrated with omics data type to rank the high-potential SNPs and to highlight the causal genes for development of genetic markers and functional genomics studies. The integration of SNPs and co-expressed genes through construction of biological network offers a comprehensive interpretation of genetic variation at the biological system level. Integrative approach links biologically meaningful sets of genes to reveal the molecular basis of trait variation.

In this study, integration of SNPs and co-expressed genes was performed to prioritize the functional SNPs that can be suggested as a candidate for the development of molecular markers and to highlight the causal genes that might contribute in the flavonoid biosynthesis process. The co-expressed genes can be suggested as a functional target for functional validation in the future study to unravel their potential in rice breeding improvement.

Integration of SNPs and co-expressed genes was performed using Cytoscape v3.7 [67]. Flavonoid biosynthetic pathway templates in BioPAX, KGXML and SIF formats was retrieved from KEGG, Plant Reactome and RiceCyc databases. To merge the SNP and co-expressed genes, gene identifier (ID) was used as a matching identity. Every gene in the network consists of gene ID, chromosome, start and stop position based on physical genome coordinates (bp). Similarly, SNPs consists of the position in genome locations (bp) and gene ID.

Analysis on network integration between SNPs and co-expressed genes highlight that co-expressed genes can be integrated by multiple numbers of SNPs and will reveal, which SNPs appear to play an essential role in the flavonoid biosynthesis process. The co-expressed genes connected to SNPs can be prioritized as candidate genes. The high false-positive rate in SNPs also can be reduced by incorporating putative functional co-expressed genes information [68]. Several biological questions can be asked from the integration of SNPs and co-expressed genes into pathway-network, such as i) how many functional SNPs and co-expressed genes important to the expression of black and red pigmentation; ii) any regulatory genes regulate the biosynthetic pathway?; iii) and which biological process are underlying this trait.

### 4.2 Pathway-network analysis

There are different ways in bioinformatics approaches that could be applied to the pathway-network analysis. In pathway-network analysis, the description of connected genes can be interpreted into biologically meaningful information and provide insights into biological processes, molecular function and cellular components. This analysis is known as gene ontology (GO) enrichment and pathway enrichment analysis.

To date, several bioinformatic tools are available to perform GO and pathway enrichment analysis in the network. For example, ClueGO [69] and BiNGO [70]. Both plugins are available in the Cytoscape and are user-friendly. Statistical values such as Hypergeometric testing and Bonferroni method is used to calculate the p-value. Parameter such as p-value less and equal than 0.05 (p-value <= 005) and a minimum number of mapping entries > = 2 can be used to select the significant or enrich genes in the pathway-network.

**Tables 1** and 2 provide list of bioinformatics tools and databases that are used for data mining and data integration in search for potential SNPs involved in the flavonoid biosynthesis.

| Bioinformatic tools | Description | References |
|---|---|---|
| BLASTX | Sequence similarity search analysis | [22] |
| eggNog | Annotation of orthologous and paralogous sequences | [23] |
| BWA | Genome mapping | [45] |
| TopHat2 v2.3 | Transcriptome mapping | [46] |
| Picard v0.7.12 | Post mapping processes | [47] |
| GATK v3.6 (HaplotypeCaller) | SNPs discovery | [47] |
| SnpEff v4.1 | SNPs annotation | [48] |
| R package (corrr) | Gene co-expression analysis | [62] |
| ShinyGO | Gene ontology enrichment analysis | [63] |
| Cytoscape v3.7 | Data integration | [66] |
| Cytoscape v3.7 plugin ClueGO | Gene ontology enrichment analysis | [68] |
| R packages (dplyr, tidyr, sqldf) | Data cleaning and filtering | [71] [72] [73] |

**Table 1.**
*Summary of the bioinformatic tools used in search for potential SNPs involved in the flavonoid biosynthetic pathways.*

| Bioinformatic databases | Descriptions | URL | References |
|---|---|---|---|
| Rice Annotation Project Database (RAP-DB) | Is a rice genome database that has been developed by the International Rice Genome Sequencing. Information provided are genome sequences, chromosom, gene annotation dan description. | https://rapdb.dna.affrc.go.jp/ | [74] |
| Kyoto Encyclopedia of Genes and Genomes (KEGG) | A pathway database that provides biological information related to genes, proteins, enzymes and pathways involve in biological systems. | https://www.kegg.jp/ | [75] |
| PlantReactome | A plant pathway database that provides user genes, proteins, enzymes and reactions involve in the specific biological systems. | https://plantreactome.gramene.org/ | [76] |
| RiceCyc | A rice metabolic pathway database that has been developed to provide predicted biochemical pathways in rice. Several biological information, such as genes, proteins, enzymes and reactions have been displayed in diagram and all the data can be downloaded by user. | http://pathway.gramene.org/gramene/ricecyc.shtml | [77] |

**Table 2.**
*Summary of the biological databases used in search for potential SNPs involved in the flavonoid biosynthetic pathways.*

## 5. Repository of omics data

Continuously increased number of biological data is in need for a database to systematically store, organize and manage them. To date, several rice databases are available for the application in rice breeding programme, such as rice genome

**Figure 2.**
*The homepage of* MyNutRice*Base (http://www.mynutricebase.org).*

database [74], SNP databases [53, 54, 78] and pathway databases [75, 76]. Each of these databases have their uniqueness and specific target users.

In this study, the flavonoid related gene, co-expressed gene and SNP are stored in a one-stop database that is specifically developed as a genetics and genomics repository to keep all information related to nutritional traits in rice known as *MyNutRice*Base (http://www.mynutricebase.org) (**Figure 2**). It provides a platform for data mining of SNPs and genes, data visualization and sequence similarity search analysis. *MyNutRice*Base aims to accelerate the genomics and genetic analysis by enabling the rice geneticist and breeders to mine and export the biological information for the application in rice breeding improvement.

## 6. Conclusions

To date, several bioinformatics tools are available for the researcher to use and connect in an analysis pipeline. Criteria and parameter are the essential part that always carefully revised and look into while performing the bioinformatics analysis. This review highlights bioinformatics workflow used in the identification of SNPs in genomic and transcriptomic data, gene co-expression network analysis, omics data integration. This result facilitates the interpretation of SNPs into comprehensive biological information with the identification of potential. By integrating these two types of data, it may shed some light on the roles of various genes and SNPs that may play an essential role in the accumulation of flavonoid content in rice.

## Acknowledgements

This work was carried out at the Centre for Bioinformatics Research, Institute of Systems Biology (INBIOSIS), Universiti Kebangsaan Malaysia and Malaysian Agricultural Research & Development Institute (MARDI). The open access publishing fees are funded by GP-2020-K007217 and GP-2019-K021204 grants.

## Conflict of interest

The authors declare no conflict of interest.

## Author details

Rabiatul-Adawiah Zainal-Abidin[1,2] and Zeti Azura Mohamed-Hussein[1,3]*

1 Institute of Systems Biology (INBIOSIS), Universiti Kebangsaan Malaysia, Bangi, Selangor, Malaysia

2 Malaysian Agricultural Research and Development Institute (MARDI), Serdang, Malaysia

3 Department of Applied Physics, Faculty of Science and Technology, Universiti Kebangsaan Malaysia, Bangi, Selangor, Malaysia

*Address all correspondence to: zeti.hussein@ukm.edu.my

## References

[1] Wang L, Zhong M, Li X, Yuan D, Xu Y, Liu H, et al. The QTL controlling amino acid content in grains of rice (Oryza sativa) are co-localized with the regions involved in the amino acid metabolism pathway. Mol Breed. 2008;21(1):127-37. DOI: 10.1007/s11032-007-9141-7.

[2] Ye X, Salim A-B, Kloti A, Zhang J, Lucca P, Beyer P, et al. Engineering the Provitamin A (Beta-Carotene) Biosynthetic Pathway into (Carotenoid-Free) Rice Endosperm. Science (80). 2000;287:303-5. DOI: 10.1126/science.287.5451.303.

[3] Stangoulis JCR, Huynh BL, Welch RM, Choi EY, Graham RD. Quantitative trait loci for phytate in rice grain and their relationship with grain micronutrient content. Euphytica. 2007;154(3):289-94. DOI: 10.1007/s10681-006-9211-7.

[4] Maeda H, Yamaguchi T, Omoteno M, Takarada T, Fujita K, Murata K, et al. Genetic dissection of black grain rice by the development of a near isogenic line. Breed Sci. 2014;64(2):134-41. DOI: 10.1270/jsbbs.64.134.

[5] Aung MS, Masuda H, Kobayashi T, Nakanishi H, Yamakawa T, Nishizawa NK. Iron Biofortification of Myanmar Rice. Front Plant Sci. 2013;4:1-14. DOI: 10.3389/fpls.2013.00158.

[6] Nachimuthu VV, Robin S, Sudhakar D, Rajeswari S, Raveendran M, Subramanian KS, et al. Genotypic variation for micronutrient content in traditional and improved rice lines and its role in biofortification programme. Indian J Sci Technol. 2014;7(9):1414-25. DOI: 10.1.1.1028.9732.

[7] Anukul N, Ramos RA, Mehrshahi P, Castelazo AS, Parker H, Diévart A, et al. Folate polyglutamylation is required for rice seed development. Rice. 2010;3(2-3):181-93. DOI: 10.1007/s12284-010-9040-0.

[8] Dipti SS, Bergman C, Indrasari SD, Herath T, Hall R. The potential of rice to offer solutions for malnutrition and chronic diseases. Rice. 2012;5:1-18. DOI: 10.1186/1939-8433-5-16.

[9] Petrussa E, Braidot E, Zancani M, Peresson C, Bertolini A, Patui S, et al. Plant flavonoids-biosynthesis, transport and involvement in stress responses. Int J Mol Sci. 2013;14(7):14950-73. DOI: 10.3390/ijms140714950.

[10] Tohge T, De Souza LP, Fernie AR. Current understanding of the pathways of flavonoid biosynthesis in model and crop plants. J Exp Bot. 2017;68(15):4013-28. DOI: 10.1093/jxb/erx177.

[11] Lepiniec L, Debeaujon I, Routaboul J-M, Baudry A, Pourcel L, Nesi N, et al. Genetics and Biochemistry of Seed Flavonoids. Annu Rev Plant Biol. 2006;57(1):405-30. DOI: 10.1146/annurev.arplant.57.032905.105252.

[12] Oikawa T, Maeda H, Oguchi T, Yamaguchi T, Tanabe N, Ebana K, et al. The birth of a black rice gene and its local spread by introgression. Plant Cell. 2015; 27(9):2401-2414. DOI: 10.1105/tpc.15.00310.

[13] Sweeney MT, Thomson MJ, Pfeil BE, McCouch S. Caught red-handed: Rc encodes a basic helix-loop-helix protein conditioning red pericarp in rice. Plant Cell. 2006;18(2):283-94. DOI: 10.1105/tpc.105.038430.

[14] Sakamoto W, Ohmori T, Kageyama K, Miyazaki C, Saito A, Murata M, et al. The Purple leaf (Pl) locus of rice: the Pl(w) allele has a complex organization and includes

two genes encoding basic helix-loop-helix proteins involved in anthocyanin biosynthesis. Plant Cell Physiol. 2001;42(9):982-91. DOI: 10.1093/pcp/pce128.

[15] Saitoh K, Onishi K, Mikami I, Thidar K, Sano Y. Allelic diversification at the C (*OsC1*) locus of wild and cultivated rice: Nucleotide changes associated with phenotypes. Genetics. 2004;168(2):997-1007. DOI: 10.1534/genetics.103.018390.

[16] Goufo P, Trindade H. Rice antioxidants: phenolic acids, flavonoids, anthocyanins, proanthocyanidins, tocopherols, tocotrienols, γ -oryzanol, and phytic acid. Food Sci Nutr. 2014;2(2):75-104. DOI: 10.1002/fsn3.86.

[17] Shih CH, Chu H, Tang LK, Sakamoto W, Maekawa M, Chu IK, et al. Functional characterization of key structural genes in rice flavonoid biosynthesis. Planta. 2008;228(6):1043-54. DOI: 10.1007/s00425-008-0806-1.

[18] Dong X, Chen W, Wang W, Zhang H, Liu X, Luo J. Comprehensive profiling and natural variation of flavonoids in rice. J Integr Plant Biol. 2014;56(9):876-86. DOI: 10.1111/jipb.12204.

[19] Clotault J, Peltier D, Soufflet-Freslon V, Briad M, Geoffriau E. Differential selection on carotenoid biosynthesis genes as a function of gene position in the metabolic pathway: A study on the carrot and dicots. PLoS One. 2012;7(6):1-13.

[20] Manrique-Carpintero NC, Tokuhisa JG, Ginzberg I, Holliday J a, Veilleux RE. Sequence diversity in coding regions of candidate genes in the glycoalkaloid biosynthetic pathway of wild potato species. Genes|Genomes|Genetics. 2013;3(9):1467-79. 10.1371/journal.pone.0038724.

[21] Jin L, Xiao P, Lu Y, Shao Y, Shen Y, Bao J. Quantitative Trait Loci for Brown Rice Color , Phenolics , Flavonoid Contents , and Antioxidant Capacity in Rice Grain. Cereal Chem. 2009;86(6):609-15. DOI: 10.1007/s00122-010-1505-4.

[22] Altschul SF, Gish W, Miller W, Myers EW, Lipman DJ. Basic local alignment search tool. J Mol Biol. 1990;215(3):3-10.DOI:10.1016/S0022-2836(05)80360-2.

[23] Huerta-Cepas J, Szklarczyk D, Heller D, Hernández-Plaza A, Forslund SK, Cook H, et al. EggNOG 5.0: A hierarchical, functionally and phylogenetically annotated orthology resource based on 5090 organisms and 2502 viruses. Nucleic Acids Res. 2018;47(1): 9-14. DOI: 10.1093/nar/gky1085.

[24] Qu C, Zhao H, Fu F, Wang Z, Zhang K, Zhou Y, et al. Genome-Wide Survey of Flavonoid Biosynthesis Genes and Gene Expression Analysis between Black- and Yellow-Seeded *Brassica napus*. Front Plant Sci. 2016;7(1). DOI:10.3389/fpls.2016.01755.

[25] Dequigiovanni G, Ritschel PS, Maia JDG, Quecini V. In silico SNP detection for anthocyanin metabolism genes in Vitis. Acta Hortic. 2014;1046:341-8. DOI: 10.17660/ActaHortic.2014.1046.46.

[26] Singh N, Singh B, Rai V, Sidhu S, Singh AK, Singh NK. Evolutionary Insights Based on SNP Haplotypes of Red Pericarp, Grain Size and Starch Synthase Genes in Wild and Cultivated Rice. Front Plant Sci. 2017;8(1):1-9. DOI:10.3389/fpls.2017.00972.

[27] McNally KL, Childs KL, Bohnert R, Davidson RM, Zhao K, Ulat VJ, et al. Genomewide SNP variation reveals relationships among landraces and modern varieties of rice. Proc Natl Acad Sci U S A. 2009;106(30):12273-8. DOI: 10.1073/pnas.0900992106.

[28] Zhao K, Tung C, Eizenga GC, Wright MH, Ali ML, Price AH, et al. Genome-wide association mapping reveals a rich genetic architecture of complex traits in *Oryza sativa*. Nat Commun. 2011;2(467):1-10. DOI: 10.1038/ncomms1467.

[29] Edwards D, Forster JW, Chagné D, Batley J. What are SNPs? In: Association mapping in plants. New York: Springer; 2005. p. 41-52. DOI: 10.1007/978-0-387-36011-9_3.

[30] Kono TJY, Fu F, Mohammadi M, Hoffman PJ, Liu C, Stupar RM, et al. The Role of Deleterious Substitutions in Crop Genomes. Mol Biol Evol. 2016;33(9):2307-17. DOI: 10.1093/molbev/msw102.

[31] Huq A, Akter S, Nou S, Kim HT, Jung YJ, Kang KK. Identification of functional SNPs in genes and their effects on plant phenotypes. J Plant Biotechnol. 2016;43(1):1-11. DOI: 10.5010/JPB.2016.43.1.1.

[32] Günther T, Schmid KJ. Deleterious amino acid polymorphisms in *Arabidopsis thaliana* and rice. Theor Appl Genet. 2010;121:157-68. DOI: 10.1007/s00122-010-1299-4.

[33] Pagani F, Raponi M, Baralle FE. Synonymous mutations in CFTR exon 12 affect splicing and are not neutral in evolution. Proc Natl Acad Sci. 2005;102(18):6368-72. DOI: 10.1073/pnas.0502288102.

[34] Kimchi-Sarfaty C, Oh JM, Kim IW, Sauna ZE, Calcagno AM, Ambudkar S V., et al. A "silent" polymorphism in the MDR1 gene changes substrate specificity. Science (80). 2007;315(5811):525-8. DOI: 10.1126/science.1135308.

[35] Sauna ZE, Kimchi-Sarfaty C. Understanding the contribution of synonymous mutations to human disease. Nat Rev Genet. 2011;12(10):683-91. DOI: 10.1038/nrg3051.

[36] Tatarinova T V, Chekalin E, Nikolsky Y, Bruskin S, Chebotarov D, McNally KL, et al. Nucleotide diversity analysis highlights functionally important genomic regions. Sci Rep. 2016;6(1):1-12. DOI: 10.1038/srep35730.

[37] Shabalina SA, Spiridonov NA, Kashina A. Sounds of silence: Synonymous nucleotides as a key to biological regulation and complexity. Nucleic Acids Res. 2013;41(4):2073-94. DOI: 10.1093/nar/gks1205.

[38] Yuan JS, Galbraith DW, Dai SY, Griffin P, Stewart CN. Plant systems biology comes of age. Trends Plant Sci. 2008;13(4):165-71. DOI: 10.1016/j.tplants.2008.02.003.

[39] Seol Y, Won SY, Shin Y, Lee J, Chun J, Kim Y, et al. A Multilayered Screening Method for the Identification of Regulatory Genes in Rice by Agronomic Traits. Evol Bioinforma. 2016;12:253-62. DOI: 10.4137/EBO.S40622.

[40] Ficklin SP, Feltus FA. A Systems-Genetics Approach and Data Mining Tool to Assist in the Discovery of Genes Underlying Complex Traits in *Oryza sativa*. PLoS One. 2013;8(7):1-7. DOI: 10.1371/journal.pone.0068551.

[41] Wang J, Chen L, Wang Y, Zhang J, Liang Y, Xu D. A Computational Systems Biology Study for Understanding Salt Tolerance Mechanism in Rice. PLoS One. 2013;8(6):1-12. DOI: 10.1371/journal.pone.0064929.

[42] Butardo VM, Anacleto R, Parween S, Samson I, de Guzman K, Alhambra CM, et al. Systems Genetics Identifies a Novel Regulatory Domain of Amylose Synthesis. Plant Physiol. 2017;173(1):887-906. DOI:10.1104/pp.16.01248.

[43] Zainal-Abidin R-A, Zainal Z, Mohamed-Hussein Z-A, Abu-Bakar N, Ab Razak S, Simoh S, et al. RNA-seq data from whole rice grains of pigmented and non-pigmented Malaysian rice varieties. Data Br. 2020;30:105432. DOI: 10.1016/j.dib.2020.105432.

[44] Zainal-Abidin R-A, Zainal Z, Mohamed-Hussein Z-A, Sew YS, Simoh S, Ab Razak S, et al. Data on genome resequencing of pigmented and non-pigmented Malaysian rice varieties. Data Br. 2020;31:105806. DOI: 10.1016/j.dib.2020.105806.

[45] Li H, Durbin R. Fast and accurate short read alignment with Burrows-Wheeler transform. Bioinformatics. 2009;25(14):1754-60. DOI: 10.1093/bioinformatics/btp324.

[46] Kim D, Pertea G, Trapnell C, Pimentel H, Kelley R, Salzberg SL. TopHat2 : accurate alignment of transcriptomes in the presence of insertions , deletions and gene fusions. Genome Biol. 2013;14(4):1-13. DOI: 10.1186/gb-2013-14-4-r36.

[47] Van der Auwera GA, Carneiro MO, Hartl C, Poplin R, del Angel G, Levy-Moonshine A, et al. From fastQ data to high-confidence variant calls: The genome analysis toolkit best practices pipeline. Curr Protoc Bioinforma. 2013;1-33. DOI: 10.1002/0471250953.bi1110s43.

[48] Cingolani P, Platts A, Wang LL, Coon M, Nguyen T, Wang L, et al. A program for annotating and predicting the effects of single nucleotide polymorphisms, SnpEff: SNPs in the genome of *Drosophila melanogaster* strain w1118; iso-2; iso-3. Fly (Austin). 2012;6(2):80-92. DOI: 10.4161/fly.19695.

[49] McLaren W, Gil L, Hunt SE, Riat HS, Ritchie GRS, Thormann A, et al. The Ensembl Variant Effect Predictor. Genome Biol. 2016;17(1):1-14. DOI: 10.1186/s13059-016-0974-4.

[50] Vergara IA, Frech C, Chen N. CooVar: Co-occurring variant analyzer. BMC Res Notes. 2012;5(615):1-7. DOI: 10.1186/1756-0500-5-615.

[51] Wang K, Li M, Hakonarson H. ANNOVAR: Functional annotation of genetic variants from high-throughput sequencing data. Nucleic Acids Res. 2010;38(16):1-7. DOI: 10.1093/nar/gkq603.

[52] Alexandrov N, Tai S, Wang W, Mansueto L, Palis K, Fuentes RR, et al. SNP-Seek database of SNPs derived from 3000 rice genomes. Nucleic Acids Res. 2015;43:1023-7. DOI: 10.1093/nar/gku1039.

[53] Zhao H, Yao W, Ouyang Y, Yang W, Wang G, Lian X, et al. RiceVarMap: A comprehensive database of rice genomic variations. Nucleic Acids Res. 2015;43(1):18-22. DOI: 10.1093/nar/gku894.

[54] Zhang Z, Hu S, He H, Zhang H, Chen F, Zhao W, et al. Information Commons for Rice (IC4R). Nucleic Acids Res. 2016;44:1172-80. DOI: 10.1093/nar/gkv1141.

[55] Hunt SE, McLaren W, Gil L, Thormann A, Schuilenburg H, Sheppard D, et al. Ensembl variation resources. Database. 2018;2018:1-12. DOI: 10.1093/database/bay119.

[56] Usadel B, Obayashi T, Mutwil M, Giorgi FM, Bassel GW, Tanimoto M, et al. Co-expression tools for plant biology: Opportunities for hypothesis generation and caveats. Plant, Cell Environ. 2009;32(12):1633-51. DOI: 10.1111/j.1365-3040.2009.02040.x.

[57] Li Y, Pearl SA, Jackson SA. Gene Networks in Plant Biology: Approaches in Reconstruction and Analysis. Trends Plant Sci. 2015;20(10):664-75. DOI: 10.1016/j.tplants.2015.06.013.

[58] van Dam S, Võsa U, van der Graaf A, Franke L, de Magalhães JP. Gene

co-expression analysis for functional classification and gene–disease predictions. Briefing in Bioinformatics. 2017;1-18. DOI: 10.1093/bib/bbw139.

[59] Obayashi T, Aoki Y, Tadaka S, Kagaya Y, Kinoshita K. ATTED-II in 2018: A Plant Coexpression Database Based on Investigation of the Statistical Property of the Mutual Rank Index. Plant Cell Physiol. 2017;59(1):1-7. DOI: 10.1093/pcp/pcx191/4690683.

[60] Barrett T, Wilhite SE, Ledoux P, Evangelista C, Kim IF, Tomashevsky M, et al. NCBI GEO: Archive for functional genomics data sets - Update. Nucleic Acids Res. 2013;41(1):91-95. DOI: 0.1093/nar/gks1193.

[61] Leinonen R, Sugawara H, Shumway M. The sequence read archive. Nucleic Acids Res. 2011;39(1):19-21. DOI: 10.1093/nar/gkq1019.

[62] Kuhn M, Jackson S and Cimentada J. corrr: Correlations in R. R package version 0.4.2. (2020). https://CRAN.R-project.org/package=corrr.

[63] Ge SX, Jung D. ShinyGO: a graphical enrichment tool for animals and plants. bioRxiv. 2018;(315150):2. DOI: 10.1101/315150v1.

[64] Bhardwaj A, Dhar YV, Asif MH, Bag SK. In Silico identification of SNP diversity in cultivated and wild tomato species: Insight from molecular simulations. Sci Rep. 2016;6(1):1-13. DOI: doi.org/10.1038/srep38715.

[65] Sholikhah A, Khasna EN, Dahlia, Listyorini D. Polymorphism of gene encoding granule bound starch synthase i (GBSSI) involved in starch biosynthesis in local rice from Banyuwangi. AIP Conf Proc. 2018;2002. DOI: 10.1063/1.5050100.

[66] Cirillo E, Kutmon M, Hernandez MG, Hooimeijer T, Adriaens ME, Eijssen LMT, et al. From SNPs to pathways: Biological interpretation of type 2 diabetes (T2DM) genome wide association study (GWAS) results. PLoS One. 2018;13(4):1-19. DOI: 10.1371/journal.pone.0193515.

[67] Shannon P, Markiel A, Ozier O, Baliga NS, Wang JT, Ramage D, Amin N, Schwikowski B, Ideker T. Cytoscape: a software environment for integrated models of biomolecular interaction networks. Genome Research. 2003 Nov;13(11):2498-504. doi: 10.1101/gr.1239303.

[68] Schaefer RJ, Michno J-M, Jeffers J, Hoekenga O, Dilkes B, Baxter I, et al. Integrating Coexpression Networks with GWAS to Prioritize Causal Genes in Maize. Plant Cell. 2018;30(12):2922-42. DOI: 10.1105/tpc.18.00299.

[69] Bindea G, Mlecnik B, Hackl H, Charoentong P, Tosolini M, Kirilovsky A, et al. ClueGO: A Cytoscape plug-in to decipher functionally grouped gene ontology and pathway annotation networks. Bioinformatics. 2009;25(8):1091-3. DOI: 0.1093/bioinformatics/btp101.

[70] Maere S, Heymans K, Kuiper M. BiNGO: A Cytoscape plugin to assess overrepresentation of Gene Ontology categories in Biological Networks. Bioinformatics. 2005;21(16):3448-9. DOI: 10.1093/bioinformatics/bti551.

[71] Hadley Wickham, Romain François, Lionel Henry and Kirill Müller. dplyr: A Grammar of Data Manipulation. R package version 0.7.6. (2018). https://CRAN.R-project.org/package=dplyr.

[72] Hadley Wickham and Lionel Henry. tidyr: Tidy Messy Data. R package version 1.0.2. (2020). https://CRAN.R-project.org/package=tidyr

[73] G. Grothendieck. sqldf: Manipulate R Data Frames Using SQL. R package version 0.4-11. (2017). https://CRAN.R-project.org/package=sqldf.

[74] Matsumoto T, Wu J, Itoh T, Numa H, Antonio B, Sasaki T. The Nipponbare genome and the next-generation of rice genomics research in Japan. Rice. 2016;9(33). DOI: 10.1186/s12284-016-0107-4.

[75] Kanehisa M, Sato Y, Kawashima M, Furumichi M, Tanabe M. KEGG as a reference resource for gene and protein annotation. Nucleic Acids Res. 2016;44:457-62. DOI: 10.1093/nar/gkv1070.

[76] Naithani S, Preece J, D'Eustachio P, Gupta P, Amarasinghe V, Dharmawardhana PD, et al. Plant Reactome: A resource for plant pathways and comparative analysis. Nucleic Acids Res. 2017;45(1):1029-1039. DOI: 10.1093/nar/gkw932.

[77] Dharmawardhana P, Ren L, Amarasinghe V, Monaco M, Thomason J, Ravenscroft D, McCouch S, Ware D, Jaiswal P. A genome scale metabolic network for rice and accompanying analysis of tryptophan, auxin and serotonin biosynthesis regulation under biotic stress. Rice. 2013;6(1):1-15. DOI: 10.1186/1939-8433-6-15.

[78] Hawkins C, Caruana J, Schiksnis E, Liu Z. Genome-scale DNA variant analysis and functional validation of a SNP underlying yellow fruit color in wild strawberry. Sci Rep. 2016;6(July):1-11. DOI: 10.1038/srep29017.

Chapter 6

# Current Applicable DNA Markers for Marker Assisted Breeding in Rice (*Oryza sativa* L.)

*Nor'Aishah Hasan, Mohd Rafii Yusoff, Abdul Rahim Harun and Faiz Ahmad*

## Abstract

Rice, (*Oryza sativa* L.) account as the second cereal most cultivated in the world. Unfortunately, global rice production is rendered by significant number abiotic and biotic stresses. Breeding for resistant variety through conventional breeding is an economical method; generally, it takes at least 10 years to release a new rice variety. Advance technology in molecular marker had revolutionized and irreversibly changes the disciplines of plant genetic and breeding. Integration of DNA-based markers in selection process enhances the effectiveness and accuracy of conventional plant breeding. It offers a novel tool for discovering and tagging alleles and genes specifically in plant. Ubiquitous of DNA marker-trait associations for diverse crops species are available with the findings of many quantitative trait loci (QTLs) mapping studies. The linkage drags, and time-consuming in conventional breeding can minimize with the application of DNA markers in plant breeding. The utilization of DNA marker in QTL mapping, MAS and gene pyramiding has been investigated. In this chapter, we discussed the recent utilizing markers in rice breeding program against abiotic and biotic stresses. In a few decades, molecular marker assisted breeding (MAB) provide a boundless task for breeders in attaining an important impact on crop development.

**Keywords:** abiotic stress, biotic stress, marker-assisted selection, rice, resistant

## 1. Introduction

*Oryza sativa* is an essential food crop contributing 30% of calories for at least 62.8% of the world population. Presently, rice was cultivated over 167.13 m ha within 114 countries with annual production of 486.62 mt (on milled basis) [1]. Majority of population in Asian consumed rice as a basic staple crop with over than 90% of rice were produced in Asia [2]. Revolution in Green technology played an important role in adoption of major increasing in rice production over last four decades. Approximately 450 million tons to 650 million tons of world rice consumption is projected to surge by 2011 and 2050, respectively [3].

Scientist and plant breeder are facing a new challenge in crop improvement due to climate change issue including limitation of water and cultivated land, new emerging pathogen and pest resulted serious yield losses. Among the major constraints, blast disease, bacterial blight, sheath blight, tungro viruses, brown spot and a number of grain discoloration diseases [4] accounted for serious rice losses.



IntechOpen

Several amendments can be implemented to overcome these limitations such as breeding high yielding varieties, durable resistance to diseases and tolerance to abiotic stresses [5]. Rice yield potential can be enhanced by using numerous approaches such as conventional hybridization, ideotype breeding, heterosis breeding, wide hybridization and molecular breeding [6]. In molecular rice breeding, there are two main strategies can be used by applying the biotechnology concept which are MAS and Genetically Engineering. MAS like genetic engineering is a technique used to introgress targeted gene however its offers tools for specific selection for further breeding in existing plant material. The field of plant genetics and breeding has irreversibly changed with the development of DNA (or molecular markers) [7].

Since past two decade, revolutionized in molecular marker to enhance the effectiveness in breeding and to significantly shorten the development time of varieties bring about the concept of molecular marker assisted selection (MAS) as an efficient plant selection [8–10]. A genetic marker is any noticeable character or otherwise assayable phenotype, for which alleles at individual loci segregate in a Mendelian manner. The genetic markers covered include (1) morphological markers (2) biochemical markers (alloenzymes and other protein markers) and (3) molecular markers (based on DNA-DNA hybridization) [11, 12]. In this chapter, we present the latest DNA-based markers in a few studies related to abiotic and biotic stress genes using the MAS application in rice breeding programs. These markers have proven significantly useful from different findings published in screening, fine mapping and gene pyramiding approaches.

## 2. Molecular marker

Rediscovery Mendelian theory in the early part of 20th century, reveal that inheritance (genes) linked together in chromosome. Molecular marker defines as individual genes flanking within a defined close interval. It can be found at a specific location of the genome and linked with inheritance of a trait or gene [13]. Thottappilly et al. [14] refers molecular marker as polymorphism identified between species including proteins and nucleic acids. Collard et al. [15] had divided molecular marker into 3 major groups which are morphological marker, biochemical marker and DNA marker (**Figure 1**). However, Xu [16] had categorized markers into two

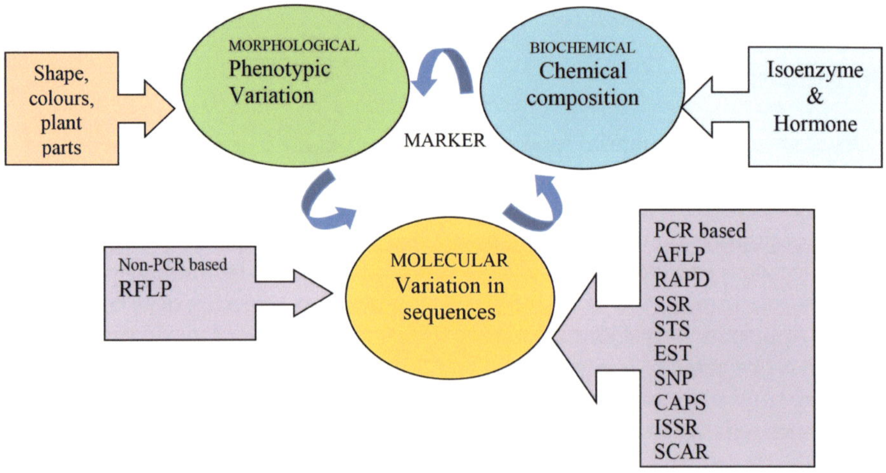


**Figure 1.**
*Different types of molecular marker available.*

broad groups based on detecting method; (i) classical marker and (ii) DNA marker. Classical marker was initiated with the mutations in the genomic loci controlling plant morphology [17] (**Table 1**). Advent in polymorphism detecting methods such as Polymerase Chain Reaction, Southern Blotting and Sequencing had invented the development of DNA markers (**Table 2**). Variation were detected at DNA level such as nucleotides changes ie; substitution (point) mutation, rearrangement (insertion and deletions) or errors in replication of tandemly repeated DNA [18].

| Types of marker | Morphological marker | Biochemical marker |
|---|---|---|
| Advantages | Readily available | Highly reproducible |
| | Usually require only simple equipment | Applicable for measure a wide range of population genetic parameter |
| | The most direct measure of phenotype | Inexpensive |
| Disadvantages | Requires expertise on crop or species | Limited biochemical assay for detection |
| | Subjected to environmental influences | Phenotype-based analysis |
| | Limited in number | |
| Application | Conventional plant breeding program | Genetic diversity<br>Map construction |

**Table 1.**
*Comparison between classical markers.*

| PCR based markers | Advantages | Disadvantages | Application |
|---|---|---|---|
| RFLP | Reliable and rigid | Harmful and tedious | Map construction<br>Hybrid fixation<br>Genetic diversity |
| | Easy to conduct across laboratory | Required a larger amount of DNA | |
| RAPD | A small amount of DNA required | Not transferable across laboratory | Genetic diversity<br>Saturation mapping |
| | Simple, rapid and lowest cost | Irreproducibility | |
| | Single primer generates multiple loci | | |
| SSR | Reliable, powerful and easy | Tedious and costly for initial establishment | Linkage and QTL mapping<br>Marker-assisted selection<br>Hybrids fixation |
| | Transferable between populations | Required polyacrylamide gel electrophoresis | |
| AFLP | Multiple loci | Required several DNA | Saturation mapping<br>Genetic diversity |
| | | Complicated | |
| SNP | Multiplexing large number of markers | Costly to assay | Fine mapping<br>Map-based cloning |
| | Degraded DNA can be used | Marker has less allele | |

**Table 2.**
*Comparison of different PCR based marker.*

Although there are many characteristics determine the suitable markers to be considered in MAS development; there are five main characteristics need to be considered [13, 19–22].

1. **Reliability:** Attribute of the desirable marker need to close proximity with the target loci, preferably less than 5 cm genetic distance or one marker every 10 cm [23].

2. **Repeatability and reproducible:** In plant breeding, a lot number of plants are usually screened for desirable marker pattern and result is required instantly. Therefore, level of simplicity in terms of time required and quick detection method are highly desirable for rapid selection. Statistical and bioinformatics software are significant for this purpose.

3. **Cost effective:** Marker assay should be not expensive feasible for selection process.

4. **Level of polymorphism:** Marker should discriminate between two similar accessions of species especially in parent breeding material.

5. **DNA quality and quantity:** A few markers require a complex DNA extraction technique and it too laboratories. In addition, it adds cost to the process.

## 3. Successful application of molecular marker in rice

Recently, MAB has become a well-known approach develop a superior genotype in rice. The aims of the breeding programs were to develop a rice variety with high yield potential, resistant to diseases and insects, tolerance to adverse environments, and acceptable grain quality. Since the first successful experiment stated for rice by Chen et al. [24] with introducing resistance to bacterial blight (BB) disease into Chinese hybrid parents. Till date, many varieties had been developed thru the application of molecular marker with backcrossing method. Varieties developed were improved from the wild type by incorporate desirable gene depending for biotic and abiotic stresses. Recently, in a few years there is a trend in producing more rice varieties that resistant in abiotic stress such as drought tolerance, submergence tolerance and salt tolerance. This demand comes from the periodic natural disaster in many regions of the rice producing country specifically in Asian country [25]. Resistant varieties against pathogenic disease were continuously developed. This scenario can be seen in blast disease where there are increasing number of resistant varieties were developed due to instability of the fungus pathogen.

Numerous molecular markers have been utilized in marker-assisted breeding however microsatellites are the preferable markers for plant breeding applications. The characteristic as codominant markers which inherited in a Mendelian fashion have emerged as a best choice compare to other markers. Additionally, microsatellite identify as preferable markers used in plant breeding programs due to ubiquitous, high polymorphism rates and wide range of distribution throughout the genome [26, 27]. Other markers possess major drawbacks including Restriction fragment length polymorphism (RFLP) markers which are not integrated with high-throughput technique; Random amplification of polymorphic DNA (RAPD) assays which are often not reproducible or immobile. AFLP analysis is not straightforward and commonly produces multiple fragments with the use of large genomic templates [28]. Application of molecular marker in rice breeding was demonstrated in **Table 3** to develop superior varieties in rice against abiotic and biotic stresses.

| Stress | Traits | Gene/QTLs | Foreground and background selection | | Reference |
|---|---|---|---|---|---|
| Abiotic | Deep roots | QTLs on chromosomes 1, 2, 7 and 9 | RFLP and SSR | SSR | [29] |
| | Quality | waxy | RFLP | AFLP | [30] |
| | Root traits and aroma | QTLs on chromosomes 2, 7, 8, 9 and 11 | RFLP and SSR | RFLP and SSR | [31] |
| | Submergence tolerance | *Sub1* QTL | Phenotyping and SSR | SSR | [32] |
| | | *Sub1* QTL | SSR | SSR | [33] |
| | | *Sub1*QTL | SSR | STS | [34] |
| | | *Sub1* QTL | SSR | SSR | [35] |
| | | *Sub*1 | SSR | SSR | [36] |
| | | *Sub1* | SSR | SSR | [37] |
| | | *Sub1* | SSR | SSR | [38] |
| | | *Sub1* | SSR | SSR | [39] |
| | Salt tolerance | *Saltol* | SSR | SSR | [40] |
| | | *Saltol* QTL | SSR | SSR | [41] |
| | | *Saltol* QTL | SSR | SSR | [42] |
| | | *Saltol* QTL | SNP | SSR | [43] |
| | Early maturation | *Hd2* | SSR | SSR | [44] |
| | High yield and drought tolerance | *qDTY1.1, qDTY2.2, qDTY3.1, qDTY3.2, qDTY6.1*, and *qDTY12* | SSR | SSR | [45] |
| | | *qDTY1.1 + qDTY2.1 + qDTY3.1 + qDTY11.1 and two QTLs qDTY1.1 + qDTY11.1* | SSR | SSR | [46] |
| | Heading time | QTL (*Hd1, Hd4, Hd5* and *Hd6*) | RFLP, STS, SSR, CAPS, dCAPS | RFLP, STS, SSR, CAPS, dCAPS | [47] |
| | Phosphorus tolerance | *Pup1* | SSR | SSR | [48] |
| | Drought tolerance | QTL | SSR | SSR | [42] |
| | | MQTL1.1 | SSR | SSR | [49] |
| | | QTL | SSR | SSR | [50] |

| Stress | Traits | Gene/QTLs | Foreground and background selection | | Reference |
|---|---|---|---|---|---|
| | | (qDTY1.1, qDTY2.1) | SSR | SSR | [51] |
| | | QTL | SSR | SSR | [52] |
| Biotic | Bacterial blight | *xa21* | STS | RFLP | [24] |
| | | *xa21* | STS | AFLP | [53] |
| | | *xa5, xa13* and *xa21* | STS, CAPS | Not performed | [54] |
| | | *xa5, xa13* and *xa21* | STS | Not performed | [55] |
| | | *xa33* | SSR | SSR | [56] |
| | | *xa34(t)* | SSR | SSR | [57] |
| | | *xa35(t)* | SSR | SSR | [58] |
| | | *xa38* | SSR | SSR | [59] |
| | | *xa21* | STS | STS | [60] |
| | | *xa5* | CAPS | CAPS | [61] |
| | | *xa13, xa21* | STS | SSR | [62] |
| | | *xa5* and *xa13* | CAPS | STS | [63] |
| | | *xa13* and *xa21* | CAPS | STS | [62, 64] |
| | | *xa13+ xa21* and *xa5 + xa21* | SSR | SSR | [65] |
| | | *xa13 and xa21* | SSR | SSR | [66] |
| | Blast resistance | *Pi1, Pi2, Pi33* and *Pi54* | SSR | SSR | [67] |
| | | *Pi* genes, *xa5* | SSR | SSR | [68] |
| | | *Pi54* and *Pi5* | SSR | SSR | [69] |
| | | *Pi1* and *Pi2* | SSR | SSR | [70] |
| | | *Pi1* | SSR | ISSR | [71] |
| | | *Pikh and pi7(t)* | SSR | SSR | [9] |
| | | *QTLs on chromosomes 1, 2, 11 and 12* | SSR | SSR | [72] |

| Stress | Traits | Gene/QTLs | Foreground and background selection | | Reference |
|---|---|---|---|---|---|
| | Submergence tolerance, disease resistance, quality | *Subchr9* QTL, *xa21*, *Bph* and blast QTLs and quality loci | SSR and STS | Not performed | [34] |
| | Thermo-sensitive genic male sterile (TGMS) and blast resistance | TGMS and *Pi* gene | SNP | SNP | [73] |
| | Brown planthopper resistance | *Bph14* and *Bph15* | InDEL | InDEL | [74] |
| | | *Bph18* | SSR | SSR | [75] |
| | Gall midge resistance | *Gm8* | SSR | SSR | [76] |
| | Rice stripe resistance | *Stv-bi* | SSR | SSR | [77] |

**Table 3.**
*Application of molecular marker in MAB in rice.*

## 4. Abiotic stresses

### 4.1 Deep root

Four QTL (chromosome 1, 2, 7 and 9) was introgressed to NIL developed from crossed between Azucena (donor parent) and IR64 (recipient parent) [29]. Twenty-nine NIL successfully developed in which three NIL carrying target 1 demonstrated significantly increased trait over IR64 while 3 out of 8 NIL consist combination target 1 and 7 exhibited higher root mass and depth. Four NIL carrying targets 9 showed increased maximum number of root length and 2 NIL showed no significant difference on root phenotypic compare to IR64. Results suggest the introgression of genes linked to root improvement is an ideal system compared to change the morphological characters of plant.

### 4.2 Quality

Waxy locus (*wx*-MH) from Mighui 63 was successfully introgressed into Zhenshan 97 which having poor agronomically quality including amylose content (AC), gel consistency (GC), gelatinization temperature (GT) and endosperm [30]. MABC scheme had been used to develop 3 backcross and 1 selfing generation. Improved variety of Minghui 63 (Shanyou 63) carrying *wx*-MH gene at 6.1 cm length have similar agronomy characteristic as Minghui 63 with low AC, high GC and GT with low grain opacity. Result indicated that waxy region gives significant effect on quality trait.

### 4.3 Root trait and aroma

MABC strategy had adopted by high yielding India upland rice variety, Kalinga III to insert 5 segments of QTL including 4 QTL carrying root length and thickness and 1 QTL control recessive aroma trait from Azucena donor variety which come from Philippines japonica rice [31]. This experiment was performed until 8 generation which using 3000 SSR markers for almost 8 years to develop NIL. Twenty-two NIL successfully develop which carrying QTL on root length at chromosome number 9 linked with RM242-RM201 and RM248 on chromosome 7 control for delayed flowering.

### 4.4 Submergence tolerance

Ten improves lines from OM 1490/IR 64 -Sub1 was developed from Marker-assisted backcrossing and having 90–99% survival under field submergence treatment [78]. Foreground selection was carried out using microsatellite, RM23805 marker which identified allele as 240 and 230 bp bands, respectively from OM1490, susceptible parent and IR64-*Sub*1, tolerant parent. Entire lines of OM1490/IR64-*Sub*1 introgression possess an increase rate of survival from original parent. In India, high yielding mega variety, Swarna had been incorporated with *Sub*1 gene by using three selections namely foreground, recombinant, and background selection, respectively. Findings indicated that superior variety Swarna successfully transformed into submergence tolerant variety within 2–3 years with three backcross generations [33]. Popular rice variety, AS996 successful incorporated *sub1* gene from rice variety IR64 as a donor parent through MABC. Insertion of *sub1* was confirmed with foreground SSR marker, ART5 and SC3 [36]. Background analysis was screened with 53 polymorphic SSR markers to assess $BC_2F_1$ and $BC_3F_1$ generations. The highest recurrent background was achieved 100% donor parent. Results suggest the development a new submergence tolerant rice variety

ASS996-SUB1 to stand with climate change. Superior varieties from India (Samba Mahsuri and CR1009), IRRI, Philippines (IR64), Laos (Thadokkham 1) and Bangladesh (BR11) successfully introgress *sub1* gene exhibited an improve survival rate compare to parental variety by applying MABC approach. Septiningsih et al. [79] studied the expression level of sub genes revealed that *sub1C*-1 was dominant over the *sub1A* allele however no reduction of expression observed when intolerant *sub1C* allele merged with the tolerant *sub1A*-1 allele. Survival rates of heterozygous plant with *sub1A* expression higher compared to homozygous tolerant parent. Existence of *sub*1A-2 will not significantly function in intolerant plant with turn off expression of sub*1C*-1. Study by Ahmad et al. [38] developed 30 newly lines of $BC_2F_3$ population from backcross of MR219 and Swarna-Sub1 with 95.37% recurrent parent genome (RPG) recovery. Rahman et al. [39] stated a marker-assisted backcross breeding (MABB) approach to incorporate the *Sub1* locus into popular rice varieties of southern India, CO 43, from tolerant FR13A variety. An elite NILs of CO 43 harboring *Sub1* locus possessing 95.78% of the recurrent parental CO 43 genome produced from genotyping and phenotyping analysis.

### 4.5 Salt tolerance

ASS996 a high yielding and widely grown cultivar in Vietnam had been cross with FL478, a highly salt tolerant rice variety using MABC strategy [40]. Four *Saltol*-AS996 introgression lines, QF3–1, QF3–2, QF 4–3-3, QF6–4 was developed with 3 foreground SSR markers, AP3206, RM3412 and RM10793 and 63 polymorphic markers were applied across rice genome for background analysis. The promising lines showed significant salt tolerance or similar compare to recurrent parent, FL478. MABC approach had been applied for introgressed *Saltol* QTL from donor parent, FL478 into Bacthom 7 recipient rice cultivar. Eighty-nine polymorphic SSR markers including 8 foreground markers to screen heterozygous plant carrying *Saltol* locus in each backcrossed generation. Background analysis demonstrated the restoration of recipient genome up to 96.8–100% in the best plant of $BC_3F_1$ generation. Finding concluded that improve version of salt tolerance-Bacthom 7 is relevant for cultivation in coastal areas of the Vietnamese Deltas [41]. De leon et al. [43] reported the utilization of SSR and SNP markers to characterize the ILs (Pokkali x 'Bengal) and identify 14 QTLs traits related to salinity tolerance.

### 4.6 Early maturation

In Indonesia, *Hd*2 gene located at chromosome 7 from Nipponbare was transferred to Code variety [44]. Two markers linked with early maturation trait, RM1362 and RM7601 was used to identify heterozygous carrying this gene in 2 backcross and 1 selfing generation. Improved lines showed a similar agronomic characteristic with code variety except for heading time which 73–85 days in improved version compared to recipient parents, 89 days. Maturation days for improved version of Code carrying *Hd*2 gene were 103–104 day which is at considerable maturation days. Result concluded the availability to use these improved lines in Indonesia for increasing rice production.

### 4.7 Heading time

Koshihikari and Kasalath were crossed by using MABC strategy to introgress 3 QTL segment which code for *Qdth*3, *Qdth*6 and *Qdth*8 [47]. Genotyping analysis was performed using five different types of DNA marker including RFLP, STS, SSR, CAPS and dCAPs. Three NIL successfully produced namely Wakei 367, Wakei 371 and Kanto IL1 which carrying Hd2 gene coded for heading time. All NIL exhibited same morphological trait with Koshihikari with early heading time for Kanto IL1 (12 days), Wakei 367 (10 days) and Wakei 371 (11 days) compare to 95 days on Koshihikari.

### 4.8 Phosphorus tolerance

MABC approach had been successfully adapted to introgress a major quantitative trait locus (QTL), *Phosphorus uptake1* (*Pup1*), positioned on rice (*Oryza sativa*) chromosome 12 into two types of Indonesian rice varieties, irrigated and upland. SNP and CAPs had been used as foreground marker to identify desirable traits and background selection had been conducted using 47 to 61 polymorphic SSRs markers and restored 89–93% of the recipient genome in $BC_2F_3$ population. Phenotyping finding suggests that *Pup1* is effective to enhance grain yield under field treatment [48].

### 4.9 Drought tolerance

Crossing by hand emasculation and artificial pollination was performed between IR20 (*japonica*) and lowland (*indica*) cultivar to develop near isogenic lines (NILs) [42]. Foreground and background selection were conducted using 100 SSR markers that distributed entire genome. Five NILs exhibited high yield performance compare to IR20 in rain-fed and irrigated conditions. NIL-212 and 297 successfully incorporated with three and two root QTLs with NIL-297 demonstrated maximum value of nodal root and surface area whereas increase number of nodal roots than recurrent parent, IR20 observed in NIL 212. Backcrossed population ($BC_2F_3$) was developed through Marker Assisted Selection (MAS) to transfer QTLs for drought tolerant characters into Thai famous variety KMDL105 using 3 different donor parents, DH212, DH103 and DH126, respectively. Hundred and three KMDL 105 introgression lines were developed carrying selected target [80]. Genome analysis claimed that donor segment was found in non-carrier chromosomes and minimum proportion of recurrent genome was determined in carrier chromosome. Results suggest that background analysis is required to restore the recurrent genome segments. Sandhu et al. [46] enhance the grain yield of Samba Mahsuri under reproductive-stage drought (RS) through marker-assisted selection (MAS) and marker-assisted recurrent selection (MARS). Findings recorded four *qDTYs* ($qDTY_{1.1} + qDTY_{2.1} + qDTY_{3.1} + qDTY_{11.1}$) and two QTLs ($qDTY_{1.1} + qDTY_{11.1}$). Presence of gene MQTL1.1 responsible for the drought tolerance was identified from the cross (Sarjoo52 × Nagina- 22) in the study reported by Awasthi and Lal [49]. Ramchander et al. [50] study 101 Backcross Inbred Lines (BILs) of rice ($BC_1F_5$) for drought tolerance by develop a cross between Norungan//TKM9/Norungan. Marker aided selection used in this study efficiently pyramiding favorable QTL alleles to enhance drought tolerance in elite background of famous rice varieties using 15 polymorphic SSR markers. Muthu et al. [51] successfully developed abiotic stress-tolerant rice genotypes of the famous rice variety, Improved White Ponni (IWP) through a marker assisted backcross breeding approach by introgression major effect quantitative trait loci (QTLs) conferring tolerance against drought (*qDTY1.1*, *qDTY2.1*), salinity (*Saltol*), and submergence (*Sub1*). Backcrossed inbred lines (BILs) developed in this study showed minor effect caused by drought, salinity, and submergence. Root trait genes were introgressed through marker-assisted backcross breeding (MABC) between aerobic rice variety, AERON1 and MRQ74, a high yielding rice variety of Malaysia, [52]. Six best rice lines in $BC_2F_1$ with highest recovery percentage and resemble similar phenotypic appearance were identified with RM242 and RM263 foreground and 57 background markers.

### 4.10 Grain fragrance

Ahn et al. [81] was first discover the molecular marker linked with grain aroma. Lorieux [82] was used RFLP marker, RG28, which linked with aromatic located at chromosome 8 with genetic distance of 4.5 cm. The uniqueness of RG28 was interchangeable into STS marker and can be used to distinguish between fragrance and non-fragrance rice cultivars [83–85]. Improved version Pusa Bashmati and improved versions of PRR78 was successfully introgressed with quality trait from Basmati 370 as a donor with Manawthukha cultivar as a recipient using MABC scheme [86, 87].

### 4.11 Semi-dwarfing

*Sd*1 gene coded for semi-dwarf trait creates an interest among breeder as it enhances crop production. Advent in MAB had made it possible to track the introgression of *sd*1 gene by using SSR markers in *two traditional japonica varieties*, 'IR36' and 'Allorio' (IRGC284) [88].

## 5. Biotic stresses

### 5.1 Bacterial blight resistance

Marker assisted backcrossing coupled with stringent phenotypic screening was adapted to transfer BB resistance gene, *Xa*23 into Minghui63, YR293 and Y1671 which famous recipient lines. Foreground selection was screened with RM206 located at 1.9 cm near to the *Xa*23 locus [89]. The improved version of BB resistant was identical with original recipient for agronomic traits [30]. NSIC Rc142 and NSIC Rc154 donor cultivar for *Xa*4 + *Xa*5 + *Xa*21 genes were incorporated into susceptible cultivar IR64 in Philippines [61]. Currently, *Xa*13 and *Xa*21 had been successfully inserted into recurrent line PRR78 through MABC [90]. In Thailand, three backcrossed generation required to insert *Xa*21 (BB resistance gene) from IR1188 variety into famous KDML105 variety using MABC approach together with phenotypic selection. Similar agronomic characteristic was observed between developed introgressed lines having BB resistant gene (*Xa*21) with local variety KDML105 [91]. Foreground selection was screened with six SSR markers namely RM302, RM212, RM210, RM149, RM287, RM224 at different chromosome while 66 rice SSR markers was applied for background genotype. In China, popular restorer line, Yihui1577 was incorporated with *Xa*7 and *Xa*21 from Huahui20 thru MABC strategy [92]. Foreground marker were applied using RM20582 in existence of *Xa*7 to map a 0.14 cM interval at chromosome 6 between RM20582 and RM20593 [93]. Huang et al. [40] in his study successfully employed of *Xa*7, *Xa*21, *Xa*22 and *Xa*23 from Jinke 1 a cytoplasmic male sterile line into Huahui 1035 line using MABC. High yielding $F_1$ hybrid was used to develop more hybrid rice in China. Samba Mahsuri a mega high-yielding, fine grain-type variety was selected as donor for *Xa*21and *Xa*13 to incorporate into susceptible varieties, Taraori Basmati and Basmati 386 through MABC scheme [94]. Molecular backcross breeding was used to pyramided *Xa*21 + *xa*13 and *Xa*21 + *xa*5 combinations genes in a background of lowland cultivar, Jalmagna [65]. Twenty pyramided lines with targeted gene combinations showed resistant against BB pathogen. Balachiranjeevi et al. [66] reported the development of durable BB resistance through marker-assisted backcross breeding (MABB) into background of elite rice, DRR17B by introgress *Xa21* and *Xa33* dominant genes.

Six promising pyramiding lines of DRR17B with durable resistance and similar or superior agro-morphological attributes to DRR17B have been identified.

### 5.2 Blast resistance

C101A51 and Tetep contribute for *Piz*-5 and *Pi*54 blast gene was introgressed into PRR78 variety [69]. Two improved lines were developed, Pusa1602 (Piz5) and Pusa1603 (*Piz*54) through MABC breeding strategy by two backcross series. Foreground selection was performed by using AP5930 and RM206 while background analysis using SSR polymorphic markers revealed 89.01 and 87.88% in Pusa1602 and Pusa1603 lines recovery of RPG, respectively. Hybrid develop carrying blast resistance gene produced exhibited similar performance with parental Pusa RH10 in terms of yield, grain and cooking quality traits. Multiple resistance gene against leaf blast (*Pi*1), neck blast (*Pi*2) (donor BL122) and bacterial (*Xa*23) (donor CBB23) was successfully introgressed into Ronfeng B, a mega variety with early maturating line thru MABC scheme [70]. Improved lines D521, D524 was developed using three foreground markers (MRG4766, AP22 and RM206) and 131 polymorphic background markers. Proportional recipient genome for D521 and D524 were 96.18 and 96.56%, respectively after 4 backcrossed generations. The size of lesions exhibited from 0.77 to 1.18 cm with 96.7–100% resistance frequencies values. Rongfeng 3A was developed as an improve version of cytoplasmic male sterile line from Rongfeng by a series of conducive backcross breeding with *Pi*1, *Pi*2 and *Xa*23 gene. A series of superior lines with resistance to bacterial blight and blast diseases were developed by using MAB [30, 90, 95]. Hybrid Pusa RH10 is likely susceptible to BB and blast therefore by employing the bacterial and blast resistance gene, it will be improved adjustability to disease endemic parts and also retain the rice production. In Iranian, two crosses were made between susceptible, Tarom Mahali (TAM) and resistance cultivar, Khazar (KHZ) to increase blast resistance. Screening with 74 polymorphic markers on 192 $F_{2:3}$ families demonstrated both parental exhibited maximum genetic variation between the two varieties on entire chromosomes except chromosome 6, 7, 8 and 12 [96]. In Malaysia, improved version of MR219 was developed by introgression of *Pi-z* and *Pikh* and *Pi-b* by using two different donor, Pongsu Seribu 1 and 2 respectively [10, 97]. Hasan et al. [98] had successfully introgressed *Pi7(t)* and *Pikh* in MR264 from Pongsu Seribu 2 by using MABC approach. Thailand rice, the popular Sakon Nakhon (SKN) rice had been introgressed with QTLs on chromosomes 1, 2, 11 and 12 by using marker-assisted backcrossing. Srichant et al. [72] in his study presented 2 lines (SKN 39-10-19-29-12 and SKN 39-10-19-29-13) with high resistance to leaf and neck blast; and having similar agronomic traits as the SKN.

### 5.3 Brown plant hopper resistance

*Bph*3, gene coded for resistance against brown plant hopper had been introgressed into the famous Thai variety 'Khao Dawk Mali 105' [99]. Backcrossed population had been screened with RM589 and RM190, highly linked to *Bph*3 and *Wx*-RH loci. Fifty NIL were developed from background selection using 75 polymorphic SSR markers disperse 12 rice chromosomes. Cross had been made between Junambyeo, an elite japonica cultivar with IR65482-7-216-1-2 a donor indica rice by a series of backcrossing through MAB [75]. *Bph18* gene was transferred and 7312T4A, the most closely polymorphic STS marker was applied to screen the existence of the *Bph*18 gene in BPH-introgress backcrossed lines [100]. Two hundred and sixty SSR marker were used to screen the RPG percentage in the four NIL breeding lines.

### 5.4 Gall midge resistance

Gm8 resistance gene code for rice gall midge had been employed from donor parent Aganni into Samba Mashuri having g*ene Xa21* using MABC. Four plants identified in selfing population (BC2F2) carrying *Gm*8 and *Xa*21 genes and further tested for phenotypic selection against BB and gall midge [76]. Exploration of gm gene had been a great achievement in modern crop however resistant variety carrying 1 major gene easily lost the resistancy [101]. Till date, 11 Gm genes in the plant [102] and 7 pathogens have been identified [103].

## 6. Conclusion

Advent in marker technology has offer a remarkable progress in crop improvement specifically in rice. Present paper reviews the application of DNA markers in abiotic and biotic stresses in rice using MAS. Interval between molecular markers and gene/QTLs linked with target traits play a significant role in efficiency of MAS technique. Generally, marker will be validated in fine mapping study. Thus, the development of important marker associated with abiotic and biotic stresses resistance is proficient by QTL mapping experiments. MAB greatly increase the effectiveness and efficiency of breeding. Desirable individual in germplasm collection with genes or QTLs can be characterized in genotyping analysis by using the DNA markers for target genomic regions rather than phenotyping analysis. Pyramiding tolerance genes and QTLs will develop a resistant cultivar against abiotic and biotic stresses. MAS refers to selection by DNA markers linked to QTLs or target genes. Nowadays, DNA-based genetic markers play a significant part in the forthcoming of MAB and molecular genetics analysis for establishment of stress-tolerance and resistance in plants through molecular linkage maps. Knowledge of markers for abiotic and biotic tolerant traits presented here will serve a fundamental guide and help rice breeders in their MAB.

## Acknowledgements

The authors are grateful to Universiti Teknologi MARA, Universiti Putra Malaysia and Malaysian Nuclear Agency for the facilities, support and financial aids during the research work. The author also would like to extend appreciation to the Ministry of Higher Education (MOHE) for providing financial aids during this research.

## Author details

Nor'Aishah Hasan[1,2*], Mohd Rafii Yusoff[1,3], Abdul Rahim Harun[4] and Faiz Ahmad[4]

1 Institute of Tropical Agriculture and Food Security, Universiti Putra Malaysia, Serdang, Selangor, Malaysia

2 Faculty of Applied Science, Universiti Teknologi MARA, Cawangan Negeri Sembilan, Kampus Kuala Pilah, Negeri Sembilan, Malaysia

3 Department of Crop Science, Universiti Putra Malaysia, Serdang, Selangor, Malaysia

4 Agrotechnology and Bioscience Division, Malaysian Nuclear Agency, Kajang, Selangor, Malaysia

*Address all correspondence to: aishahnh@uitm.edu.my

## References

[1] Shahbandeh M. World rice acreage 2010-2018. 2020. Available from: https://www.statista.com/statistics/271969/world-rice-acreage-since-2008/

[2] Bandumula N. Rice production in Asia: Key to global food security. Proceedings of the National Academy of Sciences, India Section B: Biological Sciences. 2017;**88**:1323-1328. DOI: 10.1007/s40011-017-0867-7

[3] Rejesus RM, Mohanty S, Balagtas JV. Forecasting Global Rice Consumption. Working Paper. North Carolina State University; 2012

[4] Singh PK, Nag A, Arya P, Kapoor R, Singh A, Jaswal R, et al. Prospects of understanding the molecular biology of disease resistance in Rice. International Journal of Molecular Sciences. 2018;*19*:1141

[5] Oladosu Y, Rafii MY, Samuel C, Fatai A, Magaji U, Kareem I, et al. Drought resistance in Rice from conventional to molecular breeding: A review. International Journal of Molecular Sciences. 2019;**20**:3519

[6] Khan MH, Dar ZA, Dar SA. Breeding strategies for improving Rice yield—A review. Agricultural Sciences. 2015;**6**:467-478. DOI: 10.4236/as.2015.65046

[7] Collard BCY, Mackill DJ. Marker-assisted selection: An approach for precision plant breeding in the twenty-first century. Philosophical Transactions of the Royal Society B. 2007;**363**:557-572

[8] Chukwu SC, Rafii MY, Ramlee SI, Ismail SI, Hasan MM, Oladosu YA, et al. Bacterial leaf blight resistance in rice: A review of conventional breeding to molecular approach. Molecular Biology Reports. 2019;**46**:1519-1532

[9] Hasan N, Rafii MY, Abdul Rahim H, Nusaibah SA, Mazlan N, Abdullah S. Genetic analysis and identification of SSR markers associated with rice blast disease in a $BC_2$ $F_1$ backcross population. Genetics and Molecular Research. 2015b;**16**(1):1-11

[10] Miah G, Rafii MY, Ismail MR, Puteh AB, Rahim HA, Asfaliza R, et al. Blast resistance in rice: A review of conventional breeding to molecular approaches. Molecular Reproduction and Development. 2013;**40**:2369-2388

[11] Nadeem MA, Nawaz MA, Shahid MQ, Doğan Y, Comertpay G, Yıldız M, et al. DNA molecular markers in plant breeding: Current status and recent advancements in genomic selection and genome editing. Biotechnology & Biotechnological Equipment. 2018;**32**(2):261-285

[12] Shabir G, Aslam K, Khan AR, Shahid M, Manzoor H, Noreen S, et al. Rice molecular markers and genetic mapping: Current status and prospects. Journal of Integrative Agriculture. 2017;**16**(9):1879-1891

[13] Jiang GL. Plant marker-assisted breeding and conventional breeding: Challenges and perspectives. Advances in Crop Science and Technology. 2013;**1**:105

[14] Thottappilly G, Magonouna HD, Omitogun OG. The use of DNA markers for rapid improvement of crops in Africa. African Crop Science Journal. 2000;**8**(1):99-108

[15] Collard BCY, Jahufer MZZ, Brouwer JB, Pang ECK. An introduction to markers, quantitative trait loci (QTL) mapping and marker-assisted selection for crop improvement: The basic concepts. Euphytica. 2005;**142**:169-196

[16] Xu Y. Molecular Plant Breeding. Wallington, UK: CABI International; 2010. p. 734. ISBN 978-184593-392-0

[17] Xu Y, Crouch JH. Marker-assisted selection in plant breeding: From publications to practice. Crop Science. 2008;**48**:391-407

[18] Modi D, Pandey B. Potential of molecular markers in plant biotechnology. Plant Omics. 2009;**2**:141-162

[19] Farooq S, Azam F. Molecular markers in plant breeding-II. Some pre-requisites for use. Pakistan Journal of Biological Sciences. 2002;**5**:1141-1147

[20] Mackill DJ, Ni J. Molecular mapping and marker assisted selection for major-gene traits in rice. In: Khush GS, Brar DS, Hardy B, editors. Proceeding Fourth International Rice Genetics Symposium. Los Baños, The Philippines: International Rice Research Institute; 2000. pp. 137-151

[21] Mohler V, Singrun C. General considerations: Marker-assisted selection in biotechnology in agriculture and forestry. Molecular Marker Systems. 2004;**55**:305-317

[22] Sabina A, Shafiq MA, Wani A, Bhat KA, Chalkoo S, Mir MR, et al. Marker assisted selection in rice. Journal of Phytopathology. 2010;**2**(10):66-81

[23] Boopathi NM. Marker-assisted selection (MAS). In: Genetic Mapping and Marker Assisted Selection. Singapore: Springer; 2020

[24] Chen S, Lin XH, Xu CG, Zhang QF. Improvement of bacterial blight resistance of 'Minghui 63', an elite restorer line of hybrid rice, by molecular marker-assisted selection. Crop Science. 2000;**40**:239-244

[25] Aryal JP, Sapkota TB, Khurana R, Chhetri AK, Rahut DB, Jat ML. Climate change and agriculture in South Asia: Adaptation options in smallholder production systems. Environment, Development and Sustainability. 2019;**21**:1-31. DOI: 10.1007/s10668-019-00414-4

[26] Kalia RK, Rai MK, Kalia S, Dhawan KD. Microsatellite markers: An overview of the recent progress in plants. Euphytica. 2011, 2011;**177**(3):309-334. DOI: 10.1007/s10681-010-0286-9

[27] Morgante M, Hanafey H, Powell W. Microsatellites are preferentially associated with nonrepetitive DNA in plant genome. Nature Genetics. 2002;**30**(22):194-200

[28] Paun O, Schönswetter P. Amplified fragment length polymorphism (AFLP) – An invaluable fingerprinting technique for genomic, transcriptomic and epigenetic studies. Methods in Molecular Biology. 2012;**862**:75-87

[29] Shen ZK, Wang M, Li Y, Jackson D, Yin A, Dong D, et al. Crustal deformation associated with the Altyn Tagh fault system, western China, from GPS. Journal of Geophysical Research. 2001;**106**(30):607-621

[30] Zhou WC, Kolb FL, Bai GH, Domier LL, Boze LK, Smith NJ. Validation of a major QTL for scab resistance with SSR markers and use of marker-assisted selection in wheat. Plant Breeding. 2003;**122**:40-46

[31] Steele KA, Price AH, Shashidhar HE, Witcombe JR. Marker-assisted selection to introgress of rice QTLs controlling root traits and aroma into an Indian upland rice variety. Theoretical Application Genetics. 2006;**112**:208-221

[32] Mackill DJ, Collard BCY, Neeraja CN, Rodriguez RM, Heuer S, Ismail AM. QTLs in rice breeding: Examples for abiotic stresses. In: Brar DS, Mackill DJ, Hardy B, editors. Rice Genetics 5: Proceedings of the International Rice Genetics Symposium. Manila: International Rice Research Institute; 2006. pp. 155-167

[33] Neeraja C, Maghirang-Rodriguez R, Pamplona A, Heuer S, Collard B,

Septiningsih E. A marker-assisted backcross approach for developing submergence-tolerant rice cultivars. Theoretical and Applied Genetics. 2007;**115**:767-776

[34] Toojinda T, Tragoonrung S, Vanavichit A, Siangliw JL, Pa-In N, Jantaboon J, et al. Molecular breeding for rainfed lowland rice in the Mekong region. Plant Production Science. 2005;**8**:330-333

[35] Iftekharuddaula M, Ahmed HU, Ghosal S, Moni ZR, Md SA. Development of new submergence tolerant Rice variety for Bangladesh using marker-assisted backcrossing. Rice Science. 2015;**22**(1):201

[36] Cuc LM, Huyen LTN, Hien PTM, Hang VTT, Dam NQ, Mui PT, et al. Application of marker assisted backcrossing to introgress the submergence tolerance QTL *Sub*1 into the Vietnam elite rice variety-AS996. American Journal of Plant Sciences. 2012;**3**:528-536

[37] Ling ZZ, Mew T, Wang JL, Lei CL, Huang L. Development of Chinese near-isogenic lines of rice and their differentiating ability to pathogenic races of *Pyricularia grisea*. Science in Agriculture. 2000;**33**:1-8

[38] Ahmed F, Rafii MY, Ismail MR, Juraimi AS, Rahim HA, Latif MD, et al. The addition of submergence-tolerant Sub1 gene into high yielding MR219 rice variety and analysis of its BC2F3 population in terms of yield and yield contributing characters to select advance lines as a variety. Biotechnology & Biotechnological Equipment. 2016;**30**(5):853-863. DOI: 10.1080/13102818.2016.1192959

[39] Rahman H, Dakshinamurthi V, Ramasamy S, Manickam S, Kaliyaperumal AK, Raha S, et al. Introgression of submergence tolerance into CO 43, a popular Rice variety of India, through marker-assisted backcross breeding. Czech Journal of Genetics and Plant Breeding. 2018;**54**(3):101-108

[40] Huyen LTN, Cuc LM, Ismail AM, Ham LH. Introgression the salinity tolerance QTLs *Saltol* into AS996, the elite rice variety of Vietnam. American Journal of Plant Sciences. 2012;**3**:981-987

[41] Vu HTT, Le DD, Ismail AM, Le HH. Marker-assisted backcrossing (MABC) for improved salinity tolerance in rice (Oryza sativa L.) to cope with climate change in Vietnam. Australian Journal of Crop Science. 2012;**6**(12):1649-1654

[42] Suji KK, Prince KSJ, Mankhar PS, Kanagaraj P, Poornima R, Amutha K, et al. Evaluation of rice (*Oryza sativa* L.) near iso-genic lines with root QTLs for plant production and root traits in rainfed target populations of environment. Field Crops Research. 2012;**137**:89-96

[43] De Leon TB, Linscombe S, Subudhi PK. Identification and validation of QTLs for seedling salinity tolerance in introgression lines of a salt tolerant rice landrace 'Pokkali'. PLoS One. 2017;**12**(4):e0175361. DOI: 10.1371/journal.pone.0175361

[44] Fatimah MA, Emmy FA, Kusairi M, Muhammad T. Food security: Self-sufficient of rice in Malaysia. International Journal of Management. 2011;**18**(2):83-100

[45] Kumar A, Dixit S, Ram T, Yadaw RB, Mishra KK, Mandal NP. Breeding high-yielding drought-tolerant rice: Genetic variations and conventional and molecular approaches. Journal of Experimental Botany. 2014;**65**:6265-6278. DOI: 10.1093/jxb/eru363

[46] Sandhu N, Dixit S, Swamy BPM, Vikram P, Venkateshwarlu C, Catolos M, et al. Positive interactions of major-effect

QTLs with genetic background that enhances rice yield under drought. Scientific Reports. 2018;**8**:1626. DOI: 10.1038/s41598-018-20116-7

[47] Takeuchi N, Ebitani T, Yamamato T, Sato H, Ohta H, Nemoto H, et al. Development of isogenics of rice cultivar Koshihikari with early and late heading by marker-assisted selection. Breeding Science. 2006;**56**:405-413

[48] Chin JH, Gamuyao R, Dalid C, Bustamam M, Prasetiyono J, Moeljopawiro S, et al. Developing rice with high yield under phosphorus deficiency: Pup1 sequence to application. Plant Physiology. 2011;**156**:1202-1216

[49] Awasthi S, Lal JP. Application of molecular markers to improve drought tolerance and yield enhancement in upland rice. Annals of plant sciences. Annals of Plant Sciences. 2015;**4**(04):1051-1056

[50] Ramchander S, Raveendran M, Robin S. Mapping QTLs for physiological traits associated with drought tolerance in rice (Oryza Sativa L.). Journal of Investigative Genomics. 2016;**3**(3):56-61. DOI: 10.15406/jig.2016.03.00052

[51] Muthu V, Abbai R, Nallathambi J, Rahman H, Ramasamy S, Kambale R, et al. Pyramiding QTLs controlling tolerance against drought, salinity, and submergence in rice through marker assisted breeding. PLoS One. 2020;**15**(1):e0227421. DOI: 10.1371/journal.pone.0227421

[52] Jalil MA, Juraimi AS, Yusop MR, Uddin MK, Hakim MA. Introgression of root trait genes for drought tolerance to a Malaysian rice variety by marker-assisted backcross breeding. International Journal of Agriculture and Biology. 2018;**20**:119-126

[53] Chen S, Xu CJ, Lin XH, Zhang QQ. Improving bacterial blight resistance of '6078', an elite restorer line of hybrid rice, by molecular marker-assisted selection. Plant Breeding. 2001;**120**(2):133-137

[54] Sanchez AC, Brar DS, Huang N, Li Z, Khush GS. Sequence tagged site marker-assisted selection for three bacterial blight resistance genes in rice. Crop Science. 2000;**40**:792-797

[55] Singh S, Sidhu JS, Huang N, Vikal Y, Li Z, Brar DS, et al. Pyramiding three bacterial blight resistance genes (*xa5*, *xa13* and *Xa21*) using marker-assisted selection into *indica* rice cultivar PR106. Theoretical and Applied Genetics. 2001;**102**:1011-1015

[56] Natrajkumar P, Sujatha K, Laha GS, Srinivasarao K, Mishra B, Viraktamath BC, et al. Identification and fine mapping of *Xa33*, a novel gene for resistance to *Xanthomonas oryzae* pv. *oryzae*. Phytopathology. 2012;**102**:222-228

[57] Chen S, Liu X, Zeng L, Ouyang D, Yang J, Zhu X. Genetic analysis and molecular mapping of a novel recessive gene xa34(t) for resistance against *Xanthomonas oryzae* pv. Oryzae. Theoretical and Applied Genetics. 2011;**122**:1331-1338

[58] Guo S, Zhang D, Lin X. Identification and mapping of a novel bacterial blight resistance gene Xa35(t) originated from Oryza minuta. Scientia Agricultura Sinica. 2010;**43**:2611-2618

[59] Bhasin H, Bhatia D, Raghuvanshi S, Lore JS, Sahi GK, Kaur B, et al. New PCR-based sequence-tagged site marker for bacterial blight resistance gene Xa38 of rice. Molecular Breeding. 2012;**30**:607-611

[60] Ronald PC, Albano B, Tabien R, Abenes L, Wu KS, McCouch S, et al. Genetic and physical analysis of the rice bacterial blight disease resistance locus, Xa27. Gene. 1992;**236**:113-120

[61] Toenniessen GH, O'Toole JC, DeVries J. Advances in plant biotechnology and its adoption in developing countries. Current Opinion in Plant Biology. 2003;**6**:191-198

[62] Joseph M, Gopalakrishnan S, Sharma RK, Singh VP, Singh AK, Mohapatra T. Combining bacterial blight resistance and basmati quality characteristics by phenotypic and molecular marker-assisted selection in rice. Molecular Breeding. 2004;**13**:377-387

[63] Sundaram RM, Priya MRV, Laha GS, Rani NS, Rao PS, Balachandran SM, et al. Introduction of bacterial blight resistance into Triguna, a high yielding, mid-early duration rice variety by molecular marker assisted breeding. Biotechnology Journal. 2009;**4**:400-407

[64] Gopalkrishnan RK, Rajkumar AK, Josheph M, Singh VP, Singh AK, Bhat KV, et al. Integrating marker assisted background analysis with foreground selection for identification of superior bacterial blight resistant recombinants in basmati rice. Plant Breeding. 2008;**127**:131-139

[65] Pradhan SK, Pandit E, Pawar S, et al. Development of flash-flood tolerant and durable bacterial blight resistant versions of mega rice variety 'Swarna' through marker-assisted backcross breeding. Scientific Reports. 2019;**9**:12810. DOI: 10.1038/s41598-019-49176-z

[66] Balachiranjeevi CH, Bhaskar NS, Abhilash KV, Harika G, Mahadev Swamy HK, Sk H, et al. Marker-assisted pyramiding of two major, broad-spectrum bacterial blight resistance genes, *Xa*21 and *Xa*33 into an elite maintainer line of rice, DRR17B. PLoS One. 2018;**13**(10):e0201271

[67] Divya B, Robin S, Rabindran R, Senthil S, Raveendran M, John Joel A. Marker assisted backcross breeding approach to improve blast resistance in Indian rice (*Oryza sativa*) variety ADT43. Euphytica. 2014;**200**:61-77

[68] Pinta W, Toojinda T, Thummabenjapone P, Sanitchon J. Pyramiding of blast and bacterial leaf blight resistance genes into rice cultivar RD6 using marker assisted selection. African Journal of Biotechnology. 2013;**12**(28):4432-4438

[69] Singh VK, Singh A, Singh SP, Ellur RK, Choudhary V, Sarkel S, et al. Incorporation of blast resistance into "PRR78", an elite basmati rice restorer line, through marker assisted backcross breeding. Field Crops Research. 2012;**128**:8-16

[70] Fu CY, Wu T, Liu WG, Wang F, Li JH, Zhu XY, et al. Genetic improvement of resistance to blast and bacterial blight of the elite maintainer line Rongfeng B in hybrid rice (*Oryza sativa* L.) by using marker-assisted selection. African Journal of Biotechnology. 2012;**11**:13104-13124. DOI: 10.5897/AJB12.1465

[71] Liu DW, Oard SV, Oard JH. High transgene expression levels in sugarcane (*Saccharum officinarurn* L) driven by rice ubiquitin promoter RUBQ2. Plant Sciencc. 2003;**165**:743-750

[72] Srichant N, Chankaew S, Monkham T, Thammabenjapone P, Sanitchon J. Development of Sakon Nakhon rice variety for blast resistance through marker assisted backcross breeding. Agronomy. 2019;**9**(2):67

[73] Jiang J, Yang D, Ali J, Mou T. Molecular marker-assisted pyramiding of broad-spectrum disease resistance genes, *Pi2* and *Xa23*, into GZ63-4S, an elite thermo-sensitive genic male-sterile line in rice. Molecular Breeding. 2015;**35**:1-12

[74] Yu K, Park SJ, Poysa V. Marker-assisted selection of common beans for

resistance to common bacterial blight: Efficacy and economics. Plant Breeding. 2000;**119**:411-415

[75] Suh JP, Yang SJ, Jeung JU, Pamplona A, Kim JJ, Lee JH, et al. Development of elite breeding lines conferring Bph18 gene-derived resistance to brown plant hopper (BPH) by marker-assisted selection and genome-wide background analysis in japonica rice (*Oryza sativa* L.). Field Crops Research. 2011;**120**:215-222

[76] Sama VSAK, Himabindu K, Naik SB, Sundaram RM, Viraktamath BC, Bentur JS. Mapping and marker-assisted breeding of a gene allelic to the major Asian rice gall midge resistance gene *Gm8*. Euphytica. 2012;**187**:393-400

[77] Xu J. Pyramiding of two BPH resistance genes and *Stv-bi* gene using marker-assisted selection in japonica rice. Crop Breeding and Applied Biotechnology. 2013;**13**:99-106

[78] Lang NT, Phuoc NT, Ha PTT, Toan TB, Buu BC, Reinke R, et al. Development of submergence tolerant breeding lines for Vietnam. Sabrao Journal of Breeding and Genetics. 2015;**47**:448-459

[79] Septiningsih EM, Pamplona AM, Sanchez DL, Neeraja CN, Vergara GV, Heuer S, et al. Development of submergence-tolerant rice cultivars: The Sub1 locus and beyond. Annals of Botany. 2008;**103**:151-160

[80] Siangliw JL, Jongdee B, Pantuwan G, Toojinda T. Developing KDML105 backcross introgression lines using marker-assisted selection for QTLs associated with drought tolerance in rice. Science Asia. 2007;**33**:207-214

[81] Ahn SN, Bollich CN, Tanksley SD. RFLP tagging of a gene for aroma in rice. Theoretical Application Genetic. 1992;**84**:825-828

[82] Lorieux M, Petrov M, Huang N, Guiderdoni E, Ghesquière A. Aroma in rice: Genetic analysis of a quantitative trait. Theoritical Application Genetic. 1996;**93**:1145-1151

[83] Lang NT, Buu BC. Initial marker-assisted selection in rice breeding at Cuu Long Delta Rice Research Institute. Omonrice. 2010;**17**:8-21

[84] Jin Q, Waters D, Cordeiro GM, Henry RJ, Reinke RF. A single nucleotide polymorphism (SNP) marker linked to the fragrance gene in rice (*Oryza sativa* L.). Plant Science. 2003;**165**:359-364

[85] Mahdav MS, Pandey MK, Kumar PR, Sundaram RM, Prasad GSV, Sudarshan L, et al. Identification and mapping of tightly linked SSR marker for aroma trait for use in marker assisted selection in rice. Rice Genetics Newsletter. 2009;**25**:38-39

[86] Yi M, Nwe KT, Vanavichit A, Chai-arree W, Toojinda T. Markerassisted backcross breeding to improve cooking quality traits in Myanmar ricecultivar Manawthukha. Field Crops Residue. 2009;**113**:178-186

[87] Singh AK, Gopalakrishnan S, Singh VP, Prabhu KV, Mohapatra T, Singh NK, et al. Marker assisted selection: A paradigm shift in Basmati breeding. Indian Journal Genetic Plant Breeding. 2011;**71**(2):120-128

[88] Negrão S, Palaniappan J, Maroco J, Lourenço T, Mackill D, Oliveira MM. Bridging sd1 molecular knowledge with recent breeding strategies for the improvement of traditional rice varieties: A japonica case-study. African Journal Biotechnology. 2010;**9**:2192-2200

[89] Pan HJ, Wang CL, Zhao KJ, Zhang Q, Fan YL, Zhou SC, et al. Molecular mapping by PCR-based markers and marker-assisted selection of Xa23 a bacterial blight resistance

gene in rice. Acta Agronomica Sinica. 2003;**29**(4):501-507

[90] Basavaraj SH, Singh VK, Singh A, Singh A, Singh A, Yadav S, et al. Marker-assisted improvement of bacterial blight resistance in parental lines of Pusa RH10, a superfine grain aromatic rice hybrid. Molecular Breeding. 2010;**26**:293-305

[91] Win KM, Korinsak S, Jantaboon J, Siangliw M, Lanceras-Siangliw J, Sirithunya P, et al. Breeding the Thai jasmine rice variety KDML105 for non-age-related broad-spectrum resistance to bacterial blight disease based on combined marker-assisted and phenotypic selection. Field Crops Residue. 2012;**137**:186-194

[92] Xu J, Jiang J, Dong X, Ali J, Mou T. Introgression of bacterial blight (BB) resistance genes Xa7 and Xa21 into popular restorer line and their hybrids by molecular marker-assisted backcross (MABC) selection scheme. African Journal of Biotechnology. 2012;**11**:8225-8233

[93] Chen S, Huang Z, Zeng L, Yang J, Liu Q, Zhu X. High-resolution mapping and gene prediction of *Xanthomonas Oryzae* pv. Oryzae resistance gene Xa7. Molecular Breeding. 2008;**22**:433-441. DOI: 10.1007/s11032-008-9187-1

[94] Pandey MK, Shobharani N, Sundaram RM, Laha GS, Madhav MS, Rao KS, et al. Improvement of two traditional Basmati rice varieties for bacterial blight resistance and plant stature through morphological and marker-assisted selection. Molecular Breeding. 2013;**31**:239-246

[95] Basavaraj SH, Singh VK, Atul S, Devender S, Nagarajan M, Mohapatra T, et al. Marker aided improvement of Pusa 6B, the maintainer parent of rice hybrid Pusa RH10, for resistant to bacterial blight. Indian Journal of Genetics. 2009;**69**(1):10-16

[96] Sabouri H, Sabouri A, Jafarzadeh MR, Mollashahi M. Detection of QTLs controlling field blast resistance in rice (*Oryza sative* L.). Plant Omics. 2011;**4**:1-5

[97] Tanweer FA, Rafii MY, Sijam K, Rahim HA, Ahmed F, Ashkani S, et al. Introgression of Blast Resistance Genes (Putative Pi-b and Pi-kh) into Elite Rice Cultivar MR219 through Marker-Assisted Selection. Frontier Plant Science. 2015;**6**:1002. DOI: 10.3389/fpls.2015.01002

[98] Hasan MM, Rafii MY, Ismail MR, Mahmood M, Rahim HA, Alam MA. Marker-assisted backcrossing: A useful method for rice improvement. Biotechnology Equipment. 2015a;**29**:237-254

[99] Jairin J, Teangdeerith S, Leelagud P, Kothcharerk J, Sansen K, Yi M, et al. Development of rice introgression lines with brown planthopper resistance and KDML105 grain quality characteristics through marker-assisted selection. Field Crop Residue. 2009;**110**:263-271

[100] Jena KK, Jeung JU, Lee JH, Choi HC, Brar DS. High-resolution mapping of a new brown planthopper (BPH) resistance gene, Bph18(t), and marker-assisted selection for BPH resistance in rice (*Oryza sativa* L.). Theoretical Application Genetics. 2006;**112**:288-297

[101] Bentur JS, Srinivasan TE, Kalode MB. Occurrence of a virulent gall midge (GM), Orseolia oryzae Wood-Mason biotype in Andhra Pradesh, India. International Rice Research Newsletter. 1987;**12**(3):33-34

[102] Himabindu K, Suneetha K, Sama VSAK, Bentur JS. A new rice gall midge resistance gene in the breeding line CR57-MR1523, mapping with flanking markers and development of NILs. Euphytica. 2010;**174**:179-187

[103] Vijaya LP, Amudhan S, Himabindu K, Cheralu C, Bentur JS. A new biotype of the Asian rice gall midge Orseolia oryzae (Diptera: Cecidomyiidae) characterized from the Warangal population in Andhra Pradesh, India. International Journal Insect Science. 2006;**26**:207-211

Chapter 7

# Pathogens Transmitted through Contaminated Rice

*Leka Lutpiatina*

## Abstract

Rice can be a source of food poisoning because it can be contaminated with dangerous pathogens. Pathogens that often transmitted through rice are *Bacillus cereus* and *Staphylococcus aureus*. This chapter aims to explain the dangers of pathogens transmitted through contaminated rice, modes of transmission, contamination cases, and precautions. The method used in writing is to review articles. It is known that pathogens transmitted through contaminated rice can cause food poisoning, which occurs due to consuming rice containing pathogenic bacteria. Several cases of contamination of *Bacillus cereus* and *Staphylococcus aureus* in rice occurred in Indonesia, Pakistan, India, Malaysia, Belgium, America, Australia, Korea, Iran, China, and Nigeria. In general, prevention is by proper handling of raw materials, controlling the temperature of cooking and storing rice, and personal hygiene of food handlers.

**Keywords:** rice, pathogens transmitted, food poisoning, *Bacillus cereus*, *Staphylococcus aureus*

## 1. Introduction

Most of the world's population, especially Asia, use rice as the primary source of carbohydrates in their daily menu. Rice, as a staple food, is usually served with side dishes to complement the taste and also complement one's nutritional needs. Rice can process with other food ingredients into new dishes, such as fried rice, yellow rice, or uduk rice.

Rice can be a source of food poisoning because it contaminated with dangerous pathogens. In general, food poisoning can cause by contaminant bacteria such as *Bacillus cereus*, *Staphylococcus aureus*, *Salmonella* group (except *Salmonella typhi*), *Shigella, Vibrio, Escherichia coli, Campylobacter, Yersinia enterocolitis, Clostridium* [1]. Rice-based food poisoning more often caused by *Bacillus cereus* and *Staphylococcus aureus* based on several cases in the world [2–15]. Besides, *Bacillus cereus* and *Staphylococcus aureus* are significant sources of microbiological harm from cereal grains and related products [16].

*Bacillus cereus* can found in soil, plants, and the intestinal tract of insects and mammals. In poor environmental conditions, bacteria can turn into spore forms. *Bacillus cereus* in spore form can found hiding in raw rice. The bacteria move from the soil to the paddy fields, their spores persist for years, even surviving during cooking due to their resistance to extreme temperatures. However, if rice left at room temperature, in warm and humid conditions, the spores can turn into bacteria and produce toxins that can cause vomiting and diarrhea [17].

*Staphylococcus aureus* is naturally present in the human body, so these bacteria are one of the essential agents causing food poisoning that often occurs in society. The most significant cause of *Staphylococcus aureus* entry into the food chain (which then causes staphylococcal poisoning) is the low sanitation of workers handling food [18].

According to the Food Standards Agency (FSA), there are nearly 900,000 food poisoning cases each year. The lifestyle that has changed in recent years has resulted in an increasing dependence on ready-to-eat food, eating out more than cooking, busyness results in having less time to prepare and cook food. This habit is the reason that increases the number of cases of food poisoning [19]. Apart from that, environmental factors also influence the level of contamination. Food prepared under unfavorable conditions and environment implies a higher incidence of food poisoning than others [18]. Food poisoning occurs more frequently in developing countries than in developed countries. This situation is due to differences in the level of sanitation between developed and developing countries [18].

## 2. Pathogenic *Bacillus cereus* transmitted through rice

### 2.1 Characteristics and diseases caused by *Bacillus cereus*

*Bacillus cereus* is a spore-forming bacteria that produces a toxin that causes vomiting or diarrhea. Symptoms are generally mild and short-lived (up to 24 h). *Bacillus cereus* commonly found in the environment (e.g., soil) and various foods. Spores are able to withstand harsh environments, including average cooking temperatures. *Bacillus cereus* is a Gram-positive, motile (flagellated), spore-forming, rod-shaped bacterium belonging to the genus Bacillus. Species in this genus include *Bacillus anthracis, Bacillus cereus, Bacillus mycoides, Bacillus thuringiensis, Bacillus pseudomycoides, and Bacillus weihenstephanensis* [17, 20].

*Bacillus cereus* is widespread and easy to find in the soil, where it adopts a saprophytic life cycle, germinate, grow, and sporulate in this environment [21]. Spores are more resistant to environmental stress than vegetative cells because of their metabolic dormancy and hard physical properties [22].

*Bacillus cereus* causes two types of diseases, namely emetic syndrome and diarrhea syndrome. Emetic syndrome causes by emetic toxins produced by bacteria during the growth phase in food, a diarrheal syndrome caused by diarrheal toxins produced during bacterial growth in the small intestine [23].

*Bacillus cereus* has a mesophilic or psychrotrophic strain. Mesophilic strains grow well at 37°C but do not grow below 10°C, whereas psychrotrophic strains grow well at cold temperatures but grow poorly at 37°C [24]. All *Bacillus cereus* isolates associated with emetic toxin production have found to be mesophilic [25]. *Bacillus cereus* growth is optimal in the presence of oxygen, although it continues to grow under anaerobic conditions. *Bacillus cereus* cells grown in aerobic conditions were less resistant to heat and acid than *Bacillus cereus* cells that grew anaerobically or microaerobically [26]. *Bacillus cereus*' mesophilic strains have shown to have higher acid resistance than psychotropic strains [25].

Spores are more resistant to dry heat than humid heat, with heat resistance usually higher in foods with lower water activity. Spores are also more resistant to radiation than vegetative cells [22]. Nisin is a preservative that used to inhibit germination and spore growth. Antimicrobials that inhibit the growth of *Bacillus cereus* include benzoic, sorbic, and ethylenediaminetetraacetic acids [22].

Symptoms of *Bacillus cereus* disease cause two types of foodborne illness, namely emetic syndrome (vomiting) and diarrhea. Vomiting syndrome is poisoning caused by the ingestion of a cyclic peptide toxin called cereulide that has been pre-formed in food during growth by *Bacillus cereus*. This syndrome has a short incubation period and recovery time. Symptoms of nausea, vomiting, and stomach cramps occur within 1–5 h after ingestion, with recovery usually within 6–24 h [27].

The diarrheal syndrome caused by enterotoxins produced by *Bacillus cereus* in the host body, the incubation period before the onset of the disease is 8–16 h, and infection usually lasts 12–14 h. However, it can continue for several days. Symptoms are generally mild, with stomach cramps, watery diarrhea, and nausea [28]. In a small number of cases, both toxins produced, and vomiting and diarrhea develop [17].

No form of the disease is considered life-threatening for normal healthy individuals, with few fatal cases reported [22]. *Bacillus cereus* has been associated with the non-food-related illness, although this is rare. These bacteria have been found in postoperative and traumatic wounds and can cause opportunistic infections, especially in individuals with immune system disorders, such as septicemia, meningitis, and pneumonia. *Bacillus cereus* has also known to occasionally cause localized eye infections in humans [29].

The pathogenic mechanisms for *Bacillus cereus* emetic disease are well known. Emetic toxins (cereulide) cause vacuole formation in HEp-2 cells in the laboratory [29]. Cereulide in experimental animals caused vomiting, potentially by binding to 5-HT3 receptors in the stomach/small intestine to stimulate the vagus nerve and brain [30]. Cereulide produced by the Non-Ribosomal Peptide Synthetase Complex (NRPS) [31]. All NRPS clusters have characterized [23], resulting in a precise method for detecting cereulide-producing *Bacillus cereus* strains [32]. Emetic toxin production has shown to occur in skim milk in the temperature range of 12–37°C, with more toxins produced at 12 and 15°C than at higher temperatures [33]. Emetic toxins are highly resistant to environmental factors, exhibiting stability from a pH of 2–11 and heating to 100°C for 150 min (pH 8.7–10.6) [22].

Three types of enterotoxins are associated with a form of diarrhea syndrome, namely three components of the enterotoxin Hemolysin BL (HBL), three parts of Non-Hemolytic Enterotoxin (NHE) and one element of cytotoxin K. Enterotoxins are released into the small intestine by the surviving vegetative cells of *Bacillus cereus* [34]. Diarrheal enterotoxins are stable at pH 4–11 and deactivated by heating to 56°C for 5 min [22].

Up to 26% of the vegetative cells of *Bacillus cereus* can survive as they travel through the stomach. Diarrheal enterotoxins are unstable at low pH in the stomach and degraded by digestive enzymes. Any previously formed enterotoxins in food are destroyed during passage through the stomach so that they do not cause disease if ingested [22]. In contrast, *Bacillus cereus* spores can pass unaffected by the gastric barrier. Spores need to be triggered by nutrients and intestinal epithelial cells to initiate germination. In the small intestine, spores germinate, grow and produce enterotoxins [35].

A vital virulence factor required to cause diarrhea symptoms is the ability of vegetative cells and *Bacillus cereus* spores to adhere to the small intestine's epithelial cell walls. Spore and cell adhesion efficiency show to be low, around 1% [35]. Enterotoxins' ability to damage tissue and damage the plasma membrane of small intestinal epithelial cells plays a role in causing diarrhea [27].

### 2.2 Mode of transmission *Bacillus cereus* food poisoning

The pattern of transmission *Bacillus cereus* food poisoning can be caused by ingesting large numbers of bacterial cells and spores in contaminated food (diarrhea type) or by ingesting food contaminated with pre-formed toxins (emetic type). Transmission of this disease caused by the consumption of contaminated food, improper handling/storage of food, and inadequate cooling of cooked food [36].

| Year | Food | Country | Findings | Article title |
|---|---|---|---|---|
| 2009 | Brown rice and glutinous rice | Korea | 15 (37%) of 83 samples of brown rice, 23 (37%) of 63 samples of glutinous rice | Prevalence, Genetic diversity, and Antibiotic Susceptibility of *Bacillus cereus* Strains Isolated from Rice and Cereals Collected in Korea [2] |
| 2009 | Raw rice | Amerika | *Bacillus* species spores found in 94 (52.8%) rice samples with an average concentration of 32.6 CFU/g | Detection of Toxigenic *Bacillus cereus* and *Bacillus thuringiensis* in US Rice [3] |
| 2012 | Cooked rice (white rice, fried rice) | Belgia | The concentration of cereulide found in rice dishes is around four ng/g | Prevalence and Levels of *Bacillus cereus* Emetic Toxin in Rice Dishes Randomly Collected from Restaurants and Comparison with the Levels Measured in a Recent Foodborne Outbreak [4] |
| 2012 | Cooked and raw rice | Pakistan | All rice samples showed the presence of *Bacillus cereus*, the highest number: 3.34 × 101 CFU/ml | Microbial Assessment of Uncooked and Cooked Rice Samples Available in Local Markets of Lahore [5] |
| 2013 | Baby food (made from rice) | Iran | *Bacillus cereus* and its enterotoxigenic genes have found in infant diets in Iran | *Bacillus cereus* in Infant Foods: Prevalence Study and Distribution of Enterotoxigenic Virulence Factors in Isfahan Province, Iran [6] |
| 2018 | Local unhulled (coarse) rice | Malaysia | The number of *Bacillus cereus* bacteria in all samples found to be more than 1100 MPN/g | Presence of *Bacillus cereus* from Local Unhusked (Rough) Rice Samples in Sarawak, Malaysia [7] |
| 2019 | Cooked rice (yellow rice) | Indonesia | 21% of yellow rice contaminated with *Bacillus cereus* | *Staphylococcus aureus* and *Bacillus cereus* in Yellow Rice [8] |
| 2020 | Rice/ noodles | China | 59 out of 119 rice/noodle samples (50%) were positive for *Bacillus cereus* | A Study on Prevalence and Characterization of *Bacillus cereus* in Ready-to-Eat Foods in China [9] |

**Table 1.**
*Cases of* Bacillus cereus *contamination in rice (rice-based food).*

## 2.3 Case contamination and precautions for *Bacillus cereus* in rice

Cases of contamination of *Staphylococcus aureus* on rice, either in the form of raw rice or cooked rice and other processed rice products, are found in several countries. Further explanations can see in **Table 1**.

Precautions for contamination of *Bacillus cereus* in rice:

1. Processing (thoroughly cooked and quickly cooled) is one of the easiest ways to prevent foodborne illness associated with Bacillus spp. [37].

2. Hot foods should store at 140°F/60°C or higher [37].

3. Reheating cooked food should be stored at 165°F/74°C [37].

4. If frozen food is allowed to thaw, it must remain at 41°F/5°C or lower [37].

5. Steaming under pressure, roasting, frying, and grilling foods will destroy the vegetative cells and spores if temperatures within foods are ≥ 145°F/63°C [38].

6. Foods that contain emetic toxins need to be heated to 259°F/126°C for more than 90 min—reheating food until steaming is not sufficient to kill emetic toxins [38].

## 3. Pathogenic *Staphylococcus aureus* transmitted through rice

### 3.1 Characteristics and diseases caused by *Staphylococcus aureus*

*Staphylococcus aureus* is one of the bacteria that cause food poisoning. *Staphylococcus aureus* is commonly found in the environment (soil, water, and air) and located on humans' nose and skin. *Staphylococcus aureus* is a spherical, Gram-positive, non-spore bacteria. The genus Staphylococcus divided into 32 species and subspecies. *Staphylococcus aureus* causes food poisoning by producing Staphylococcal Enterotoxin (SE) [39, 40].

*Staphylococcus aureus*'s growth and survival depend on several environmental factors such as temperature, water activity (aw), pH, presence of oxygen, and food composition. These physical growth parameters varied for different strains of *Staphylococcus aureus* [41]. The temperature range for *Staphylococcus aureus* growth is 7–48°C, with an optimum temperature of 37°C. *Staphylococcus aureus* is resistant to freezing and does well in foods stored below −20°C; however, viability is reduced at −10 to 0°C. *Staphylococcus aureus* easily killed during pasteurization or cooking. *Staphylococcus aureus* growth occurs in the pH range 4.0–10.0, with an optimum of 6–7 [41].

*Staphylococcus aureus* is a facultative anaerobe so it can grow in both aerobic and anaerobic conditions. However, growth occurs at a much slower rate under anaerobic conditions [41]. For non-sporing mesophilic bacteria, *Staphylococcus aureus* has relatively high heat resistance [41]. A highly heat resistant *Staphylococcus aureus* strain (D-value at 60°C > 15 min in broth) has identified from foodborne outbreaks in India [42].

Several chemical preservatives, including sorbate and benzoate, inhibit the growth of *Staphylococcus aureus*. The effectiveness of this preservative increases as the pH decreases. Methyl and propyl parabens are also useful [41, 43].

Symptoms of staphylococcal food poisoning generally have a rapid onset, appearing approximately 3 h after ingestion (range 1–6 h). Common symptoms include nausea, vomiting, stomach cramps, and diarrhea. The individual may not show all the signs associated with the disease. In severe cases, headaches, muscle cramps, and temporary changes in blood pressure and pulse may occur. Recovery is usually between 1 and 3 days [39, 41]. Death is rare (0.03% for the general population) but occasionally reported in children and the elderly (death rate 4.4%) [40]. *Staphylococcus aureus* can cause various health problems not related to food such as skin inflammation (e.g., ulcers and style), mastitis, respiratory tract infections, wound sepsis, and toxic shock syndrome [40, 41].

Staphylococcal food poisoning caused by the ingestion of foods containing pre-formed SE [44], there are several types of SE; enterotoxin A is most commonly associated with staphylococcal food poisoning. Enterotoxins D, E, and H, and to a lesser extent B, G, and I have also associated with staphylococcal food poisoning [45, 46].

SE produced during the exponential phase of *Staphylococcus aureus* growth in a strain-dependent quantity. Typically, the disease-inducing dose of SE occurs when at least 105–108 CFU/g of *Staphylococcus aureus* are present [45, 40]. Most of the genes for SE located in plasmid or prophage elements. Thus, transfer between strains can occur, modifying the ability of *Staphylococcus aureus* strains to cause disease and contributing to pathogen evolution [44, 46].

As the temperature decreases, the SE production rate also decreases. However, SE remained stable under frozen storage. SE is highly resistant to heating and can withstand the processes used to sterilize low-acid canned foods. SE production can occur in the pH range 4.5–9.6, with an optimum of 7–8. SE production can occur in anaerobic and aerobic environments; however, toxin production is optimal under aerobic conditions [41].

## 3.2 Mode of transmission. *Staphylococcus aureus* food poisoning

Staphylococcal food poisoning occurs when the food consumed contains SE produced by *Staphylococcus aureus*. Food handlers carrying enterotoxin-producing *Staphylococcus aureus* in their nose or hands are considered a significant source of food contamination through direct contact or respiratory secretions [44].

| Year | Food | Country | Findings | Article title |
|---|---|---|---|---|
| 2001 | Rames rice | Indonesia | The number of *Staphylococcus aureus* in the rice sample: 3.21 Log CFU/g | Study of Microbiological Safety of Snack Food at the FATETA-IPB Canteen, Bogor [10] |
| 2003 | Rice at the restaurant | Brazil | Rice containing *Staphylococcus aureus*: 100 CFU/g | An outbreak of staphylococcal food poisoning in the Municipality of Passos, MG, Brazil [11] |
| 2004 | Rice cake | Korea | 19.3% of rice cakes were contaminated with *Staphylococcus aureus* | Occurrence of Toxigenic *Staphylococcus aureus* in Ready-to-Eat Food in Korea [12] |
| 2010 | Uduk rice | Indonesia | The frequency of isolation of coagulase-positive *Staphylococcus aureus* in uduk rice samples was 6.67%, and not all were found in sufficient amounts to form enterotoxins. | Risks of *Staphylococcus aureus* in Traditional Ready-to-Eat Food and Evaluation of Its Presence in uduk Rice [13] |
| 2014 | Kerala matta rice | India | Kerala matta rice samples contained coagulase-positive *Staphylococcus aureus*. | Outbreak of Staphylococcal Food Poisoning [14] |
| 2019 | Yellow rice | Indonesia | 7% of yellow rice contaminated with *Staphylococcus aureus* | *Staphylococcus aureus* and *Bacillus cereus* in yellow rice [8] |
| 2019 | Jollof rice | Nigeria | *Staphylococcus aureus* found in jollof rice samples at the campus cafeteria. | Identification and anti-bacterial Testing of *Staphylococcus aureus* Isolated from Jollof Rice sold at selected Cafeterias in Federal University [15] |

**Table 2.**
*Cases of* Staphylococcus aureus *contamination in rice (rice-based food).*

It estimated that in the US, *Staphylococcus aureus* accounts for 2.6% of foodborne diseases caused by 31 significant pathogens [47]. The incidence of staphylococcal food poisoning is seasonal. Most cases occur in late summer when temperatures are warm, and food is stored incorrectly [40].

Foods associated with the staphylococcal food poisoning outbreak include meat and meat products, poultry and egg products, milk and dairy products, salads, cream sandwich products, and sandwich stuffing. Foods that require extensive handling during preparation and stored above refrigeration temperature (4°C) for a long time after development frequently implicated in staphylococcal food poisoning [39]. Foods high in starch (such as rice) and protein believed to support SE production [41].

### 3.3 Case contamination and precautions for *Staphylococcus aureus* in rice

Cases of contamination of *Staphylococcus aureus* on rice, either in the form of raw rice or cooked rice and other processed rice products, are found in several countries in the world. Further explanation can see in **Table 2**.

Precautions for contamination of *Staphylococcus aureus* in rice:

1. The permissive temperature for growth and toxin production by *Staphylococcus aureus* is between 6 and 46°C. Thus, the ideal cooking and cooling temperatures should be above 60°C and below 5°C, respectively, is below the recommended temperature [48].

2. Serving food quickly when stored at room temperature, wearing gloves, masks, hairpins during food handling and processing, washing hands frequently, maintaining personal hygiene for food handlers can help prevent *Staphylococcus aureus* contamination [49].

3. Other precautions such as raw material control, proper handling and processing, adequate cleaning, and disinfection of equipment used in food processing and preparation must take [50].

4. Environmental factors that can play an essential role in the proliferation of bacteria and the production of *Staphylococcus aureus* enterotoxins are the storage of rice at room temperature for an extended period between preparation and consumption [14].

## 4. Conclusions

It knows that pathogens transmitted through contaminated rice can cause food poisoning, which occurs due to consuming rice containing pathogenic bacteria. Several cases of contamination of *Bacillus cereus* and *Staphylococcus aureus* in rice occurred in Indonesia, Pakistan, India, Malaysia, Belgium, America, Australia, Korea, Iran, China, Nigeria. In general, prevention by proper handling of raw materials, controlling the temperature of cooking and storing rice, and personal hygiene of food handlers.

## Acknowledgements

Thanks go to students and lecturers of the Medical Laboratory Technology Poltekkes Kemenkes Banjarmasin, Indonesia who have supported the writing of this manuscript and to all parties who did not directly play a role in the writing process.

## Conflict of interest

The authors declare no conflict of interest.

## Acronyms and abbreviations

| | |
|---|---|
| HBL | hemolysin BL |
| NHE | non-hemolytic enterotoxin |
| SE | staphylococcal enterotoxin |

## Author details

Leka Lutpiatina
Medical Laboratory Technology, Poltekkes Kemenkes Banjarmasin, Indonesia

*Address all correspondence to: leka.zns@gmail.com

## References

[1] Bardale R. Principles of Forensic Medicine & Toxicology. 2nd ed. New Delhi-110002: Jaypee Brothers Medical Publishers Ltd; 2017. p. 606

[2] Park Y-B, Kim J-B, Shin S-W, Kim J-C, Cho S-H, Lee B-K, et al. Prevalence, genetic diversity, and antibiotic susceptibility of *Bacillus cereus* strains Isolated from rice and cereals collected in Korea. Journal of Food Protection. 2009;**72**(3):612-617. DOI: 10.4315/0362-028x-72.3.612

[3] Ankolekar C, Rahmati T, Lebbe RG. Detection of toxigenic *Bacillus cereus* and *Bacillus thuringiensis* in US rice. International Journal of Food Microbiology. 2009;**128**:460-466

[4] Delbrassinne L, Andjelkovic M, Dierick K, Denayer S, Mahillon J, Van Loco J. Prevalence and levels of *Bacillus cereus* emetic toxin in rice dishes randomly collected from restaurants and comparison with the levels measured in a recent foodborne outbreak. Foodborne Pathogens and Disease. 2012;**9**(9):809-814

[5] Tahir A, Hameed I, Aftab M, Mateen A. Microbial assessment of uncooked and cooked rice samples available in local markets of Lahore. Pakistan Journal of Botany. 2012;**44**:267-270

[6] Rahimi E, Abdos F, Momtaz H, Torki Baghbadorani Z, Jalali M. *Bacillus cereus* in infant foods: Prevalence study and distribution of enterotoxigenic virulence factors in Isfahan Province, Iran. The Scientific World Journal. 2013;**2013**:1-6

[7] Bilung LM, Tesfamariam F, Andriesse R, San FYK, Ling CY, Tahar AS. Presence of *Bacillus cereus* from local Unhusked (rough) rice samples in Sarawak, Malaysia. Journal of Sustainability Science and Management. 2018;**13**(1):181-187

[8] Leka L, Nudia K, Dewi DR, Wahdah N. *Staphylococcus aureus* and *Bacillus cereus* in yellow rice. Indian Journal of Public Health Research & Development. 2019;**10**(8):2104-2108. DOI: 10.5958/0976-5506.2019.02166.1

[9] Yu S, Yu P, Wang J, et al. A study on prevalence and characterization of *Bacillus cereus* in ready-to-eat foods in China. Frontiers in Microbiology. 2020;**10**:1-11. DOI: 10.3389/fmicb.2019.03043

[10] Hartini PB. Study of microbiological safety of snack food at the FATETA-IPB Canteen, Bogor [thesis]. Bogor: Institut Pertanian Bogor; 2001

[11] do Carmo LS, Dias RS, Linardi VR, de Sena MJ, dos Santos DA. An outbreak of staphylococcal food poisoning in the municipality of Passos, MG, Brazil. Brazilian Archives of Biology and Technology. 2003;**46**(4):581-586. DOI: 10.1590/S1516-89132003000400012

[12] Su Kyung O, Lee N, Cho YS, Shin D-B, Choi SY, Koo M. Occurrence of toxigenic *Staphylococcus aureus* in ready-to-eat food in Korea. Journal of Food Protection. 2007;**70**(5):1153-1158. DOI: 10.4315/0362-028X-70.5.1153

[13] Erza Apriyadi T. Risks of *Staphylococcus aureus* in traditional ready-to-eat food and evaluation of its presence in uduk rice [thesis]. Bogor: Fakultas Teknologi Pertanian Institut Pertanian Bogor; 2010

[14] Basavegowda M, Rajegowda RM, Bettappa P, Shivanna UR, Channabasappa AN, Vijaykumar GS. Outbreak of staphylococcal food poisoning. International Journal of Medicine and Public Health. 2014;**4**(3):257-259

[15] Radji M, Adeleye A, Amoo A, Barde G, Mohammed S. Identification

and anti-bacterial testing of *Staphylococcus aureus* isolated from jollof rice sold at selected cafeterias in Federal University. UJMR. 2019;**4**(1):104-109

[16] FDA. Analysis of microbial hazards related to time/temperature control of foods for safety. Comprehensive Reviews in Food Science and Food Safety. 2003;**2**:33-41. DOI: 10.1111/j.1541-4337.2003.tb00049.x

[17] Montville TJ, Matthews KR. Food Microbiology: An Introduction. Washington D.C: ASM Press; 2005

[18] Ray B. Fundamental Food Microbiology Second Edition. Boca Raton/London/New York, Washington D.C: CRC Press; 2001

[19] Parashnath M, Indranil C, Food Poisoning: Illness ranges from relatively mild through to life threatening. Journal of Medical and Health Sciences. 2016;**5**(4):1-19

[20] Rajkowski KT, Bennett RW. *Bacillus cereus*. Ch 3. In: Miliotis MD, Bier JW, editors. International Handbook of Foodborne Pathogens. New York: Marcel Dekker; 2003. pp. 27-39

[21] Vilain S, Luo Y, Hildreth M, Brözel V. Analysis of the life cycle of the soil saprophyte *Bacillus cereus* in liquid soil extract and in soil. Applied and Environmental Microbiology. 2006;**72**:4970-4977

[22] Jenson I, Moir CJ. *Bacillus cereus* and other *Bacillus* species. Ch 14. In: Hocking AD, editor. Foodborne Microorganisms of Public Health Significance. 6th ed. Sydney: Australian Institute of Food Science and Technology (NSW Branch); 2003. pp. 445-478

[23] Ehling-Schulz M, Guinebretière M, Monthan A, Berge O, Fricker M, Svensson B. Toxin gene profiling of enterotoxic and emetic *Bacillus cereus*. FEMS Microbiology Letters. 2006;**260**(2):232-240

[24] Wijnands LM, Dufrenne JB, Zwietering MH, van Leusden FM. Spores from mesophilic *Bacillus cereus* strains germinate better and grow faster in simulated gastrointestinal conditions than spores from psychrotrophic strains. International Journal of Food Microbiology. 2006a;**112**(2):120-128

[25] Wijnands LM, Dufrenne JB, Rombouts FM, in 't Veld PH, van Leusden FM. Prevalence of potentially pathogenic *Bacillus cereus* in food commodities in the Netherlands. Journal of Food Protection. 2006;**69**(11):2587-2594

[26] Mols M, Pier I, Zwietering MH, Abee TJ. The impact of oxygen availability on stress survival and radical formation of *Bacillus cereus*. International Journal of Food Microbiology. 2009;**135**(3):303-311

[27] Senesi S, Ghelardi E. Production, secretion and biological activity of *Bacillus cereus* enterotoxins. Toxins. 2010;**2**:1690-1703

[28] Granum PE. *Bacillus cereus*. Ch 20. In: Doyle MP, Beuchat LR, editors. Food Microbiology: Fundamentals and Frontiers. 3rd ed. Washington D.C: ASM Press; 2007. pp. 445-455

[29] Schoeni JL, Wong ACL. *Bacillus cereus* food poisoning and its toxins. Journal of Food Protection. 2005;**68**(3):636-648

[30] Agata N, Ohta M, Mori M, Isobe M. A novel dodecadepsipeptide, cereulide, is an emetic toxin of *Bacillus cereus*. FEMS Microbiology Letters. 1995;**129**:17-20

[31] Horwood PF, Burgess GW, Oakey HJ. Evidence for non-ribosomal

peptide synthetase production of cereulide (the emetic toxin) in *Bacillus cereus*. FEMS Microbiology Letters. 2004;**236**(2):319-324

[32] Fricker M, Misselhausser U, Busch U, Scheres S, Ehling-Schulz M. Diagostic real time PCR assays for the detection of emetic food-borne *Bacillus cereus* in foods and recent food-borne outbeaks. Applied and Environmental Microbiology. 2007;**73**:1892-1898

[33] Finlay WJJ, Logan NA, Sutherland AD. *Bacillus cereus* produces most emetic toxin at lower temperatures. Letters in Applied Microbiology. 2000;**31**:385-389

[34] Wijnands LM, Pielaat A, Dufrenne JB, Zwietering MH, van Leusden FM. Modelling the number of viable vegetative cells of *Bacillus cereus* passing through the stomach. Journal of Applied Microbiology. 2009;**106**:258-267

[35] Wijnands LM. *Bacillus cereus* associated food borne disease: Quantitative aspects of exposure assessment and hazard characterization [thesis]. The Netherlands: Wageningen University; 2008

[36] Schneider KR, Parish ME, Goodrich RM, Cookingham T. Preventing Foodborne Illness: *Bacillus cereus* and *Bacillus anthracis*. Florida Cooperative Extension Service, Institute of Food and Agricultural Sciences, University of Florida: 2004. pp. 1-5

[37] Nazrul Islam M. Fried Rice syndrome, a disease of fast world: Scientific analysis. American Journal of Biomedical Science & Research. 2019;**5**(6):512-514. DOI: 10.34297/AJBSR.2019.05.000979

[38] Keith RS, Renée GS, Rachael S, Ploy K, Bruna B. Preventing Foodborne Illness: *Bacillus cereus*. Florida Cooperative Extension Service. University of Florida: Institute of Food and Agricultural Sciences; 2015. pp. 1-5

[39] FDA. Bad Bug Book: Foodborne Pathogenic Microorganisms and Natural Toxins Handbook. 2nd ed. Silver Spring: US Food and Drug Administration; 2012. pp. 87-92

[40] Montville TJ, Matthews KR. Food Microbiology: An Introduction. 2nd ed. Washington D.C: ASM Press; 2008

[41] Stewart CM. *Staphylococcus aureus* and staphylococcal enterotoxins. Ch 12. In: Hocking AD, editor. Foodborne Microorganisms of Public Health Significance. 6th ed. Sydney: Australian Institute of Food Science and Technology (NSW Branch); 2003. pp. 359-380

[42] Nema V, Agrawal R, Kamboj DV, Goel AK, Singh L. Isolation and characterization of heat resistant entertoxigenic *Staphylococcus aureus* from a food poisoning outbreak in Indian subcontinent. International Journal of Food Microbiology. 2007;**117**:29-35

[43] Davidson PM, Taylor TM. Chemical preservatives and natural antimicrobial compounds. Ch 33. In: Doyle MP, Beuchat LR, editors. Food Microbiology: Fundamentals and Frontiers. 3rd ed. Washington D.C: ASM Press; 2007. pp. 713-745

[44] Argudin MA, Mendoza MC, Rodicio MR. Food poisoning and *Staphylococcus aureus* enterotoxins. Toxins. 2010;**2**(7):1751-1773

[45] Seo KS, Bohach GA. *Staphylococcus aureus*. Ch 22. In: Doyle MP, Beuchat LR, editors. Food Microbiology: Fundamentals and Frontiers. 3rd ed. Washington D.C: ASM Press; 2007. pp. 493-518

[46] Pinchuk IV, Beswick EJ, Reyes VE. Staphylococcal enterotoxins. Toxins. 2010;**2**:2177-2197

[47] Scallan E, Hoekstra RM, Angulo FJ, Tauxe RV, Widdowson M, Roy SL, et al. Foodborne illness acquired in the United States—Major pathogens. Emerging Infectious Diseases. 2011;**17**(1):7-11

[48] James SJ, Evans J, James C. A review of the performance of domestic refrigerators. Journal of Food Engineering. 2008;**87**(1):2-10

[49] Aycicek H, Cakiroglu S, Stevenson TH. Incidence of *Staphylococcus aureus* in ready-to-eat meals from military cafeterias in Ankara, Turkey. Food Control. 2005;**16**(6):531-534

[50] Hennekinne J-A, de Buyser M-L, Dragacci S. *Staphylococcus aureus* and its food poisoning toxins: Characterization and outbreak investigation. FEMS Microbiology Reviews. 2012;**36**:815-836

Section 4

# Rice Biology under Stress

Chapter 8

# Nitrogen Use Efficiency in Rice under Abiotic Stress: Plant Breeding Approach

*Satyen Mondal, Jamil Hasan, Priya Lal Biswas, Emam Ahmed, Tuhin Halder, Md. Panna Ali, Amina Khatun, Muhammad Nasim, Tofazzal Islam, Evangelina S. Ella and Endang M. Septiningsih*

## Abstract

Nitrogenous fertilizer has remarkably improved rice (*Oryza sativa* L.) yield across the world since its discovery by Haber-Bosch process. Due to climate change, future rice production will likely experience a wide range of environmental plasticity. Nitrogen use efficiency (NUE) is an important trait to confer adaptability across various abiotic stresses such as flooding, drought and salinity. The problem with the increased N application often leads to a reduction in NUE. New solutions are needed to simultaneously increase yield and maximize the NUE of rice. Despite the differences among flooding, salinity and drought, these three abiotic stresses lead to similar responses in rice plants. To develop abiotic stress tolerant rice varieties, speed breeding seems a plausible novel approach. Approximately 22 single quantitative trait loci (QTLs) and 58 pairs of epistatic QTLs are known to be closely associated with NUE in rice. The QTLs/genes for submergence (*SUB1A*) tolerance, anaerobic germination (*AG*, *TPP7*) potential and deepwater flooding tolerance (*SK1, SK2)* are identified. Furthermore, phytochrome-interacting factor-like14 (*OsPIL14*), or loss of function of the slender rice1 (*SLR1*) genes enhance salinity tolerance in rice seedlings. This review updates our understanding of the molecular mechanisms of abiotic stress tolerance and discusses possible approaches for developing N-efficient rice variety.

**Keywords:** abiotic stress, crop establishment, climate change, QTLs, food security

## 1. Introduction

Nitrogen is the most abundant (78%) of the atmosphere in gaseous form as an $N_2$ molecule. But it is not directly available to the plants for their growth and development [1]. It is the foremost important major essential nutrient element involved in the physiological processes in plants. Globally, nitrogen deficiency is a crucial growth-limiting factor for plants, especially under abiotic stresses. The nitrogen use efficiency (NUE) is defined as the output of any crop plant per unit of nitrogen applied under a specific set of soil and climatic conditions [2]. Agronomist usually considers the amount of rough rice produced per unit of nitrogen applied as the efficiency of nitrogen, but physiologist defined it as the amount of rough rice produced per unit of nitrogen absorbed [3, 4]. The latter is also termed as N utilization efficiency. Apparent N recovery is based on N uptake measurement in the above-ground plant parts and assumes that fertilized and control crops absorb the same amount of soil N. On the other hand, physiological and agronomic efficiencies are based on grain yield rather than total dry matter production. However, the enhancement of NUE under an abiotic stressful environment has paramount importance to the future rice breeder.

Rice (*Oryza sativa* L.) is grown in a wide range of ecosystems from the tropic to the temperate regions, but productivity is severely tormented by various abiotic stresses [5, 6]. Farmers may encounter flooding or waterlogging if heavy rain occurs immediately after seeding before or after transplanting. The flooding can cause complete crop failure because of the high sensitivity of rice to anaerobic conditions caused by flooding during germination [7, 8]. There are various forms of flooding caused by directly from heavy rains and/or flooding from adjacent rivers, leading to drastic reductions in rice yield, ranging from 0.5 to 2.0 t $ha^{-1}$ [9]. Flash floods are relatively short durations, prevailing some days to a couple of weeks. Apart from this, stagnant flooding (30–50 cm water depth) may occur at any time of the monsoon. Sometimes, the stagnant flooding may have coincidence with the flash flood resulting in severe impacts on rice production. In deepwater areas, stagnant water present from 0.5 m to a few meters in the field, usually for 4-6 months. The depth of water in some of these deepwater areas can exceed 4 m as in floating-rice areas. Rice breeders have been trying to develop a unique rice variety having specific adaptive traits to tackle these types of floods [10–12]. Recently, the Bangladesh Rice Research Institute has developed a deepwater rice variety, BRRI dhan91, for the deepwater ecosystem. However, the application of nitrogenous fertilizer is very challenging to the deepwater rice field and the NUE of this ecosystem is not yet been well investigated.

Another one among the most important abiotic stresses is drought. Despite the importance of drought as a major factor in yield reduction in rainfed ecosystems, few efforts have been made to develop high-yielding drought-tolerant rice variety. Impending rice production will experience a range of drought stress. The root architectural plasticity is taken into accounts as a very important characteristic to confer tolerance to drought stress [13]. Deciphering the genetic and molecular mechanisms controlling root phenotypic plasticity is important for effective screening, selection and rice breeding efforts. Despite the likely genetic complexity behind the regulation of trait expression in line with environmental conditions, phenotypic plasticity is heritable and selectable. The QTLs have been identified incur for plasticity in aerenchyma development and lateral root growth in response to drought stress in rice [14]. These QTLs can be used in advanced breeding for the development of a drought-tolerant rice variety. Due to global climate change, rice crops will face diverse stresses, including prolonged drought stress, poor soil fertility, and unpredictable rainfall. Rice establishment, either by transplanting or direct seeding, depends upon the rainfall pattern. Therefore, the identification of root phenotypic plasticity traits suitable for adaptation to the particular range of conditions faced by rice crops, as well as the genetic regions responsible for those plasticity traits, may facilitate selection for wide adaptation of rice genotypes to variable conditions to confer sustainable yield. Quantification of root architectural plasticity possesses significant value to detect which root traits may play the pivotal roles in rice adaptability to drought. It is reported that the most plastic genotypes in root traits may show the most yield stability under various dynamics of drought stress [11]. In this regard, many drought-tolerant cultivars, like N22 and Moroberekan, have been selected from rainfed ecosystems through traditional processes. These cultivars harbor genes for tolerance to abiotic stresses, including a wide range of drought [15]. But due to their low yield potential and poor grain quality, farmers and consumers are reluctant to prefer these cultivars. This provides a unique opportunity for rice breeders to develop high-yielding drought-tolerant varieties.

Salinity is another major abiotic stress that is globally distributed in both irrigated and non-irrigated areas [16, 17]. On a global basis, salinity stress ranked second after the drought [18]. Salt stress affects many aspects of rice growth and development, especially during seed germination and seedling growth [19]. It is one of the most prevalent environmental threats to global agricultural productivity, especially in arid and semi-arid climates, where population growth, water shortage and land degradation are major concerns [1, 20]. Salt-affected soils are identified by high electrical conductivity (EC), sodium adsorption ratio (SAR) and pH, calcareousness, poor organic matter, less biological activity and imbalance in physical soil conditions. Salinity causes toxicities of ions like $Na^+$ and $Cl^-$, osmotic stress and ionic imbalance to the root zone or in the soil body, including soil impermeability [21], resulting in nutrient uptake problems in rice plant. Salt stress is the osmotic stress expressed on seedling to the reproductive stage when they are growing under high saline conditions. The N is the essential element for the synthesis of chlorophyll, amino acids, nucleic acids, and proteins. Reduction in plant dry matter is sometimes observed under severe NaCl salt stress and N deficiency. This phenomenon possibly happens because of the decrease in sugar or starch accumulation [1, 22]. The NUE of nitrogenous fertilizers in saline soil depends upon its mineralization pattern, soil salinity levels, soil texture, temperature, freshwater irrigation and soil pH [23]. As NUE for rice plants under salt-affected soils is relatively lower than those on normal soils, the judicious use of nitrogenous fertilizer application in saline soil is needed. Breeders involved in salinity tolerant rice, it is groundbreaking news that the overexpression of PHYTOCHROME-INTERACTING FACTOR-LIKE14 (*OsPIL14*), or loss of function of the DELLA protein SLENDER RICE1 (*SLR1*), accelerate mesocotyl and root growth under salt stress and minimize the sensitivity to NaCl-induced hindrance of seedling growth in rice [17].

## 2. Crop establishment methods under abiotic stress

Crop establishment under abiotic stress is crucial for farmers, even though farmers are coping with this stress condition. There are many more abiotic stresses; out of those, we will discuss only flooding, drought and salinity stress.

### 2.1 Crop establishment under flooding stress

Proper rice establishment is significantly important in flood-prone areas because of its sensitivity to flooding during germination (**Figure 1**) and early seedling stage relative to other growth stages [24, 25]. In most areas of Asia, irrigated rice is established by transplanting of seedlings into puddle soil [26, 27], after which the fields are flooded for a prolonged time and recession of water is done before harvesting. Puddling gives some advantages such as it reduces water loss by percolation, assists weed control through destroying weeds, burying weed seeds and maintaining anaerobic conditions that impede weed germination, and makes the soil soft for transplanting [28, 29]. In many rainfed areas of Bangladesh and the eastern part of India, water deposits in the field to around 30 cm or more within a few days after the onset of the rainy season, making the farmers to transplant taller and older seedlings being their only viable option in their hand [30]. Many variations in direct-seeding are being practiced depending on water availability

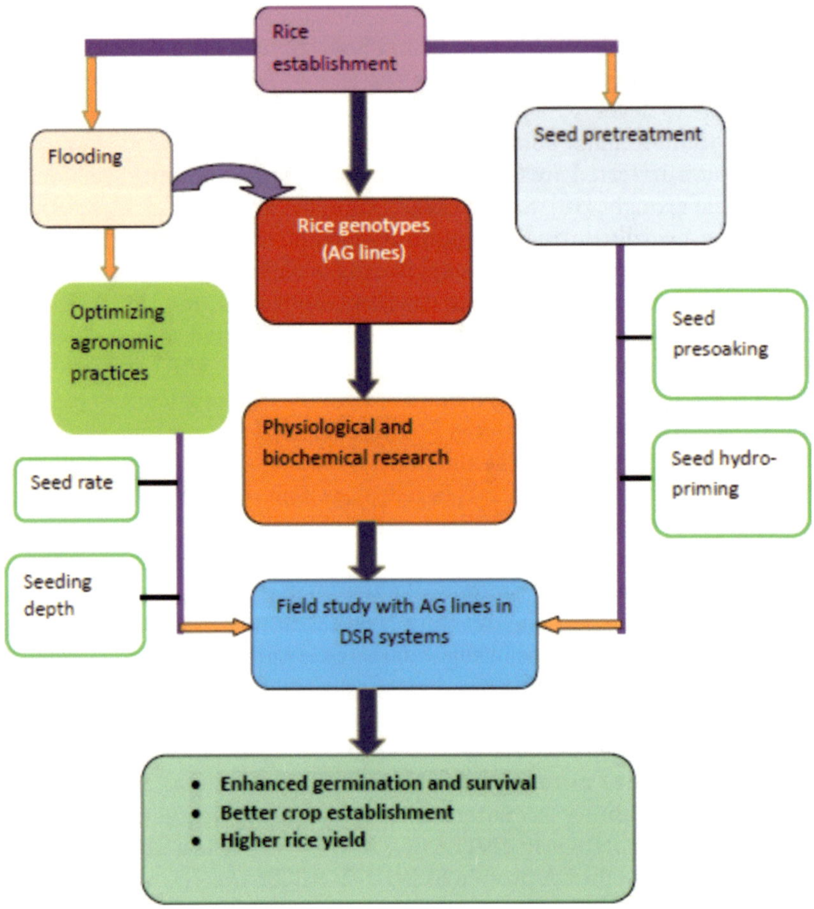


**Figure 1.**
*Crop establishment methods and seed management options under early flooding stress using anaerobic germination (AG) potential rice genotypes in direct-seeded rice (DSR) system under field condition.*

and field conditions [29]. Due to increasing labor scarcity and cost, however, the need to shift a more suitable establishment method with much lower labor requirement than manual transplanting is conducted. This can be achieved by changing to mechanical transplanting or direct-seeding, which also enable timelier planting/seeding and improved crop stand [31]. Researchers in China [32], South Asia [33], and Australia [34] reported that rice could be successfully grown using dry-seeding. Dry-seeding rice has been developed as an alternative establishment method of rice that alters labor requirements and other inputs while increasing or maintaining economic productivity and alleviating soil degradation problems in cropping systems [35, 36].

The three basic methods of direct seeding are water seeding (broadcasting seed into standing water), dry-seeding and wet-seeding [31]. In wet seeded rice (WSR), the pre-germinated seeds are broadcasted or sown in rows on the saturated soil surface, typically after puddling. Dry-seeding involves broadcasting or preferably drilling the seed into non-puddled soil, usually after dry tillage [31]. Water seeding involves pre-germinated rice seeds broadcast in standing water and is practiced in some cooler areas like in California, Central Asia and Australia [30]. The main advantage of this method is that the standing water suppresses the majority of weed species. This is common in temperate irrigated areas, but could potentially be adapted in flood-prone rainfed lowlands in the tropical area where farmers can practice early sowing without waiting for a complete recession of floodwater, to minimize the risk of delayed maturity and late-season drought [26]. Once the rice crop has been established in direct-seeded systems and based on water availability, the field is flooded to suppress weed growth and water depth is then maintained at 5–10 cm through most of the season, later water is gradually drained prior to harvest [30]. The type and degree of adoption of alternative rice crop establishment methods to puddling and manual transplanting vary across Asia. In some parts of South East Asia (Philippines, Malaysia, and Vietnam) and Sri Lanka, transplanting has been replaced in large areas by wet-seeding on puddled soil [2, 26]. In the more developed East Asian countries, like Japan and South Korea, transplanting in puddled soil using specialized machinery has been a common practice for many years, and there is now emerging interest in mechanical transplanting into either puddle or non-puddle/dry tilled soil in parts of India. In parts of South Asia, especially in the rice-wheat systems of north-west India, dry-seeding of rice is at the early stages of adoption. The same seed drill can also be used for sowing other crops; thus, dry-seeding may be more conducive to the mechanization of rice establishment than the use of a single purpose mechanical transplanters in regions where farmers also grow non-rice crops [37].

The establishment methods involving puddling have several disadvantages, including higher tillage costs, adverse effects on soil structure for upland crops grown in rotation with rice, and high water requirement for crop establishment. Irrigation cost for crop establishment can be reduced by avoiding puddling, with or without a change in the crop establishment method. For example, both mechanical transplanting and wet-seeding can be done in non-puddled soil after saturating the soil (after dry tillage or no-tillage) [38]. Dry-seeding into dry or moist soil, can further reduce the water requirement for crop establishment, with or without prior dry tillage as for transplanted and wet seeded rice. Nevertheless, direct-seeded rice in the field for 2–3 weeks is longer than transplanted rice, increasing the length of the irrigation season. It has been observed that the extraction of water is more uniform across depths with direct-seeded rice because of better root growth than with transplanted rice [39]. At the early stage of crop growth, up to 60 days after sowing (DAS) growth rate is relatively higher in DSR and WSR than transplanted rice, having more plant density per unit area compared to transplanting [40].

### 2.2 Crop establishment under drought stress

Drought is an environmental occurrence imposed by the synergistic effect of hydrological, climatic, and natural forces that result in insouciant precipitation for agricultural production over a long period [41]. Globally drought severity is one of the serious concerns because of its immense impacts [42]. The frequency and severity of global drought remain omnipresent and the incidence or extremity of drought has been increasing globally, such as in the Mediterranean region [43], Central China [44], and Africa [45]. Drought is a major constraint to rice production worldwide, as it can occur for varying lengths of time and intensity at any stage of rice growth and development. With the increasing human population and depleting water resources, the development of drought-tolerant rice is of supreme importance to minimizing rice yield losses from drought stress [46]. The major obstacle of rain-fed rice production is drought [47]. Irrigated conditions induce shallow root systems to uptake the resources from the top layer of the soil, whereas rain-fed conditions favor a deep and robust root system, needed to extract the water and nutrients from a wider volume of soil [48]. Three common types of drought can be found for rice production: early water scarcity that causes a delay in seedling establishment through transplanting, mild sporadic stress having cumulative effects, and late stress affecting long duration varieties [49].

Drought stress induces different physiological and biochemical changes in rice at various developmental stages [50]. It is reported that the plant acclimatized to drought stress through modification of its roots into thicker and longer to uptake nutrient and water from a relatively higher depth of soil and it is found that assimilates are translocated to roots instead of shoots in response to drought stress [51]. In contrast, some researchers opined that root growth in rice decreases under drought stress [52]. These findings show that the response of roots to water stress is highly dependent on the rice genotype, period and intensity of stress [53]. The impact of drought stress on rice yield also depends on the growth stages, with the seedling, tillering, flowering, but if rice plant faces severe drought at the panicle initiation stage might be the most sensitive stage resulting huge loss in yield [54].

### *2.2.1 Role of root to uptake water under drought*

As roots uptake water and nutrients from the soil; hence, the morphological and physiological characteristics of roots play a vital role in determining shoot growth, successive development and ultimate crop production [55]. The access of water to a plant is measured by its root system, root properties, root structure, and distribution of root and rootlets, so improving root traits to expedite the uptake of soil moisture and uphold the productivity under drought stress is of paramount interest [56, 57]. Herbaceous plants like rice have a root system comprised of coarse roots, which include the primary roots that originate from the taproot system and the nodal/seminal roots of fibrous root systems, easily distinguishable from the finer lateral roots [58]. Moisture deficiency can be recovered through modification of the root-shoot ratio and maintain leaf gourd cell-mediated process under drought stress [59]. The optimal dry matter partitioning theory proposes that a plant distributes the assimilates among its different parts for optimum growth and development [60]. It further suggests that the shoot ratio and some other signaling processes may change the ratio to balance the assimilates that alter plant growth even the plants produce certain root for adaptation [61]. Roots having a small diameter and a high specific root length expedite the surface area of roots in contact with soil water and also increase the influx of the xylem through the apoplastic pathway [62, 63]. Moreover, the decrease in root diameter also assists in enhancing water access and upraises the productivity of plants under drought stress [64].

Agronomic adjustments to root plasticity may occur when plant combat with multiple resources limitation [65]. Root architecture varying with rice seedling establishment methods; dry direct seeding prone to more edaphic stresses than irrigated transplanted methods [31]. Moreover, the adjustments in high yield potential among genotypes showing the highest degree of root plasticity may be due to genetic potentiality rather than functional adjustments. Undesirable traits to drought stress such as tall plant height and very early flowering have been reported previously, later in high-yielding, medium-duration drought-tolerant rice varieties developed [66, 67]. So the exact identification and fine-mapping of the QTLs governing the root plasticity traits identified [68]. The positive plasticity values noticed in response to stress indicate that the growth of that particular root trait was increased due to stress application. This response is distinct from an allometric response, in which larger root biomass is related to larger shoot size, because though root growth increased under drought stress but shoot growth down-regulated under stress [68]. The genotypes showing most root-plasticity have positive correlations for root architectural traits between and drought suggest that the most root-plastic genotypes would consistently show a plastic response in different drought environments either in transplanting or direct seeding or in other soil types [68]. The genotypes having the most root-plasticity under drought also would show a relatively greater degree of plasticity under low phosphorus content soil, depending on the soil depth [68]. Combinations of multiple root plasticity traits in response to drought and/or low-phosphorus have been related to genotypic variation for adaptation to various environments [69]. It is reported that no single functional parameter was strongly incurred to trends in root plasticity or yield [68]. In line with root architectural plasticity, traits such as root anatomy, water use efficiency, and phenology has been reported to be related to more stable plant establishment across versatile environments in various species [70, 71]. In the case of rice, phenological plasticity in response to drought may be difficult to assess because rice exhibited delayed flowering under drought, and this delay can be reduced by plasticity in root architectural traits, which improve moisture uptake. A set of QTLs has been identified related to root architectural plasticity traits and phenotypic plasticity traits in rice, resulting in getting a better understanding of rice establishment under drought stress [68].

### 2.3 Crop establishment under salinity stress

Generally, rice plants are very sensitive to salinity stress during the early stages of seedling establishment, post-germination and reproductive stage and relatively less sensitive during tillering and grain filling stages [72, 73]. Under salinity altering in the shoot to root ratio as a consequence of root length reduction was supposed to be the avoidance mechanism of the seedlings from salt stress. Salinity accumulates the toxic ion in plants, causing a mineral imbalance. The essential ions are reduced and do not meet the demand resulting in hindrance in normal physiological activities of rice plants. High salt stress impedes the seed germination process, while low salt stress induces seed dormancy [74]. To cope with such stress conditions, seeds develop a mechanism of maintaining low water potential, other specific avoidance, escaping, or tolerance mechanisms to protect the damage by salt stress [75]. Salinity limits germination in a number of ways. From reducing the osmotic potential of soil, which makes a decline in water imbibitions by seed [74] to the creation of ionic toxicity, which alters enzymatic action involved in nucleic acid metabolism. Other effects of salt stress on seed germination include changes in the metabolic process of protein [76]. Seeds are usually more sensitive to salt stress due to close association to the surface of the soil. Accumulation of NaCl to a toxic level in soil, ionic stress decreases the rate of germination [77]. Seed could not absorb water properly because of lower water potential induced by salt stress resulting in toxic effects to the developing embryo and delay in the germination process [78]. The average time of seed germination depends on salinity severity and genotype's inherent quality. There is a strong negative co-relationship between the strength of salinity stress and the rate of germination [79]. Salinity exhibits an immense effect on the germination index and seed size [80]. Small-sized seeds show a relatively higher germination index than large size seeds under salinity stress. Salinity has a negative effect on germination percentage, rate of germination and germination speed [81]. After germination, in successive growth of the seedling, salinity reduces shoot and root dry matter production in rice genotypes [82], and the magnitude of reduction increased with increasing salinity level (**Table 1**).

#### *2.3.1 Plant physiology under salinity*

Higher amounts of salt in the soil cause a serious threat to various metabolic processes of plants, which results in a reduction of crop yield. Soil salinity limits the uptake of essential ions into the plants resulting in metabolic disorder leading to

| Genotype | Salinity level (dS $m^{-1}$) | | | | | | | |
|---|---|---|---|---|---|---|---|---|
| | 0 | 5 | 10 | 15 | 0 | 5 | 10 | 15 |
| | Shoot dry weight (g/10 plants) | | | | Root dry weight (g/10 plants) | | | |
| IR20 | 0.060 | 0.05 (17) | 0.028 (53) | 0.016 (73) | 0.068 | 0.048 (29) | 0.036 (47) | 0.016 (76) |
| POKKALI | 0.134 | 0.116 (13) | 0.064 (52) | 0.044 (67) | 0.152 | 0.076 (50) | 0.026 (83) | 0.018 (88) |
| IR29 | 0.140 | 0.07 (50) | 0.036 (74) | 0.014 (90) | 0.06 | 0.048 (20) | 0.022 (63) | 0.012 (80) |
| NERICA 1 | 0.084 | 0.064 (24) | 0.024 (71) | 0.008 (90) | 0.054 | 0.038 (30) | 0.018 (67) | 0.01 (82) |
| NERICA 5 | 0.076 | 0.054 (29) | 0.032 (58) | 0.02 (74) | 0.13 | 0.056 (57) | 0.016 (88) | 0.004 (97) |
| NERICA 12 | 0.092 | 0.068 (26) | 0.046 (50) | 0.028 (70) | 0.062 | 0.04 (35) | 0.024 (61) | 0.01 (84) |
| NERICA 19 | 0.054 | 0.038 (30) | 0.014 (74) | 0.004 (93) | 0.036 | 0.028 (22) | 0.002 (94) | 0.0 (100) |
| IWAII | 0.090 | 0.068 (24) | 0.032 (64) | 0.02 (78) | 0.068 | 0.046 (32) | 0.028 (59) | 0.016 (76) |

*Values in parenthesis indicate percent reduction to respective controls.*

**Table 1.**
*Effect of salinity on shoot and root dry weight (g/10 plants) of different rice varieties [82].*

downstream in plant growth rate [83]. Excess salt concentration in the root zone of plants causes a change in plant water potential. Salinity causes a reduction in turgor pressure in plant cells due to less water uptake by the plants. Insufficient water limit cell division and regulation of stomatal aperture, which lead to low photosynthesis rate and in severe case causes plant tissues death [84]. Aside from this, reduction in turgor pressure causes stomatal closure, resulting reduction in gaseous exchange of transpiration [20]. Salinity causes other physiological disorders, like changes in membrane permeability, leading to misfolding of membrane proteins [85] and suppression of the photosynthesis [86]. Reduction in enzymatic activities and photopigments causes a lowering of photosynthesis rate [87]. Many plant physiological and biochemical processes, photosynthesis [88], water conductance through stomata [75, 89] are affected by salinity, resulting in an adverse effect on biological processes and crop yield reduction.

#### *2.3.2 Plant anatomical change under salinity stress*

Rice adopts various strategies in response to salinity through their anatomical modification, which allows them to cope with the stress. Plants with growth in high salt concentration have more thickness of leaves [90], epidermis, cell walls and cuticles. The higher the salt concentrations, the higher the mesophyll cell layers and cell size up to some extend [91], due to more elasticity in the cell wall at high turgor pressure [92]. Salinity expedites the density of stomata at the lower side of leaves [93] with increased palisade tissues [94]; however, salinity downregulates the number of cells per leaf. Salinity reduces the number of stomata on the surface of the epidermis [95], vessels number [94]. However, salinity accelerates subrinization inside the roots resulting in hindrance in nutrient uptake from soil [96]. In rice, it is reported that stem diameter was reduced [97], while trichome and stomata density increased. Salt stress reduced cell size, the epidermal thickness of leaves, apical meristem, diameter of the cortex and central cylinder [98]. Salinity induced thickening of exodermis and endodermis [99] and assist in developing sclerenchymatous tissues [98]. Once the seed has germinated, the next goal for the plant is an establishment. Salinity causes a reduction in crop establishment by reducing shoot growth, sealing leaf development and expansion, reducing the growth of internodes and inducing abscission of leaf [91, 100]. Salinity causes some complexity to plants, like osmotic stress, ion toxicity and nutrient imbalance, which are detected as the most prominent reasons for a reduction in crop growth, resulting in crop failure in severe cases. Nonetheless, different developmental stages like germination, vegetative growth, flowering, spikelet's setting and grain filling of rice behave differently with salinity. It is reported that salinity decreased biomass and leaf area in rice [101].

## 3. QTLs and genes of nitrogen use efficiency

In soil, inorganic nitrogen is available for plants as nitrate ($NO_3^-$) in aerobic upland condition and ammonium ($NH_4^+$) in flooded wetland or acidic soils. Nitrogen use efficiency (NUE) is a complex trait that is controlled by multiple genes. Many genes and/QTLs associated with NUE have been identified in rice. Studying and understanding the mechanisms of N utilization at a molecular level may help to improve rice varieties for N deficiency tolerance under different abiotic stresses. Researchers [102] identified 14 putative QTLs for NUE components and 63 QTLs for 12 physiological and agronomic characteristics with six hotspot regions using 174 recombinant inbred lines derived from the IR64/Azucena cross at the vegetative phase in the hydroponic Yoshida solution with three different N concentrations: 1X (standard), 1/4X and 1/8X. In line with this, it is reported that eight QTLs for plant height in hydroponics with two N supply levels in the Yoshida culture solution and 13 QTLs for plant height in a soil mediated experiment with two N supply treatments [103]. Twelve QTLs were detected for root weight, 14 for shoot weight and 12 for biomass from 239 rice recombinant inbreed lines (RILs) derived from a cross between two *indica* parents (Zhenshan97/Minghui63) under hydroponics medium using two N treatments [104]. In another pot experiment, seven QTLs were identified associated with nitrogen use and the yield on chromosome 3 [105]. Three candidate genes *Os05g0208000, Os07g0617800 and Os10g0189600* were identified through fine-mapping of four QTLs located on chromosomes 5, 7 and 10 accelerated yield performance under low N level [106].

Five QTLs were identified on chromosomes 1, 2, 7 and 11 for grain yield (GY) using 127 RILs derived from the cross Zhanshan 97/Minghui 63 [107]. The phenotypic and genetic associations between grain NUE and GY are positive and highly significant; thus, QTLs for both NUE and GY could be used to trigger NUE and GY in a breeding program [108]. Seven QTLs for the glutamine synthetase (GS1) protein content and six QTLs for the NADH-GOGAT protein content were detected using backcross inbred lines between Nipponbare and Kasalath. Some of these QTLs were fined mapped to get a structural gene for GS1 from chromosome 2 and chromosome 1 [109]. A QTL on chromosome 2 activates cytosolic GS1 for protein synthesis in older leaves, resulting in more active tillers during the vegetative stage and a more panicle number and total panicle weight [110]. Using 166 RIL populations, 22 single QTLs and 58 pairs of epistatic QTLs associated with physiological NUE in rice have been identified [111]. With the same mapping population, 28 main effect QTLs and 23 pairs of epistatic QTLs were detected [112]. It is reported that [113], using 38 chromosome segment substitution lines derived from a cross between "Koshihikari," a *japonica* variety, and "Kasalath," an *indica* variety, identified a major QTL *qRL6.1* on the long arm of chromosome 6 associated with root elongation under deficient and sufficient $NH4^+$ condition. The "Kasalath" allele at this QTL region promoted significant root elongation. The marker interval was C11635–P3A2 and phenotypic variance explained by this QTL was 76.4%.

A set of RILs grown in four different seasons in two locations with three nitrogen fertilization treatments was analyzed for QTL for grain yield components and two main effect QTLs were detected viz., grain yield per panicle on chromosome 4 and grain number per panicle on chromosome 12 under N zero level [114]. Four QTLs for trait differences of plant height and heading date between two N levels have been mapped on chromosomes 2 and 8 co-locating with reported QTLs for NUE [111]. In response to low nitrogen application for two years, 33 QTL have been identified in RIL population, out of which only ten QTLs were consistent under low N [115]. QTL mapping for NUE and nitrogen deficiency tolerance traits in RIL population for two years resulted in four common QTL on chromosomes 1, 3, 4 and 7 [116].

From a recombinant inbred population, 20 single QTLs (S-QTLs) and 58 pairs of epistatic loci (E-QTLs) were detected for the nitrogen concentration of grain, nitrogen concentration of straw, the nitrogen content of shoot, harvest index, grain yield, straw yield and physiological nitrogen use efficiency (PNUE) [117]. Researchers [118] identified seven chromosomal regions using 40 introgression lines (ILs) derived from a cross between "Ilpumbyeo," a temperate *japonica* variety, and "Moroberekan," a tropical *japonica* accession from seedlings grown in 0, 250 and 500 μM $NH_4^+$. Among them, the *qRW6* QTL was detected on the long arm of chromosome 6 associated with root weight in temperate *japonica*.

Recently, a group of scientists reported [119] about a main effect QTL *qRD-WN6XB* (**Table 2**) on the long arm of chromosome 6, which positively confers tolerance to N deficiency in the *Indica* rice variety XieqingzaoB, was identified using a chromosomal segment substitution line population using Zhonghui9308 and XieqingzaoB. Nine candidate genes were found in this region through fine mapping. Out of these genes, *Os06g15910* was seemed to be a strong candidate gene associated with root system improvement under low N status. However, putative

| QTLs/genes | Special traits | Chr. No | Reference |
|---|---|---|---|
| *ARE1* | High-yield under N limiting condition | 8 | [108] |
| *qRL6.1* | Root elongation under deficient and sufficient $NH_4^+$ condition | 6 | [113] |
| *qRW6* | Enhance root traits and yield potential | 6 | [118] |
| *qRDWN6XB* | Confers tolerance to N deficiency | 6 | [119] |
| *qGYLN7* | Increases grain yield under low N | 7 | [106] |
| *qGYPP-4b* | Increases grain yield per plant under low N | 4 | [114] |
| *qGNPP-12* | Increases grain number per panicle under low N | 12 | [114] |

**Table 2.**
*Major QTLs/genes associated with nitrogen use efficiency under abiotic stresses.*

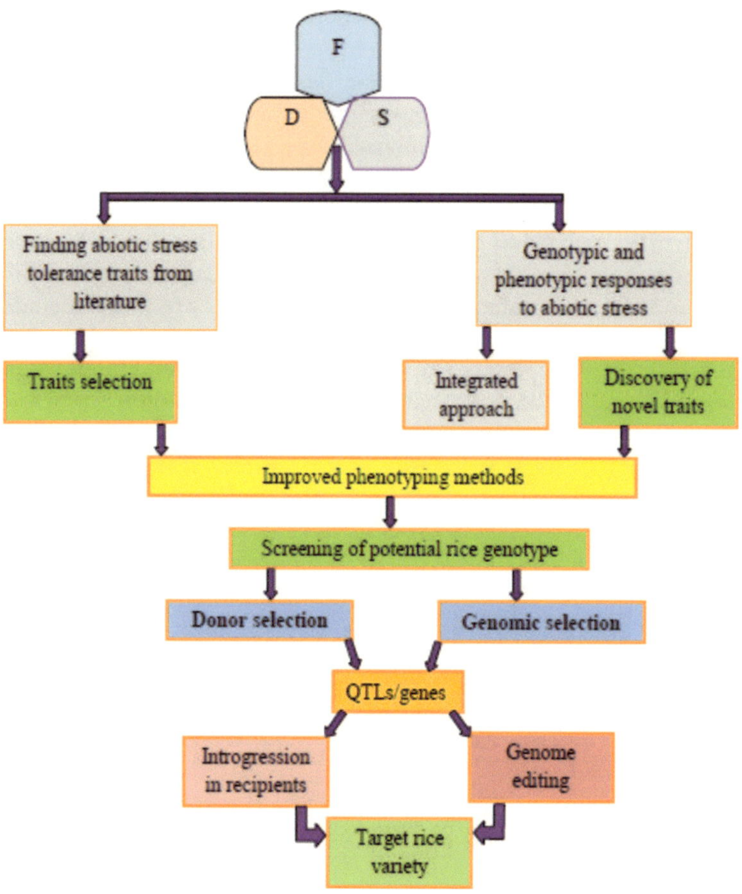


**Figure 2.**
*Holistic breeding approach for multiple abiotic stress tolerance in rice. F = flooding, D = drought, S = salinity, QTLs = quantitative trait loci.*

QTLs/genes needed for multiple abiotic stress tolerance, NUE and associated novel traits in rice could be discovered through a holistic breeding approach (**Figure 2**).

## 4. Hybrid rice production under abiotic stress

Adverse environmental conditions like abiotic factors, triggering abiotic stresses, run a key role in determining the productivity of rice yields. Biologically, abiotic stress is considered as a substantial deviation from the model environments in which plants are grown, inhibiting them from expressing their complete genetic potential regarding growth, development and reproduction [120]. Agriculture production in Bangladesh is dwindled mainly due to biotic and abiotic stresses. Abiotic stress ubiquitously affects the crop growth and development process worldwide. Hence, these are one of major areas of concern to fulfill the required food demand [121, 122]. The major abiotic stresses, drought, flooding, salinity are making the risks to food and nutritional security from tropics to temperate regions worldwide. Drought affects plants in numerous ways like it affects plant growth, yield, membrane integrity, pigment content, osmotic adjustments, water relations and photosynthetic activity [123]. Over the last three decades, the temperature of the country has increased significantly. It is estimated that by 2030, 2050 and 2100, the temperature may increase around 1, 1.4 and 2.4°C, respectively [124]. This is significant as an increased temperature reduces the yield of rice. Therefore, the country is in a risky situation in meeting future challenges concerning food security.

Bangladesh is facing salinity intrusion into the arable agricultural lands. The decline in rice yield under judiciously salt-affected soils is anticipated to be 68 percent [126]. Due to global warming, the rise in sea levels, surplus irrigation without appropriate drainage in the inland area under salt stress is growing. Flash flood and cold injury also cause rice production loss almost every year in Bangladesh. Rainfed conditions in Bangladesh are quite complex, where multiple stresses frequently prevail and even follow in quick succession within a single cropping season. Two or more abiotic stresses often coexist in many rainfed lowland and saline areas of Bangladesh. Most of the rainfed areas in Bangladesh are often occurred by multiple abiotic stresses such as flooding, drought and salinity even within the same cropping season near the coastal areas. Therefore, we need to breed new hybrid rice varieties that could tolerate more than one abiotic stress and yield high under normal favorable rainfed conditions as well.

Northern districts of Bangladesh are cold prone areas of the country. Usually, Boro (winter) rice is seriously affected by cold during the seedling and flowering stage. Seedling mortality sometimes goes up to 90%, especially in the northern part of the country. In recent years, more than 2.0 million hectares of rice crops in the cold prune area of Bangladesh have been seriously affected by extreme cold stress,

| Years | Salinity class and salt affected area (000'ha) | | | | Total (000'ha) |
|---|---|---|---|---|---|
| | S1 (2.0–4.0 dS/m) | S2 (4.1.0–8.0 dS/m) | S3 + S4 (8.1–16.0 dS/m) | S5 (>16.0 dS/m) | |
| 1973 | 287.37 | 426.43 | 79.75 | 39.9 | 833.45 |
| 2000 | 289.76 | 307.20 | 336.58 | 87.14 | 1020.75 |
| 2009 | 328.43 | 274.22 | 351.69 | 101.92 | 1056.26 |

*Source: Soil Resources Development Institute (2010).*

**Table 3.**
*Extent of soil salinity during the last four decades (1973–2009) in coastal areas of Bangladesh.*

causing partial to total yield loss, especially in the northern part of the country. In the haor areas of Bangladesh, early planted Boro rice has to face cold stress at the reproductive stages (Panicle initiation to flowering). If the mean temperature goes down below 20°C for more than 5-6 days during the reproductive stage of the hybrid rice plant associated with spikelet sterility, cause serious yield damage.

In particular, abiotic stresses significantly constrain rice production in Bangladesh and the frequency of these stresses is, unfortunately, likely to increase with climate change. Hybrid rice breeding programs around the world have pre-emptively responded by breeding stress-tolerant rice varieties. By manipulating the heritable variation present in the germplasm, we can develop abiotic stress-tolerant cultivars through breeding techniques, but it is a cumbersome and time-consuming process. The slow progress is due to the complexity of the problem involving environmental conditions and the genetic system. The development of stress-tolerant hybrid rice varieties has gained momentum among the breeders in the recent past. The development of hybrid rice with inbuilt stress tolerance is most desirable to increase the production capacity of rice under saline conditions.

Climate change has affected Bangladeshi agriculture a lot. The most pronounced effects of climate change are the heat stress, periodic drought conditions, and salinity intrusion in coastal belts due to sudden flood and flash flood in major rice-growing areas of Bangladesh. In the last couple of decades, the salinity affected area increased drastically in Bangladesh (**Table 3**). Due to periodic drought and saline water intrusion in the coastal belt, the already existing problem of high amounts of salts in the upper surface soil has intensified. In the future, efforts should be directed to develop climate-smart hybrid rice, which can perform stably under diverse environmental conditions. Nonetheless, China is now feeding 20 percent of the world's population from only 10 percent of the world's arable land where hybrid rice covers around 57 percent of the total cultivated rice area [125]. They have achieved this success by adopting research on region-based and stress-tolerant hybrid rice development. Their way of success was not so smooth, but eventually, they overcome all the obstacles. On the other hand, rice is called "the life of the people of Bangladesh." No obvious alternative crop can replace rice presently. Initially, after the introduction of hybrid rice in Bangladesh in 1998, the area under hybrid rice cultivation significantly increased, but not up to the mark. Currently, only 7.48 percent of the total rice area is under hybrid rice cultivation in Bangladesh [126]. We have released hybrid rice for a favorable environment. It is now worldwide accepted that hybrid rice can give 15-20% more yield compared to inbred high yielding rice cultivars. Therefore, the development of abiotic stress tolerant hybrid rice is the demand of the time to sustain food security.

## 5. Future outlook and conclusions

Among the essential nutrient elements, nitrogen has a paramount importance for rice growth and development in natural ecosystems. To promote optimum N nutrition for the higher rice yield, it is important to explore the possible variability in NUE in rice genotypes. Understanding the molecular mechanisms of variable NUE in rice genotypes would help to develop NUE in the elite rice variety under abiotic stressful conditions using traditional and molecular plant breeding methods, including genome editing. Global climate change plunges world rice production toward various abiotic stress. Flooding, drought and salinity are correlated to cause problems in rice production. If rice seedlings experience flooding during the vegetative stage, they may suffer from terminal drought during the reproductive stage, depending on the ecosystems. Likewise, periodic drought conditions may upregulate the existing salinity stress through intensification of a high amount of salt layer on the upper surface soil. Therefore, there is a dire need to adopt a holistic approach to address the problems of abiotic stresses for future rice breeding. Genomics and post-genomics approaches have high potentials for dissecting underlying molecular mechanisms in differential NUE in rice genotypes. With the help of molecular mapping, fine-tuning of target QTLs, genome editing of a number of major and minor QTLs associated with abiotic stress tolerance in rice have been detected in recent years. Further enhanced research endeavors are now underway toward the development of more tolerant rice varieties to abiotic stresses. The identified QTLs are valuable resources for marker-assisted selection (MAS) to develop elite rice genotypes tolerant to flood, drought and salinity. Novel approaches are needed to apply for accelerating the mitigation of the problems of abiotic stresses in rice such as marker-assisted breeding (MAB), rapid generation advance (RGA), gene editing and transgenic technology. Attempts should be taken to develop abiotic stress-tolerant rice varieties, which can perform in a sustainable manner in a wide range of environmental conditions. Identified QTLs and rice germplasms tolerant to abiotic stresses could be explored to understand the molecular genetics of flooding, drought and salt tolerance in rice. New genes involved in abiotic stress tolerance are needed to be identified. There is a need for strategic research on molecular breeding and agronomic aspects to enhance the resilience of global rice production. To achieve these goals, capacity building of rice scientists, farmers and other stakeholders involved in developing abiotic stress-tolerant rice variety might help to increase the desired NUE in rice.

## Author details

Satyen Mondal[1*], Jamil Hasan[1], Priya Lal Biswas[1], Emam Ahmed[1], Tuhin Halder[1], Md. Panna Ali[1], Amina Khatun[1], Muhammad Nasim[1], Tofazzal Islam[2], Evangelina S. Ella[3] and Endang M. Septiningsih[4]

1 Bangladesh Rice Research Institute, Gazipur, Bangladesh

2 Institute of Biotechnology and Genetic Engineering, Bangabandhu Sheikh Mujibur Rahman Agricultural University, Gazipur, Bangladesh

3 International Rice Research Institute, Manila, Philippines

4 Texas A&M University, College Station, TX, USA

*Address all correspondence to: satyen1981@gmail.com

## References

[1] Murtaza G, Azooz MM, Murtaza B, Usman Y, Saqib M. Nitrogen-use-efficiency (NUE) in plants under NaCl stress. Salt Stress in Plants: Signalling, Omics and Adaptations. 2014.p. 415-437. DOI:10.1007/978-1-4614-6108-116.

[2] De Datta SK, Gomez KA. Interpretive analysis of the international trials on nitrogen fertilizer efficiency in wetland rice. In: Fertilizer International (May issue). British Sulphur Corp., London, U.K. 1981. pp. 1-5.

[3] Yoshida S. Fundamental of Rice Crop Science. International Rice Research Institute, Los Baños, Laguna, Philippines, 269.1981.

[4] Moe K , Mg K, Win K, Yamakawa T. Combined Effect of Organic Manures and Inorganic Fertilizers on the Growth and Yield of Hybrid Rice (Palethwe-1). American Journal of Plant Sciences. 2017; **8**:1022-1042. DOI: 10.4236/ajps.2017.85068.

[5] Lal B, Gautama P, Nayaka AK, Rajaa R, Shahida M, Tripathi R, Singh S, Septiningsih EM, Ismail AM. Agronomic manipulations can enhance the productivity of anaerobic tolerant rice sown in flooded soils in rainfed areas. Field Crops Research.2018; 220: 105-116. DOI:org/10.1016/j.fcr.2016.08.026.

[6] Mondal, S, Khan, MIR., Dixit S, Cruz PCS, Ismail AM. Growth, productivity and grain quality of *AG1* and *AG2* QTLs introgression lines under flooding in direct-seeded rice system. Field Crops Research. 2020; 248: 1-12.

[7] Ismail AM, Ella ES, Vergara GV, Mackill DJ. Mechanisms associated with tolerance for flooding during germination and early seedling growth in rice (*Oryza sativa*). Annals of Botany. 2009; 103: 197-209.

[8] Mondal S, Khan MIR, Entila F, Dxit S, Cruz PCS, Ali MP, Pittendrigh B, Septiningsih EM, Ismail AM. Responses *AG1* and *AG2* QTL introgression lines and seed pre-treatment on growth and physiological processes during anaerobic germination of rice under flooding. Scientific Reports. 2020; 10: 10214.

[9] Ismail AM, Platten JD and Miro B. Physiological bases of tolerance of abiotic stresses in rice and mechanisms of adaptation. ORYZA-An International journal on Rice. 2013; 50 (2): 91-99.

[10] Lafitte RH, Ismail AM, Bennett J. Abiotic stress tolerance in tropical rice: progress and the future. Oryza. 2006; 43:171-186.

[11] Singh A, Septiningsih EM, Balyan HS, Singh NK, Rai V. Genetics, physiological mechanisms and breeding of flood-tolerant rice (*Oryza sativa* L.). Plant and Cell Physiology. 2017; 58(2):185-97.

[12] Septiningsih EM, Mackill DJ. Genetics and breeding of flooding tolerance in rice. In: Sasaki T, Ashikari M, editors. Rice Genomics, Genetics and Breeding. Springer Nature Singapore; 2018. p. 275-295. DOI:10.1007/978-981-10-7461-5_15.

[13] Sandhu N, K. Raman KA, Torres RO, Audebert A, Dardou A, Kumar A, Henry A. Rice Root Architectural Plasticity Traits and Genetic Regions for Adaptability to Variable Cultivation and Stress Conditions. 2016. DOI: https://doi.org/10.1104/pp.16.00705.

[14] Niones JM, Inukai Y, Suralta RR, Yamauchi A. QTL associated with lateral root plasticity in response to soil moisture fluctuation stress in rice. Plant Soil. 2015; 391: 63-75

[15] Dixit S, Huang BE, Sta Cruz MT, Maturan PT, Ontoy JCE. QTLs for

Tolerance of Drought and Breeding for Tolerance of Abiotic and Biotic Stress: An Integrated Approach. PLoS ONE. 2014; 9(10): e109574. doi:10.1371/journal.pone.0109574

[16] Barua R, De Ocampo M, Egdane J, Ismail AM, Mondal S. Phenotyping rice (*Oryza sativa* L.) genotypes for physiological traits associated with tolerance of salinity at seedling stage. **Scientia Agriculturae**. 2015; 12:156-162

[17] Mo W, Tang W, Du Y, Jing Y, Bu O, Lina RV. PHYTOCHROME-INTERACTING FACTOR-LIKE14 and SLENDER RICE1 interaction Vcontrols seedling growth under salt stress. Plant Physiology. 2020. DOI:10.1104/pp.20.00024.

[18] Singh BD. Plant breeding, 4th edn. Kalyani Publishers, Banaras. 2004.

[19] Park HJ, Ki, WY, Yun DJ. A new insight of salt stress signaling in Plant. Molecular. Cells. 2016; 39: 447-459.

[20] Munns R, Tester M. Mechanisms of salinity tolerance. Annual Review of Plant Biology. 2008; 59:651– 681

[21] Ashraf M, Athar HR, Harris PJC, Kwon TR. Some prospective strategies for improving crop salt tolerance. Adv Agron. 2008; 97:45-110

[22] Moradi F, Ismail AM. Responses of photosynthesis, chlorophyll fluorescence and ROS Scavenging systems to salt stress during seedling and reproductive stages in rice. Annals of Botany. 2007; 99:1161-1173

[23] Chaudhry MR, Khokhar IA, Shah MA. Rice growth and nutrient uptake (NP) as affected bydifferent salinity types. Pakistan Journal of Agricultural Sciences. 1989; 26:348-354

[24] Mackill DJ, Ismail AM, Singh US, Labios RV, Paris TR, Development and rapid adoption of submergence-tolerant (Sub1) rice varieties. Advances in. Agronomy. 2012; 115,299-352.

[25] Ella ES, Dionisio-Sese ML, Ismail A M. Proper management improves seedling survival and growth during early flooding in contrasting rice genotypes. Crop Sci. *2010;* **50**(5):1997-2008

[26] Pandey S, Velasco L. Economics of direct seeding in Asia: patterns of adoption and research priorities. *In*: Pandey S, Mortimer M, Wade L, Tuong T, Lopez K, Hardy B. Direct seeding: Research Issues and Opportunities. Proceedings of the International Workshop on Direct Seeding in Asian Rice System: Strategic Research Issues and Opportunities, International Rice Research Institute Bangkok, Thailand/Los Baños, the Philippines. 2002; 3-14.

[27] Mondal S, Hasan J, Ahamed T, Miah MG, Cruz PCS, Ismail AM. Effect of *AG1*, *AG2* QTLs on nonstructural carbohydrate and seed management options for rice seedling growth and establishment under flooding stress. Rice Science. 2020; *27(6): 515-528.*

[28] Singh R, Gajri PR, Gill KS, and Khera R. Puddling intensity and nitrogen-use efficiency of rice (*Oryza sativa*) on a sandy-loam soil of Punjab. Indian Journal of Agricultural Science. 1995; 65:749-751.

[29] Rao AN, Johnson D, Sivaprasad B, Ladha JK, Mortimer AM. Weed management in direct-seeded rice. Advances in Agronomy. 2007; 93: 153-255.

[30] Ismail AM, Johnson DE, Ella ES, Vergara GV, Baltazar AM. Adaptation to flooding during emergence and seedling growth in rice and weeds, and implications for crop establishment. AoB Plants, **2012**; pls019.

[31] Kumar V and Ladha JK. Direct-seeding of rice: recent developments

and future research needs. Advances in Agronomy. 2011; 111: 297-413.

[32] Yan J, Yu J, Tao GC, Vos J, Bouman BAM, Xie GH, Meinke H. Yield formation and tillering dynamics of direct-seeded rice in flooded and non-flooded soils in the Huai River Basin of China. Field Crops Research. 2010; 116: 252-259.

[33] Malik RK, Yadav A. Direct-seeded rice in the Indo-Gangetic Plain: progress, problems and opportunities. In: Humphreys, E., Roth, C.USD., (Eds.), Proceedings of permanent beds and rice residue management for rice–wheat systems in the Indo-Gangetic Plains. ACIAR Workshop Proceedings No. 127, ACIAR. Canberra, Australia. 2008; pp. 124-132.

[34] Beecher G, Dunn B, Mathews S, Thompson J, Singh RP, Humphreys L, Timsina J, Okeefe K, Johnston D. Permanent beds for sustainable cropping. IREC Farmers' Newsletter No. 171. Summer 2006.

[35] Chauhan BS, Awan TH, Abugho SB, Evengelista G, Yadav S. Effect of crop establishment methods and weed control treatments on weed management, and rice yield. Field Crops Research. 2015;172, 72-84.

[36] Ahmed S, Humphreys E, Chauhan BS. Optimum sowing date and cultivar duration of dry-seeded boro on the High Ganges River Floodplain of Bangladesh. Field Crops Research. 2016; 190: 91-102.

[37] Khade VN, Patil BD, Khanvilkar SA, Chavan LS. Effect of seeding rates and level of N on yield of direct-seeded (Rahu) summer rice in Konkan. Jornal of Maharastra Agricultural University. 1993; 18: 32-35.

[38] Malik R, Kamboj B, Jat M, Sidhu H, Bana A, Singh V, Sharawat H, Pundir A, Sahnawaz R, Anuradha T, Kumaran N, Gupta R. No-till and Unpuddled Mechanical Transplanting of Rice. Cereal Systems Initiative for South Asia, New Delhi, India. 2011. http://repository.cimmyt.org/xmlui/handle/10883/1296.

[39] Boling A, Tuong TP, Singh AK, Wopereis MCS. Comparative root growth and soil water extraction of dry-seeded, wet-seeded, and transplanted rice in a greenhouse experiment. Philippines Journal of Crop Science. 1998; 23: 45-52.

[40] Yadav S, Evangelista G, Faronilo J, Humphreys E, Henry A, Fernandez L. Establishment method effects on crop performance and water productivity of irrigated rice in the tropics. Field Crops Research. 2014; 166: 112-127.

[41] Kallis G, Droughts. Annual Review of Environment and Resources. **2008;** 33, 85-118.

[42] Zhao T, Dai A. The Magnitude and Causes of Global Drought Changes in the Twenty-First Century under a Low–Moderate Emissions Scenario. Journal of Climate. **2015;** 28, 4490-4512.

[43] Naumann G, Alfieri L, Wyser K, Mentaschi L, Betts RA, Carrao H, Spinoni J, Vogt J, Feyen L. Global Changes in Drought Conditions Under Di_erent Levels of Warming. Geophysical Research. Letters. **2018**; 45,3285-3296.

[44] Wang, Z.; Li, J.; Lai, C.; Zeng, Z.; Zhong, R.; Chen, X.; Zhou, X.; Wang, M. Does drought in China show a significant decreasing trend from 1961 to 2009? Science of the Total Environment. **2017**, 579, 314-324.

[45] Masih, I.; Maskey, S.; Mussá, F.E.F.; Trambauer, P. A review of droughts on the African continent: A geospatialand long-term perspective. Hydrol. Earth System Science. **2014**, 18, 3635-3649.

[46] Barnabas B, Jager K, Feher A. The e_ect of drought and heat stress on reproductive processes in cereals. Plant Cell Environment. **2008**; 31, 11-38.

[47] Cairns J, Impa S, O'Toole J, Jagadish S. Price, A. Influence of the soil physical environment on rice (*Oryza sativa* L.) response to drought stress and its implications for drought research. Field Crops Research. **2011**. 121,303-310.

[48] Ahmadi N, Audebert A, Bennett MJ, Bishopp A de Oliveira AC, Courtois B, Diedhiou A, Diévart A, Gantet P, Ghesquière A. The roots of future rice harvests. Rice. **2014**; 7, 29.

[49] Fukai S, Cooper M. Development of drought-resistant cultivars using physiomorphological traits in rice. Field Crops Research. **1995**; 40, 67-86.

[50] Tripathy J, Zhang J, Robin S, Nguyen TT, Nguyen H. QTLs for cell-membrane stability mapped in rice (*Oryza sativa* L.) under drought stress. Theoretical and Applied Genetics. **2000**; 100, 1197-1202.

[51] Yoshida S, Hasegawa S. The rice root system: Its development and function. In Drought Resistance in Crops with Emphasis on Rice; Paddyfield: Manila, Philippines. 1982; Volume 10, pp. 97-134.

[52] Tahere AS, Yamauchi A, Kamoshita A, Wade LJ. Genotypic variation in response of rainfed lowland rice to drought and rewatering: II. Root growth. Plant Production Science. **2000**; 3, 180-188.

[53] Xu W, Cui K, Xu A, Nie L. Huang, J.; Peng, S. Drought stress condition increases root to shoot ratio via alteration of carbohydrate partitioning and enzymatic activity in rice seedlings. Acta Physiologia. Plantarum. **2015**. 37, 1760.

[54] Wopereis M, Krop M, Maligaya A, Tuong T. Drought-stress responses of two lowland rice cultivars to soil water status. Field Crops Research. **1996;** 46, 21-39.

[55] Ghosh D, Xu J. Abiotic stress responses in plant roots: A proteomics perspective. Frontiers Plant Science. **2014**; 5,6.

[56] Nardini A, Salleo S, Tyree MT. Ecological aspects of water permeability of roots. In Plant Roots; CRC Press: Boca Raton, FL, USA. 2002; pp. 1069-1093.

[57] Comas LH, Becker SR, Cruz VM, Byrne PF, Dierig, D.A. Root traits contributing to plant productivity under drought. Frontiers Plant Science. **2013;** 4, 442.

[58] Fitter A. Characteristics and functions of root systems. In Plant Roots; CRC Press: Boca Raton, FL, USA. 2002;pp. 49-78.

[59] Maseda PH, Fernandez RJ. Stay wet or else: Three ways in which plants can adjust hydraulically to their environment. Journal of Experimental Botany. **2006;** 57, 3963-3977.

[60] Bloom AJ, Chapin FS, Mooney HA. Resource limitation in plants-an economic analogy. Annual Review of Ecology, Evolution and Systemetics. **1985**; 16, 363-392.

[61] Shipley B, Meziane D. The balanced-growth hypothesis and the allometry of leaf and root biomass allocation. Functional Ecology. **2002**; 16, 326-331.

[62] Comas LH, Mueller KE, Taylor LL, Midford PE, Callahan HS, Beerling DJ. Evolutionary Patterns and Biogeochemical Significance of Angiosperm Root Traits. International Journal of Plant Science. **2012;** 173, 584-595.

[63] Hernández E, Vilagrosa A, Pausas J, Bellot J. Morphological traits and water use strategies in seedlings of Mediterranean coexisting species. Plant Ecology. **2010**; 207, 233-244.

[64] Wasson AP, Richards RA, Chatrath R, Misra SC, Prasad SV, Rebetzke GJ, Kirkegaard JA, Christopher J, Watt M. Traits and selection strategies to improve root systems and water uptake in water-limited wheat crops. Journal of Experimental Botany. **2012;** 63, 3485-3498.

[65] Ho MD, McCannon BC, Lynch JP (2004) Optimization modeling of plant root architecture for water and phosphorus acquisition. Journal of Theoretical Biology 226:331-340

[66] Swamy BPM, Ahmed HU, Henry A, Mauleon R, Dixit S, Vikram P, Tilatto R, Verulkar SB, Perraju P, Mandal NP. Genetic, physiological, and gene expression analyses reveal that multiple QTL enhance yield of rice mega-variety IR64 under drought. 2013; PLoS ONE 8:e62795

[67] Vikram P, Swamy BP, Dixit S, Singh R, Singh BP, Miro B, Kohli A, Henry A, Singh NK, Kumar A. Drought susceptibility of modern rice varieties: an effect of linkage of drought tolerance with undesirable traits. Scientific Reports. 2015; 5: 14799

[68] Sandhu N, Raman KA, Torres RO, Audebert A, Dardou A, Kumar A, Henry A. Rice Root Architectural Plasticity Traits and Genetic Regions for Adaptability to Variable Cultivation and Stress Conditions. Plant Physiology. 2016. p 2562-2576. DOI: https://doi.org/10.1104/pp.16.00705

[69] Fort F, Cruz P, Catrice O, Delbrut A, Luzarreta M, Stroia C, Jouany C. Root functional trait syndromes and plasticity drive the ability of grassland Fabaceae to tolerate water and phosphorus shortage. Environmental and Experimental Botany. 2015; 110: 62-72

[70] Sadras VO, Reynolds MP, de la Vega J, Petrie PR, Robinson R. Phenotypic plasticity of yield and phenology in wheat, sunflower and grapevine. Field Crops Research. 2009; 110: 242-250.

[71] Kenney AM, McKay JK, Richards JH, Juenger TE. Direct and indirect selection on flowering time, water-use efficiency (WUE, d (13) C), and WUE plasticity to drought in *Arabidopsis thaliana*. Ecology and Evolution. 2014; 4: 4505-4521.

[72] Flowers, TJ, and Yeo AR. Variability in the resistance of sodium chloride salinity within rice (*Oryza sativa* L.) varieties. New Phytologist. 1981; 88: 363-373.

[73] Arunroj DS, Supapoj N, Toojinda T, and Vanavichit A. Relative leaf water content as an efficient method for evaluating rice cultivars for tolerance to salt stress. *Science Asia.* 2004; 30: 411-415.

[74] Khan MA, Weber DJ. Ecophysiology of high salinity tolerant plants. (Tasks for Vegetation Science). *Series*. Springer, Amsterdam. 2008.

[75] Aslam M, Ahmad K, Akhtar MA, Maqbool MA. Salinity Stress in Crop Plants: Effects of stress, Tolerance Mechanisms and Breeding Strategies for Improvement. Journal of Agriculture and Basic sciences. 2017; 2(1): 70-85

[76] Rasheed R. Salinity and extreme temperature effects on sprouting buds of sugarcane (*Saccharum officinarum* L.). Some histological and biochemical studies. Ph.D. thesis Dept. of Botany. Univ. of Agriculture. Faisalabad, Pakistan. 2009.

[77] Murillo-Amador B, Aguilar RL, Kaya C, Mayoral JL and Hernapdez AF. Comparative effects of NaCI and polyethylene glycol on germination, emergence and seedling growth of cowpea. Journal of Agronomy and Crop Science. 2002;188:235-247.

[78] Khan MA and Ungar IA. Germination responses to salinity stress in *Atriplex triangularis*. The Ohio Journal of Science. 1984; 84:13.

[79] Ditommaso A. Germination behavior of common ragweed (*Ambrosia artemisiifolia*) populations across a range of salinities. Weed Science. 2004; 52:1002-1009.

[80] Kaya M, Kaya G. Kaya MD, Atak M. Saglarn S. Khawar KM, Ciftci CY. Interaction between seed size and NaCl on germination and early seedling growth of some Turkish cultivars of chickpea (*Cicer arietinum* L.). Journal of Zhejiang University-Science. 2008; 9:371-377.

[81] Bordi A. The influence of salt stress on seed germination, growth and yield of canola cultivars. Notulae Botanicae Horti Agrobotanici Cluj-Napoca. 2010; 38:128-133.

[82] Ologundudu AF, Adelusi AA, Akinwale RO. Effect of Salt Stress on Germination and Growth Parameters of Rice (*Oryza sativa* L.). Notulae Scientia Biologicae. 2014; 6(2). 237-243.

[83] Epstein E, Norlyn JD, Rush DW, Kingsbury RW, Kelly DB, Cunningham GA, Wrona AF. Saline culture of crops: A genetic approach. Science. 1980; 210:399-404.

[84] Munns R. Comprative physiology of salt and water stress. Plant Cell Environment. 2002; 25:239-250.

[85] Gupta IK, Meena SK, Gupta S, Khandelwal SK. Gas exchange, membrane permeability, and ion uptake in two species of Indian *jajuba* differing in salt tolerance. Photosynthetica, 40:535-53.

[86] Sayed OH. Chlorophyll fluorescence as a tool in cereal crop research. Photosynthetica. 2003; 41:321-330.

[87] Saravanavel R, Ranganathan R and Anantharaman P. Effect of sodium chloride on photosynthetic pigments and photosynthetic characteristics of *Avicennia officinalis* seedlings. Recent Research on .Science and Technology. 2011; 3:177-80.

[88] Hayat S, Hasan SA, Yusuf M, Hayat Q and Ahmad A. Effect of 28-homobrassinolide on photosynthesis, fluorescence and antioxidant system in the presence or absence of salinity and temperature in *Vigna radiata*. Environmental and Experimental Botany. 2010; 69:105-112.

[89] Perez-Perez JG, Robles JM, Tovar JC and Botia P. Response to drought and salt stress of lemon 'Fino 49' under field conditions: water relations, osmotic adjustment and gas exchange. Scientia Horticulturae. 2009; 122:83-90.

[90] Waisel Y. Adaptation to salinity. In: Physiology of trees. Willey, New York 1991; pp. 359-383.

[91] Zekri M. Effects of NaCl on growth and physiology of sour orange and *Cleopatra mandarin* seedlings. Scientia Horticulturae. 1991; 47:305-315.

[92] Munns R, Termaat A. Whole plant responses to salinity. Australian journal of Plant Physiology. 1986; 13: 143-160.

[93] Raafat A, Habib SA, El-Shami IZ, El-Antably HM. The effect of salinity on the anatomical features of tomato plants. Annals of Agricultural Science. 1991;36:307-321.

[94] Hussein MM, Abo-Leila BH, Metwally SA, Leithy SZ. Anatomical

structure of *jatropha* leaves affected by proline and salinity conditions. Journal of Applied Sciences Research. 2012; 8:491-496.

[95] Cavisoglu KS, Kilic S, Kabar K. Some morphological and anatomical observations during alleviation of salinity (NaCI) stress on seed germination and seedling growth of barley by polyamines. Acta Physiologiae Plantarum. 2007. 29:551-557.

[96] Walker CD, Graham RD, Madison JT, Cary EE and Welch RM. Effects of Ni deficiency on some nitrogen metabolites in cowpea (*Vigna unguiculata* L. Walp). Plant Physiolgy. 1985; 79:474-479.

[97] PimmongkolA,TerapongtanakhonS, Udonsirichakhon K. Anatomy of salt-and non-salt-tolerant rice treated with NaCI. In: 28th Congr. On Science and Technology of Thailand, Publisher Bangkok, Thailand. 2002.

[98] Javed IH, Wahid A, Rasul E. Selection of pearl millet lines for tolerance to increased salinity. Journal of Agriculture and Plant Science. 2001; 11:18-23.

[99] Gomes MP, de SA e Melo Marques TCLL, de Oliveira Goncalves Nogueira M, de Castro EM, Soares AM. Ecophysiological and anatomical changes due to uptake and accumulation of heavy metal in *Brachiaria decumbens*. Scientia Agriculturae. 2011; 68(5):566-573.

[100] Ziska LH, Seemann JR, DeJong TM.. Salinity induced limitation on photosynthesis in *Prunus salica*, a deciduous tree species. Plant Physiol0gy. 1990, 93:864-870.

[101] Ashraf MY, Bhatti AS. Effect of salinity on growth and chlorophyll content in rice. Pakistan Journal of Scientific and Industrial Research. 2000; 43:130-131.

[102] Nguyen HTT, Dang DT, Van Pham C. and Bertin P. QTL mapping for nitrogen use efficiency and related physiological and agronomical traits during the vegetative phase in rice under hydroponics. Euphytica. 2016; 212(3), pp. 473-500.

[103] Fang P, Wu P. QTL 9 N-level interaction for plant height in rice (*Oryza Sativa* L.). Plant Soil. 2001. 236:237-242

[104] Lian X, Xing Y, Yan H. QTLs for low nitrogen tolerance at seedling stage identified using a recombinant inbred line population derived from an elite rice hybrid. Theoretical and Applied Genetics. 2005; 112, 85-96.

[105] Senthilvel S, Vinod K K, Malarvizhi P, & Maheswaran M. QTL and QTL 9 environment effects on agronomic and nitrogen acquisition traits in rice. Journal of Integrative Plant Biology. 2008; 50, 1108-1117.

[106] Rao IS, Neeraja CN, Srikanth B, Subrahmanyam D., Swamy KN, Rajesh K, Vijayalakshmi P, Kiran TV, Sailaja N, Revathi P, Rao PR. Identification of rice landraces with promising yield and the associated genomic regions under low nitrogen. Scientific reports. 2018; 8(1), pp. 1-13.

[107] Wei D, Cui K, Pan J, Ye G, Xiang J, Nie L. Huang J. Genetic dissection of grain nitrogen use efficiency and grain yield and their relationship in rice. Field Crops Research. 2011; 124(3), pp. 340-346.

[108] Wang Q, Nian J, Xie X, Yu H, Zhang J, Bai J, Dong G, Hu J, Bai B, Chen L, Xie Q. Genetic variations in ARE1 mediate grain yield by modulating nitrogen utilization in rice. Nature communications. **2018**;9(1):1-10.

[109] Yamaya T, Obara M, Nakajima H, Sasaki S, Hayakawa T, Sato T. Genetic manipulation and quantitative-trait loci

mapping for nitrogen recycling in rice. Journal of Experimental Botany. 2002; 53, 917-925.

[110] Obara M, Sato T, Sasaki S, Kashiba K, Nagano A, Nakamura I. Identification and characterization of a QTL on chromosome 2 for cytosolic glutamine synthetase content and panicle number in rice. Theoretical and Applied Genetics. 2004; 110, 1-11.

[111] Feng Y, Cao LY, Wu WM, Shen XH, Zhan XD, Zhai RR, Wang RC, Chen DB, Cheng SH. Mapping QTLs for nitrogen-deficiency tolerance at seedling stage in rice (*Oryza sativa* L.). Plant Breeding. 2010; 129:652-656

[112] Piao Z, Li M, Li P. Bayesian dissection for genetic architecture of traits associated with nitrogen utilization efficiency in rice. Journal of Biotechnology. 2009; 8, 6834-6839.

[113] Obara M, Tamura W, Ebitani T, Yano M, Sato T, Yamaya T. Fine-mapping of qRL6. 1, a major QTL for root length of rice seedlings grown under a wide range of NH 4+ concentrations in hydroponic conditions. Theoretical and Applied Genetics. 2010; *121*(3), pp. 535-547.

[114] Tong H, Chen L, Li W, Mei H, Xing Y, Yu X. Identification and characterization of quantitative trait loci for grain yield and its components under different nitrogen fertilization levels in rice (*Oryza sativa* L.). Molecular Breeding. 2011; 28, 495-509.

[115] Wei D, Cui K, Pan J, Wang Q, Wang K, Zhang X. Identification of quantitative trait loci for grain yield and its components in response to low nitrogen application in rice. Ausstralian Journal of Crop Science. 2012; 6(6), 986-994.

[116] Wei D, Cui K, Ye G, Pan J, Xiang J, Huang J. QTL mapping for nitrogen-use efficiency and nitrogen-deficiency tolerance traits in rice. Plant and Soil. 2012; 359, 281-295.

[117] Cho YI, Jiang W, Chin JH, Piao Z, Cho YG, McCouch SR, Koh HJ. Identification of QTLs associated with physiological nitrogen use efficiency in rice. Molecules & Cells (Springer Science & Business Media BV). 2007; 23(1).

[118] Kim PS, Kim DM, Kang JW, Lee HS, Ahn SN. QTL mapping of Rice root traits at different NH4+ levels in hydroponic condition. Plant Breeding and Biotechnology. 2015; 3(3):244-252.

[119] Anis GB, Zhang Y, Islam A, Zhang Y, Cao Y, Wu W, Cao L, Cheng S. RDWN6 XB, a major quantitative trait locus positively enhances root system architecture under nitrogen deficiency in rice. BMC plant biology. 2019; 19(1), p. 12.

[120] Rehman MS, Harris PJ, Ashraf M. Stress environments and their impact on crop production. In: Ashraf, M., Harris, P.J. (Eds.), Abiotic Stress—Plant Resistance through Breeding and Molecular Approaches. Food Products Press, New York, NY. 2005.

[121] Shanker A, Venkateswarlu B. Abiotic stress Response in Plants-Physiological, Biochemical and Genetic perspectives. Intech Publishers. 2011.

[122] Wani SH, Singh NB, Haribhushan A, Mir JI. Compatible solute engineering in plants for abiotic stress tolerance - role of glycine betaine. Current Genomics. 2013; 14: 157-165.

[123] Benjamin JG, Nielsen DC. Water deficit effects on root distribution of soybean, field pea and chickpea. Field Crops Research. 2006; 97:248-253.

[124] Amin MR, Zhang J, Yang M. Effects of climate change on the yield and

cropping area of major food crops: A case of Bangladesh. Sustainability. 2015; 7:898-915. DOI: 10.3390/su7010898

[125] Yuan LP. Development of hybrid rice to ensure food security. Rice Sci. 2014; 21(1):1–2.

[126] Krishi Dairy. Dhaka: Agriculture Information Services (AIS), Ministry of agriculture, the government of the people's republic of Bangladesh; 2016.

Chapter 9

# Adaptive Mechanisms of Root System of Rice for Withstanding Osmotic Stress

*Afsana Hannan, Md. Najmol Hoque, Lutful Hassan and Arif Hasan Khan Robin*

## Abstract

Being one of the major cereal crops, rice has a great effect on food security of the world population. But worldwide rice production faces severe threat due to a combination of factors like uncontrolled human rise, limited agricultural land and increasing environmental stresses. Coping with this situation is an urgent call for meeting the challenge. For overwhelming rice production by battling with this condition, scientists and researchers try their best to develop such rice varieties which can adapt to adverse climatic conditions. But, the majority of the research efforts are given on above ground parts of rice to make it stress tolerant. Root, one of the major parts of plant, remains unnoticed although it has immense possibility of adaptation under stress conditions. Fruitful and efficient utilization of limited resources are possible through healthier and competent root systems. Selection and breeding of rice genotypes with extensive root systems may contribute to more efficient use of soil nutrient resources and this ultimately influences the yield stability of rice.

**Keywords:** *Oryza sativa* L., root system, osmotic stress, adaptive mechanisms, lateral roots

## 1. Introduction

Rice is such an agricultural commodity that covers the third-highest worldwide production making it one of the most important cereal crops [1]. With its wide geographic distribution extending from 50°N to 35°S, rice is expected to be the most vulnerable cultivated crop to changing climates in future [2, 3]. Rice production is dwindled mainly because of biotic and abiotic stresses due to the complexity of interaction between the stress factors and various molecular, biochemical and physiological phenomena affecting plant growth and development [4, 5]. To battle with these situations, development of adaptive rice varieties is one of the best strategies. Since aboveground parts are often taken into consideration for making stress tolerant varieties, root study remains backward in this aspect. Roots, the hidden portion of the plant have not yet been much focused. Because exploring the root traits of the plant are much more difficult compared to its above-ground traits. But when it comes to the fact of studying the optimal developmental plasticity system and characteristic features of plant growth, the root system is given the first priority [6].



IntechOpen

Root system is the site of water and nutrient uptake from the soil, a sensor of abiotic and biotic stresses, and a structural anchor to support the shoot. The root system communicates with the shoot, and the shoot in turn sends signals to the roots [7]. Soil type, moisture and nutrients all strongly influence the architecture of the root system [8–10]. Recently it has been emphasized that root architectural traits play a decent role for the adaptation of crop varieties under different abiotic stresses [11, 12]. Root interaction with changing environment is a complex phenomenon that differs with genotypes and intensity of stress [13–17]. For that, different species and also genotypes under the same species may respond contrarily under stress conditions and show different magnitudes of tolerance or susceptibility to stress. These diversities can be exploited by plant breeders to improve stress tolerance in plants. Scientists assume that selection for yield will indirectly select for varieties with the optimum root system. But the fact is, more directed selection for specific root architectural traits could enhance yields for different soil environments [18]. As by 2035, a predicted 26% increase in rice production will be essential to feed the rising population [19], it is imperative to develop high yielding rice cultivars with efficient root systems for better exploitation of natural resources under stressed conditions.

## 2. Progress in root study of rice under osmotic stress

### 2.1 Reasons why root study has become the topic of interest

Being the hidden half of the plants, the root system performs several functions like water and nutrient acquisition, mechanical support to the plant and storage of reserve assimilates [7]. In plant, roots are the first organ for sensing the water limitation and the roots are also the signal transmitter to other plant parts through xylem sap and phytohormone which is known as one of the most important root-shoot stress signal mechanism [20–23]. Development of the root system is a major agronomic trait and proper architecture in a given environment permits plants to survive in water and nutrient deficit conditions and gives the ability to utilize minimum resources efficiently [6].

Crop loss in rice production has become severe now-a-days due to abiotic stresses. Therefore, having a clear knowledge about the architecture and development of roots of rice toward optimizing water and nutrient uptake has become crucial for exploitation and manipulation of root characteristics for enhancing yield under unfavorable conditions [24, 25]. In general, root study comprises the study of the entire root system or a large portion of the plant's root system [26, 27]. To understand the functional characteristic of root system and the necessity to exploit heterogeneous environment, root architecture study has become crucial in plant productivity as root system architecture is strongly linked with plasticity to the plant through which plant can alter its root structure according to its heterogeneous environment [26].

### 2.2 Root system architecture of rice

Elongation and branching are the mode of plant root growth. Local environmental conditions, physiological status of the plants and the type of root determine the magnitude and direction of root elongation [6]. Root system architecture (RSA) is thus the three-dimensional geometry of the root system including the primary root, branch roots, and root hairs [6, 26, 28, 29]. Topological, distributional and morphological features combine to form the root system architecture [8, 26, 30]. Topology denotes the branching pattern of individual roots including features like lengths and diameters, number of roots originating from a node, root insertion angles, magnitude and the altitude of root [31, 32]. Measures of the spatial distribution of roots simplify the dissection of root systems [26]. Root morphology refers to the external features of a root axis and may include properties of roots hairs, root diameter and trend of secondary root emergence. Acceleration or inhibition of primary root growth, increment of lateral roots (LRs) and a rise in root hairs and also the formation of adventitious roots are the ways of modification of root system architecture. The primary root is formed during embryogenesis. This primary root produces secondary roots those in turn produce tertiary roots [6, 33]. Root system architecture has proved to be a critical factor in plant survival, contributing to water and nutrient acquisition efficiency and competitive fitness in a given environment [34]. Composition of soil specially water and mineral nutrients availability and plant species have impact on root architecture [6].

Monocot cereals have a complex fibrous root system consisting of an adventitious root (ARs) bunch. Adventitious roots originate from the shoot or subterranean stem. This type of root is sometimes referred to as a nodal or crown root [35]. Root systems of rice plants (*Oryza sativa* L.) comprise numerous nodal roots of relatively short length: a mature rice plant usually has several hundreds of nodal roots, most of which are less than 40 cm in length [36]. Rice (*Oryza sativa* L.) is a model cereal crop with seminal roots that die during the growing period [36]. Thus, lateral roots and adventitious roots are the key determinants of nutrient and water use efficiency in rice [37].

Several embryonic and postembryonic roots including the radicle, the embryonic crown roots, the postembryonic crown roots, the large lateral roots (L-type), and the small lateral roots (S-type) [38] form the rice root systems (see **Figure 1**). Lateral rice roots can appear on any primary root, including embryonic and crown roots, and can be classified into two main anatomical types [39]. Numerous small lateral roots (S-type) are thin with determinate growth that can be formed from large lateral roots (L-type) and they never bear any lateral roots. Whereas large lateral (L-type) roots are few in number, thinner compared to primary roots that show indeterminate growth. Additionally, lateral elongation of small lateral roots and downward elongation of large lateral roots indicate non-responsiveness of the small lateral roots to gravity. Higher orders of branching can also be observed in the large lateral roots of the crown roots that emerge at later growth stages [40].

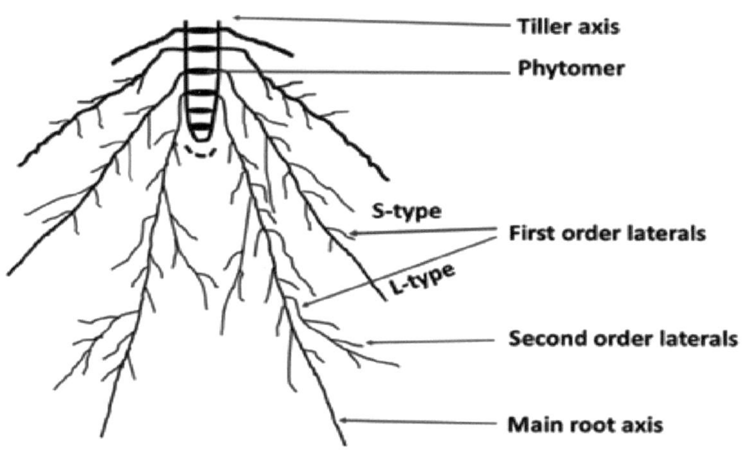


**Figure 1.**
*A typical root system architecture at the tiller axis of* Oryza sativa *L. Black disks indicate individual root bearing phytomer with progressive development chronologically from top to downward. Root hairs form on main axis and all the lateral roots [41].*

These small and large lateral roots exhibit differential growth and lateral root bearing pattern signifying unlike purposes for these two types of lateral roots [37].

### 2.2.1 Phytomer concept

The concept of a phytomer was established around 6–7 decades ago [40, 42]. Clear knowledge about phytomer is required for better understanding of plant development and architecture. Many higher plants, including rice, are composed of successive stem segments called phytomer [43–45]. Each phytomer consists of an internode of the stem with one leaf, one tiller bud and several adventitious (nodal) roots [36]. The phytomer concept has long been recognized among grass scientists [46, 47]. The coordinated development of stem, tiller bud, and adventitious roots in each phytomer corresponds to the phyllochronic time in rice [43, 44, 48]. This indicates that genotypic variation in root-and-shoot growth can be ascribed to the variation of stem and adventitious root development at the phytomer level [49].

Detailed study of root morphology and architecture at the phytomer level become more obvious with the attainment of new knowledge about segmental architecture of poaceous crops [50–53]. As the higher plant structure is formed by the repetitive unit of plant growth called phytomer [54], so phytomer formation, its growth and senescence ultimately determine development of plant canopy [47]. Therefore the phytomer components have become the interest of the plant breeder.

### 2.2.2 Lateral roots

Root axes of rice plants serve functions of anchorage and typically establish overall root system architecture [55]. The lateral roots are the functionally active part of the root system involved in nutrient acquisition and water uptake. The size, type and distribution of lateral roots eventually decide the ultimate length and surface area of an individual root and finally of a whole tiller. Understanding morphology of the lateral roots is therefore important to develop rice cultivars with an efficient root system [11, 56].

In rice, there are two types of lateral roots; long and thick roots, and short and slender roots [57–59]. It has been designated that the first type as L-type and the latter as S-type [60]. The L-type lateral roots are usually long and thick and are capable of producing higher-order lateral roots, whereas S-type ones are short, slender, and non-branching. In rice plants, these two types of lateral roots are visually distinguishable. The L-type lateral roots show basically identical tissue arrangement with seminal and nodal roots, whereas S-types are anatomically different wherein their vascular systems are simplified [35].

In rice plants, the observed average diameter of S-type lateral roots (first-order) that were produced on mature nodal roots of a one-month-old plant was 80 μm, whereas that of L-type roots was almost double that, i.e., 159 μm. Average length was 7.6 mm for S-type and about 30 mm for L-type. The S-type laterals were almost similar in length, and only very few S-type laterals exceeded 10 mm in length. The L-type laterals varied greatly in length and some of them elongated to more than 300 mm [60]. The small laterals are less effective in water and nutrient uptake than even root hairs [61].

The changes in lateral root development were triggered by changes in water status in the root zone, and these developmental changes were induced by genetic [62, 63] and environmental factors. With regard to the environmental factors, it is shown that phenotypic plasticity promoted lateral root development and that nodal root production was the key trait that ensured stable growth of rice plants grown under changing soil moisture levels [64]. As far as the literature explored, developmental morphology of the individual roots with special reference to different lateral root branches was not studied in detail, probably due to lack of the most appropriate tools and methods [11].

#### *2.2.3 Root hairs*

Root hairs are tubular-shaped cells that arise from root epidermal cells called trichoblast; they are thought to increase the absorptive capacity of the root by increasing the surface area [65]. Root hairs contribute as much as 77% of the root surface area of the cultivated crops, forming the major point of contact between the plant and the rhizosphere. Root hair is a long and narrow tube like structure originating from a single cell through tip growth (the deposition of new membrane and cell wall material at a growing tip). For being the major water and nutrient uptake site of plants, root hairs form a progressively significant model system for development studies and cell biology of higher plants [66]. Root hairs had the highest contribution toward total length and surface area of an individual root whereas main axis and first order laterals mostly contributed root volume [11].

Root hairs are localized for many water channels [67], phosphate [68], nitrogen [69], potassium [70], calcium [70], and sulfate transporters [71], all of which are beneficial to water and nutrient uptake by plants [72]. There is significant inter- and intra-specific variation exists for root hair traits, and this has been linked to P uptake. Plants with longer, denser root hairs exhibit greater P uptake and plant growth in P-deficient soils [73–75]. So, the root hair traits, especially root hair length can be exploited in breeding for improved nutrient uptake and increased fertilizer use efficiency [76]. Considerable researches support an important role for root hairs in P attainment [73–75, 77, 78]. Root hair length and root hair density (which is usually correlated with root hair length) have clear value for the acquisition of P and probably other diffusion-limited nutrients such as K and ammonium [79].

Usually root hair traits have a low heritability and their expression is influenced by soil type resulting in lack of research in this field [6, 80, 81]. It has been proposed that plasticity in root epidermis development as a response to a variety of environmental conditions might reflect a function of root hairs in sensing environmental signals, after which plants adjust themselves to the stress conditions, such as by increasing nutrient acquisition and water uptake or by helping to anchor the plant to the soil [82–87]. Root hair elongation increases root surface area. Root surface area increment is a common phenomenon when the plants are subjected to the stress condition like salinity, drought or other abiotic stresses [79, 88–91].

### 2.3 Research progress of rice root study till date under osmotic stress

Plants recurrently face several stresses like salinity, drought, submergence, low temperature, heat, oxidative stress and heavy metal toxicity while exposed to the nature. Growth and grain production in cereals is often limited by these stresses under field conditions. All these stresses either directly or indirectly impose osmotic stress to plants that ultimately affect the final yield of rice. Root is the first part which can sense these stresses better than other plant parts. So researchers prioritize the fact of understanding the root adaptive responses of plants upon osmotic stress. In the last 30 years, comprehensive studies have been performed focusing on architecture and developmental morphology of roots and their genetic and molecular basis [11]. Morphological and anatomical development of the rice root system was thoroughly reviewed [92] whereas the mystery of root length was also reviewed [93]. A recent study highlighting the growth, development and genetic reasons of root morphology and function of crop plants was provided by [94]. An outstanding study on root system architecture and its molecular and genetic background also greatly contributed to the relevant literature recently [37]. The physiological background of root branching was also studied [7, 33]. The root parameters that are focused by the studies comprising root anatomy, plant height, root-shoot ratio, length, diameter, density, surface area and volume of root, root elongation rate, root branching, expansion of root regarding tiller development, maximum root depth, distribution pattern of root in soil column, root hydraulic conductivity, hardpan penetrability, all of which possess innumerable functional implication [95]. Roots of large diameter show greater penetration ability [96–98] and branching [8, 99] because of having larger radii of xylem vessel and poorer axial resistance to water flux [100].

### 2.3.1 Plasticity of root traits under drought

Water is essential for survival and plant growth. As a sessile organism, plants constantly encounter water deficit, which is the most severe environmental stress limiting plant growth and productivity in natural and agricultural systems [101, 102]. Thus, water stress tolerance has been a fundamental scientific question in plant biology.

Plants have evolved complex adaptive mechanisms that enable them to survive drought conditions. Over more than five decades, researchers have identified osmotic adjustment, antioxidant protection, and stomatal movement as key adaptive mechanisms for survival where both osmotic adjustment and reactive oxygen species (ROS) are involved in this plastic development process [103]. To cope with the changing water status in the growing environment, plants have evolved various adaptive mechanisms by which plants can modify root allocation and root system architecture to obtain more water [104].

Numerous studies have provided evidence to show that when plants are subjected to water stress, root growth is strongly inhibited, although root development is less sensitive to water stress than that of shoots [105–107].

Root system architecture is regulated by osmotica [108]. The osmotic potential of the soil alters the depth of the root system, its overall mass, the rate of root elongation and the number of lateral roots in many plants, including *Arabidopsis* [8, 9, 107, 109, 110].

Root length, root dry weight, and root production are limited by drought stress [111, 112]. Roots are the significant plant part which increase plant adaptability power to soil water deficits by maintaining water uptake under dry conditions [113]. Root and other root components such as root hair, root-shoot ratio, and root length are found to be decreased in drought sensitive varieties. But the resistant varieties which possess tolerance capacity against drought showed increase in root hair, high root to shoot ratio and root length [114]. Roots are considered as the most efficient plant organ which helps plant to uptake water and minerals from the soil and during drought stress. Root proliferation and changes in root parts occurs to take more water from deeper regions of the soil [25]. Different types of changes are observed in root growth of drought resistant rice varieties such as a deeper and highly branched root system than drought- sensitive varieties [115]. Plant also extends its roots for more nutrients (such as phosphorus) and water uptake which results in more root to shoot ratio [116]. In recent years breeding for developing larger and more efficient root systems has become the hotspot in research in some crops such as rice, as there is a relation between root system size and tolerance to water stress [81, 117].

The change in lateral root development, i.e. in the plasticity of the root system, exhibited under water deficit conditions may play an important role in drought stress tolerance [35]. From an agronomical view, the knowledge about lateral root development is useful for breeding varieties with drought stress tolerance [118].

### 2.3.2 Modification of root system components under submergence stress

The importance of root system structure is particularly recognizable when its significance in relation to its function is clearly identified. The significance of root system structure in nutrient and water uptake was stressed in previous study [119].

Under waterlogged conditions, the plant roots have to function in anaerobic soil, and there are at least two morphological adaptations that roots exhibit in response to anaerobiosis, i.e., development of new adventitious roots [120, 121] and superficial rooting (i.e., the concentration of new root growth in the upper layers of the soil) [122]. Nodal root production (increase in number) continued to take place, however, in the sense that when adventitious roots in the lower nodal position of the plant's stem die due to waterlogging injury, new adventitious roots appear at the next highest nodal position. There appears to be a direct relationship between the death of older adventitious roots and the development of new ones. Progressively waterlogged plants generally show smaller root system size than those grown in a well-drained condition. It is considered that the turgor pressure affects the cell elongation and growth of plants [123, 124]. Aerobic cultivars of rice have greater ability for plastic lateral root production than irrigated lowland cultivars under transient moisture stresses [125].

### 2.3.3 Plasticity of root traits under salinity stress

We have a little understanding of the responses of roots and root hairs to salinity stress and their function in stress tolerance. The efficient root system can either avoid or lessen the osmotic stress. Usually, growth, morphology, and physiology of the roots alter first under salinity stress and the whole plant is then affected. Therefore, the responses and characteristics of the roots under saline conditions are of primary importance for plant salt-tolerance [126]. It is supposed that root morphology affects salt accumulation around the roots impeding uptake of water from saline areas. Modification of root morphology has a big potential to develop crop salt tolerance [127]. Root hairs have higher sensitivity to salt than other root traits and shoots [128]. Environmental factors also regulate the root hair development [128]. The development of root epidermal cells has great plasticity where the differentiation programs can be switched from one to another in response to external factors [17]. Plasticity in development of root epidermis as a response to a variety of environmental conditions might reflect a function of root hairs in sensing environmental signals, after which plants adjust themselves to the stress conditions [82, 84–87, 129].

Root hair growth and development and their physiological role in response to salt stress are largely unknown [128]. The development of root epidermis cells has great plasticity where the differentiation programs can be switched from one to another in response to external factors [17]. Root hairs have higher sensitivity to salinity than do roots and shoots [128]. Systematic study on root hair plasticity induced by salt stress and the possible role in plant adaptation/tolerance to salinity is still lacking [128]. Usually root hair traits have a low heritability and their expression is influenced by soil type resulting in lack of research in this field [6, 80, 81].

## 2.4 Varietal differences in rice root morphological characteristics

Earlier many scientists had reported root morphology and its distribution were greatly varied based on genotypes of plant species [13–16]. There is widespread evidence that root architecture and different root characteristics of many crop species varies among genotypes [14, 130–133]. In a few quite recent studies, the importance of studying root architectural traits has been emphasized for the adaptation of the crop varieties to various abiotic stress conditions. Genotypic variation has a significant role in adapting the adverse environmental and edaphic effects [14]. Inter- and intra-species variations in root architectural traits are very useful to breed the crops for root features optimum for diverse environmental conditions [134–136].

Root anatomical and morphological traits have been well studied in rice [92]. Varietal differences in root morphological characteristics such as length and thickness have been reported in cultivated rice (*Oryza sativa* L.) in various studies [11, 14, 41, 137]. In general, the roots of upland rice cultivars are thicker and penetrate more deeply into the soil than those of lowland cultivars [14]. Root distribution has also been quantitatively characterized by using several traits, including root length, volume, and density in the soil at different depths, and these characteristics differed among cultivars [92, 138–140].

## 3. Future prospects of rice root study

Understanding and improvement of root system and its genetics plays a pivotal role to become self-sufficient and to achieve sustainability in rice production. Actually more yields from the limited input rely on our capability to unambiguously manipulate the plants. And exploring the diversity of root architecture both in genetic and phenotypic basis will directly connect to this concern. Although great strides have been made to understand the root morphology but in future, more intense investigations to elucidate the functional implication of root morphological variation may aid in selection of root system with anticipated characteristics.

Future exploration of stress responses regulated by roots at cellular or tissue level will open the door of further breeding research. Besides the modern gene pools, exploration of genes and alleles in wild relatives and landraces will also provide interesting features that will be easier to transfer to cultivated rice. Further it is important to have a better understanding on the epigenetic regulation of roots and root development under stressful conditions. There will be a need for high throughput phenotyping systems coupled with automated data analysis for accelerating the development. Endorsement of approaches including both root ideotype-based screening and selection for grain yield may establish a fruitful screening system. Alongside designing new genetic screening methods based on a better knowledge of the integrated stress responses will be also appreciated. Dynamic root/soil interaction modeling will aid in integrating different functional parameters (e.g. water uptake per length of root) under a variety of environmental conditions. Overall the root system being less accessible and more complex than other agronomic traits, achieving the ambitious goal of future rice root research, coordinated effort and joint resources are required. The sensible and appropriate efforts will have a crucial role to play in future crop production in vulnerable climate and resource scarcity prioritizing the objective of serving food to 9 billion world populations by the year 2050.

## Conflict of interest

"The authors declare no conflict of interest."

## Author details

Afsana Hannan[1], Md. Najmol Hoque[2], Lutful Hassan[1] and Arif Hasan Khan Robin[1*]

1 Department of Genetics and Plant Breeding, Bangladesh Agricultural University, Mymensingh, Bangladesh

2 Department of Biochemistry and Molecular Biology, Khulna Agricultural University, Khulna, Bangladesh

*Address all correspondence to: gpb21bau@bau.edu.bd

## References

[1] Bangladesh Bureau of statistics (BBS). 45 years Agriculture Statistics of Major Crops (Aus, Amon, Boro, Jute, Potato & Wheat), Statistics and Informatics Division (SID), Ministry of Planning, Government of the People's Republic of Bangladesh; 2018. p. 1-212.

[2] Mohanty S, Wassmann R, Nelson A, Moya P, Jagadish SVK. Rice and climate change: significance for food security and vulnerability. IRRI Discussion Paper Series No. 49. Los Baños (Philippines): International Rice Research Institute; 2013. p. 14.

[3] Zhang F, Luo X, Zhou Y, Xie J. Genome-wide identification of conserved micro RNA and their response to drought stress in Dongxiang wild rice (*Oryza rufipogon* Griff.). Biotechnology Letters 2015; 38:711-721. DOI: 10.1007/s10529-015-2012-0

[4] Zhu JK. Salt and drought stress signal transduction in plants. Annual Review of Plant Physiology and Plant Molecular Biology. 2002; 53:247-273. DOI: 10.1146/annurev.arplant.53.091401.143329

[5] Razmjoo K, Heydarizadeh P, Sabzalian MR. Effect of salinity and drought stresses on growth parameters and essential oil content of Matricaria chamomila. International Journal of Agriculture and Biology. 2008; 10:451-454.

[6] Malamy JE. Intrinsic and environmental response pathways that regulate root system architecture. Plant, Cell and Environment. 2005; 28:67-77. DOI: 10.1111/j.1365-3040.2005.013e06.x

[7] Nibau C, Gibbs D, Coates J. Branching out in new directions: the control of root architecture by lateral root formation. New Phytologist. 2008; 179:595-614. DOI: 10.1111/j.1469-8137.2008.02472.x

[8] Fitter AH. The ecological significance of root system architecture: an economic approach. In: Atkinson D, editors. Plant Root Growth: An Ecological Perspective. Oxford, Blackwell Scientific Publications, 1991. p. 229-243.

[9] Kramer PJ, Boyer JS. Water Relations of Plants and Soils. Academic Press, Inc., San Diego, USA; 1995.

[10] Lopez-Bucio J, Cruz-Ramirez A, Herrera-Estrella L. The role of nutrient availability in regulating root architecture. Current Opinion in Plant Biology. 2003; 6:280-287. DOI: 10.1016/S1369-5266(03)00035-9

[11] Robin AHK, Saha PS. Morphology of lateral roots of twelve rice cultivars of Bangladesh: dimension increase and diameter reduction in progressive root branching at the vegetative stage. Plant Root. 2015; 9:34-42. DOI: 10.3117/plantroot.9.34

[12] Robin AHK, Matthew C, Uddin MJ, Bayazid KN. Salinity-induced reduction in root surface area and changes in major root and shoot traits at the phytomer level in wheat. Journal of Experimental Botany. 2016; 67:3719-3729. DOI: 10.1093/jxb/erw064

[13] Nielsen NE, Schjørring JK. Efficiency and kinetics of phosphorus uptake from soil by various barley genotypes. Plant and Soil. 1983; 72:225-230. DOI: 10.1007/978-94-009-6836-3_17

[14] O'Toole JC, Bland WL. Genotypic variation in crop plant root systems. Advances in Agronomy. 1987; 41:91-143. DOI: 10.1016/S0065-2113(08)60803-2

[15] Römer W, Augustin J, Schilling G. The relationship between phosphate absorption and root length in nine

wheat cultivars. Plant and Soil. 1988; 111:199-201.

[16] Atkinson D. Influence of root system morphology and development on the need for fertilizers and the efficiency of use. In: Baligar VC, Duncan RR, editors. Crops as Enhancers of Nutrient Use. Academic Press, San Diego, US; 1990. p. 411-451. DOI: 10.1016/B978-0-12-077125-7.50015-3

[17] Schiefelbein JW. Constructing a plant cell: The genetic control of root hair development. Plant Physiology. 2000; 124:1525-1531. DOI: 10.1104/pp.124.4.1525

[18] Wasson AP, Richards RA, Chatrath R, Misra SC, Prasad SS, Rebetzke GJ, Watt M. Traits and selection strategies to improve root systems and water uptake in water-limited wheat crops. Journal of Experimental Botany. 2012;63:3485-3498. DOI: 10.1093/jxb/ers111

[19] Seck PA, Diagne A, Mohanty S, Wopereis MCS. Crops that feed the world 7: Rice. Food Security. 2012; 4:7-24. DOI: 10.1007/s12571-012-0168-1

[20] Zhang J, Davies WJ. Increased synthesis of ABA in partially dehydrated root tips and ABA transport from roots to leaves. Journal of Experimental Botany. 1987; 38:2015-2023. DOI: 10.1093/jxb/38.12.2015

[21] Davies WJ, Zhang J. Root signals and the regulation of growth and development of plants in drying soil. Annual Review of Plant Physiology and Plant Molecular Biology. 1991; 42:55-76. DOI: 10.1146/annurev.pp.42.060191.000415

[22] Hartung W, Sauter A, Hose E. Abscisic acid in xylem: Where does it come from, where does it go? Journal of Experimental Botany. 2002; 53:27-32. DOI: 10.1093/jexbot/53.366.27

[23] Jiang F, Hartung W. Long-distance signalling of abscisic acid (ABA): The factors regulating the intensity of the ABA signal. Journal of Experimental Botany. 2008; 59:37-43. DOI: 10.1093/jxb/erm127

[24] de Dorlodot S, Forster B, Pagès L, Price A, Tuberosa R, Draye X. Root system architecture: opportunities and constraints for genetic improvement of crops. Trends in plant science. 2007; 12:474-481. DOI: 10.1016/j.tplants.2007.08.012

[25] Smith S, De Smet I. Root system architecture: insights from Arabidopsis and cereal crops. 2012. DOI: 10.1098/rstb.2011.0234

[26] Lynch, J. Root architecture and plant productivity. Plant Physiology. 1995; 109:7. DOI: 10.1104/pp.109.1.7

[27] Jung JKHM, McCouch SRM. Getting to the roots of it: genetic and hormonal control of root architecture. Frontiers in plant science. 2013; 4:186. DOI: 10.3389/fpls.2013.00186

[28] Vance CP, Uhde-Stone C, Allan DL. Phosphorus acquisition and use: critical adaptations by plants for securing a nonrenewable resource. New Phytologist. 2003; 157:423-447. DOI: 10.1046/j.1469-8137.2003.00695.x

[29] Lambers H, Shane MW, Cramer MD, Pearse SJ, Veneklaas EJ. Root structure and functioning for efficient acquisition of phosphorus: matching morphological and physiological traits. Annals of Botany. 2006; 98:693-713. DOI: 10.1093/aob/mcl114

[30] Fitter AH. Functional significance of root morphology and root system architecture. In: Fitter AH, Atkinson D, Read DJ, Usher MB, editors. Ecological Interactions in Soil, Plants, Microbes and Animals. Oxford, Blackwell Scientific Publications; 1985. p.87-106.

[31] Fitter AH. The topology and geometry of plant-root systems - influence of watering rate on root-system topology in Trifolium pratense. Annals of Botany. 1986; 58:91-101. DOI: 10.1093/oxfordjournals.aob.a087191

[32] Glimskar A. Estimates of root system topology of five plant species grown at steady-state nutrition. Plant and Soil. 2000; 227:249-256. DOI: 10.1023/A:1026531200864

[33] Osmont KS, Sibout R, Hardtke CS. Hidden branches: developments in root system architecture. Annual Review of Plant Biology. 2007; 58:93-113. DOI: 10.1146/annurev.arplant.58.032806.104006

[34] Grime JP, Crick JC, Rincon JE. The ecological significance of plasticity. In: Symposium Society of Experimental Biology; 1986. p. 5-29.

[35] Yamauchi A, Pardales JRJr, Kono Y. Root system structure and its relation to stress tolerance. In: Ito O, Katayama K, Johansen C, Kumar Rao JVDK, Adu-Gyamfi JJ, Rego TJ, editors. Roots and nitrogen in cropping systems of the semi-arid tropics. JIRCAS Publication, Tsukuba, Japan; 1996. p. 211-234.

[36] Kawata S, Yamazaki K, Ishihara K, Shibatama H, Lai KL. Studies on root system formation in rice plants in a paddy. Japanese Journal of Crop Science. 1963; 32:163-180. DOI: 10.1626/jcs.32.163

[37] Rebouillat J, Dievart A, Verdeil J, Escoute J, Giese G, Breitler J, Gantet P, Espeout S, Guiderdoni E, Périn C. Molecular genetics of rice root development. Rice. 2009; 2:15-34. DOI: 10.1007/s12284-008-9016-5

[38] Hochholdinger F, Park WJ, Sauer M, Woll K. From weeds to crops: genetic analysis of root development in cereals. Trends in Plant Science. 2004;9: 42-48. DOI: 10.1016/j.tplants.2003.11.003

[39] Sasaki O, Yamazaki K, Kawata S. The relationship between the diameters and the structures of lateral roots in rice plants. Japanese Journal of Crop Science. 1984; 53:169-175. DOI: 10.1626/jcs.50.476

[40] Kawata S, Ishihara K, Shioya T. Studies on the root hairs of lowland rice plants in the upland fields. Japanese Journal of Crop Science. 1964; 32:250-253. DOI: 10.1626/jcs.32.250

[41] Hannan A, Hassan L, Hoque MN, Arif MTU, Robin AHK . Increasing New Root Length Reflects Survival Mechanism of Rice (*Oryza sativa* L.) Genotypes under PEG-Induced Osmotic Stress. Plant Breeding and Biotechnology.2020;8:46-57. DOI: 10.9787/PBB.2020.8.1.46

[42] Arber A. The Gramineae: a Study of Cereal, Bamboo, and Grass. Cambridge University Press, Cambridge, United Kingdom; 1934. DOI: 10.1017/CBO9780511700668

[43] Morita S, Abe J. Development of Root Systems in Wheat and Rice. In: Ito O, C Johansen C, Adu-Gyamfi JJ, Katayama K, Kumar Rao JVDK, Kand TJR, editors. Roots and Nitrogen in Cropping Systems of the Semi-Arid Tropics. Japan International Research Center for Agricultural Sciences, Ohwashi, Tsukuba, Ibaraki 305, Japan; 1994. p. 185-198.

[44] Nemoto K, Morita S, Baba T. Shoot and root development in rice related to the phyllochron. Crop Science. 1995; 35:24-29. DOI: 10.2135/cropsci1995.0011183X003500010005x

[45] McMaster GS, Hargreaves JNG. CANON in D (esign): Composing scales of plant canopies from phytomers to whole-plants using the composite design

pattern. NJAS-Wageningen Journal of Life Sciences. 2009; 57:39-51. DOI: 10.1016/j.njas.2009.07.008

[46] Rickman RW, Klepper BL. The phyllochron: where do we go in the future? Crop Science. 1995; 35:44-49. DOI: 10.2135/cropsci1995.0011183X003500010008x

[47] McMaster GS. Phytomers, phyllochrons, phenology and temperate cereal development. The Journal of Agricultural Science. 2005; 143:137-150. DOI: 10.1017/S0021859605005083

[48] Fujii Y. Studies on the regularity of root growth in rice and wheat plants. Bulletin of the Faculty of Agriculture, Saga University. 1961; 12:1-117.

[49] Harada J, Yamazaki K. Science of the rice plant. In: T Matsuo T, Hoshikawa K, editors. Volume 1. Morphology. Food and Agriculture Policy Research Center, Tokyo; 1993. p.133-186.

[50] Etter AG. How Kentucky bluegrass grows. Annals of the Missouri Botanical Garden. 1951; 38:293-375. DOI: 10.2307/2394639

[51] Silsbury JH. Leaf growth in pasture grasses. Tropical Grasslands. 1970; 4:17-39.

[52] Robin AHK, Matthew C, Crush J. Time course of root initiation and development in perennial ryegrass – a new perspective. In: Proceedings of the New Zealand Grassland Association; New Zealand. 2010. p. 233-237.

[53] Robin AHK. Segmental morphology of perennial ryegrass (*Lolium perenne* L.): a study of functional implications of plant architecture [PhD thesis]. Massey University; 2011.

[54] Bossinger, G. Segments (Phytomers). eLS. 2001. DOI: 10.1002/9780470015902.a0002093.pub2

[55] Henry A, Gowda VRP, Torres RO, McNally KL, Serraj R. Variation in root system architecture and drought response in rice (*Oryza sativa*): phenotyping of the Oryza SNP panel in rainfed lowland fields. Field Crops Research. 2011;120:205-214. DOI: 10.1016/j.fcr.2010.10.003

[56] Casimiro I, Beeckman T, Graham N, Bhalerao R, Zhang H, Casero P, Sandberg G, Bennett MJ. Dissecting Arabidopsis lateral root development. Trends in Plant Science. 2003;8:165-171. DOI: 10.1016/S1360-1385(03)00051-7

[57] Juliano JB, Aldama MJ. Morphology of *Oryza sativa* Linnaeus. The Philippine Agriculturist. 1938;25:1-76.

[58] Kawata S, Shibayama S. On the lateral root primordia formation in the crown roots of rice plants. In: Proceedings of the Crop Science Society of Japan. 1965. p. 423-431 (in Japanese with English summary). DOI: 10.1626/jcs.33.423

[59] Kono Y, Igeta M, Yamada N. Studies on the rice seminal roots. In: Proceedings of the Crop Science Society of Japan. 1972. p. 192-204 (in Japanese with English summary). DOI: 10.1626/jcs.41.192

[60] Yamauchi A, Kono Y, Tatsumi J. Quantitative analysis on root system structures of upland rice and maize. Japanese Journal of Crop Science. 1987;56:608-617. DOI: 10.1626/jcs.56.608

[61] Zobel RW. Root growth and development. In: Keiserand DL, Cregan PB, editors. The rhizosphere and plant growth. Kluwer Academic, Dordrecht, The Netherlands; 1991. p. 61-71.

[62] Weaver JE, Zink E. Length of life of roots of ten species of perennial range and pasture grasses. Plant Physiology.

1955; 37:201-217. DOI: 10.1104/pp.21.2.201

[63] Hunt ERJ, Zakir NJD, Nobel PS. Water cost and water revenues for established and rain-induced roots of Agave deserti. Functional Ecology. 1987; 1125-1129. DOI: 10.2307/2389715

[64] Bañoc DM, Yamauchi A, Kamoshita A, Wade LJ, Pardales JR. Genotypic variations in response of lateral root development to fluctuating soil moisture in rice. Plant Production Science. 2000; 3:335-343. DOI: 10.1626/pps.3.335

[65] Clarkson DT. Factors affecting mineral nutrient acquisition by plants. Annual Review of Plant Physiology. 1985; 36:77-115. DOI: 10.1146/annurev.pp.36.060185.000453

[66] Parker JS, Cavell AC, Dolan L, Roberts K, Grierson CS. Genetic interactions during root hair morphogenesis in Arabidopsis. The Plant Cell. 2000; 12:1961-1974. DOI: 10.1105/tpc.12.10.1961

[67] Brechenmacher L, Lee J, Sachdev S, Song Z, Nguyen THN, Joshi T, Oehrle N, Libault M, Mooney B, Xu D, Cooper B, Stacey G. Establishment of a Protein Reference Map for Soybean Root Hair Cells. Plant Physiology. 2009: 149; 670-682. DOI: 10.1104/pp.108.131649

[68] Schünmann PHD, Richardson AE, Smith FW, Delhaize E. Characterization of promoter expression patterns derived from the Pht1 phosphate transporter genes of barley (*Hordeum vulgare* L.). Journal of Experimental Botany. 2004:55:855-865. DOI: 10.1093/jxb/erh103

[69] Loqué D, von Wirén N. Regulatory levels for the transport of ammonium in plant roots. Journal of Experimental Botany. 2004;55:1293-1305. DOI: 10.1093/jxb/erh147

[70] Lan WZ, Wang W, Wang SM, Li LG, Buchanan BB, Lin HX, Gao JP, Luan S. A rice high-affinity potassium transporter (HKT) conceals a calcium-permeable cation channel. In: Proceedings of the National Academy of Sciences. 2010. p. 7089-7094. DOI: 10.1073/pnas.1000698107

[71] Maruyama-Nakashita A, Nakamura Y, Yamaya T, Hideki T. Regulation of high-affinity sulphate transporters in plants: towards systematic analysis of sulphur signaling and regulation. Journal of Experimental Botany. 2004;55:1843-1849. DOI: 10.1093/jxb/erh175

[72] Libault M, Brechenmacher L, Cheng J, Xu D, Stacey G. Root hair systems biology. Trends in Plant Science. 2010;15:641-650. DOI: 10.1016/j.tplants.2010.08.010

[73] Gahoonia TS, Nielsen NE. Variation in root hairs of barley cultivars doubled soil phosphorus uptake. Euphytica. 1997;98:177-182. DOI: 10.1023/A:1003113131989

[74] Gahoonia TS, Nielsen NE. Barley genotypes with long root hairs sustain high grain yields in low-P field. Plant and Soil. 2004;262:55-62. DOI: 10.1023/B:PLSO.0000037020.58002.ac

[75] Brown LK, George TS, Thompson JA, Wright G, Lyon J, Dupuy L, White PJ. What are the implications of variation in root hair length on tolerance to phosphorus deficiency in combination with water stress in barley (Hordeum vulgare)? Annals of Botany. 2012;110:319-328. DOI: 10.1093/aob/mcs085

[76] Brown LK, George TS, Dupuy LX, White PJ. A conceptual model of root hair ideotypes for future agricultural environments: what combination of traits should be targeted to cope with limited P availability? Annals of Botany.

2013; 112:317-330. DOI: 10.1093/aob/mcs231

[77] Gahoonia TS, Nielsen NE. Direct evidence on partition of root hairs in phosphorus ($^{32}P$) uptake from soil. Plant and Soil. 1998;198:147-152. DOI: 10.1023/A: 1004346412006

[78] Bates TR, Lynch JP. Root hairs confer a competitive advantage under low phosphorous availability. Plant and Soil. 2001; 236:243-250. DOI: 10.1023/A: 1012791706800

[79] Haling RE, Brown LK, Bengough, AG, Young IM, Hallett PD, White PJ, George TS. Root hairs improve root penetration, root-soil contact, and phosphorus acquisition in soils of different strength. Journal of Experimental Botany. 2013; 64:3711-3721. DOI: 10.1093/jxb/ert200

[80] Cooper M, Rajatasereekul S, Immark S, Fukai S, Basnayake J. Rainfed lowland rice breeding strategies for Northeast Thailand.: I. Genotypic variation and genotype × environment interactions for grain yield. Field Crops Research. 1999; 64:131-151. DOI: 10.1016/S0378-4290(99)00056-8

[81] Tuberosa R, Salvi S, Sanguineti MC, Landi P, Maccaferri M, Conti S. Mapping QTLs regulating morpho-physiological traits and yield: Case studies, shortcomings and perspectives in drought-stressed maize. Annals of Botany. 2002; 89:941-963. DOI: 10.1093/aob/mcf134

[82] Ewens M, Leigh RA. The effect of nutrient solution composition on the length of root hairs of wheat (Triticum aestivum L.). Journal of Experimental Botany. 1985; 36:713-724. DOI: 10.1093/jxb/36.5.713

[83] Bates TR, Lynch JP. Stimulation of root hair elongation in Arabidopsis thaliana by low phosphorus availability. Plant, Cell and Environment. 1996; 19:529-538. DOI: 10.1111/j.1365-3040.1996.tb00386.x

[84] Cao XF, Linstead P, Berger F, Kieber J, Dolan L. Differential ethylene sensitivity of epidermal cells is involved in the establishment of cell pattern in the Arabidopsis root. Physiologia Plantarum. 1999;106:311-317. DOI: 10.1034/j.1399-3054.1999.106308.x

[85] Schikora A, Schmidt W. Iron stress-induced changes in root epidermal cell fate are regulated independently from physiological responses to low iron availability. Plant Physiology. 2001; 125:1679-1687. DOI: DOI: 10.1104/pp.125.4.1679

[86] Gilroy S, Jones DL. Through form to function: root hair development and nutrient uptake. Trends in Plant Science. 2000; 5:56-60. DOI: 10.1016/S1360-1385(99)01551-4

[87] Ma Z, Bielenberg DG, Brown KM, Lynch JP. Regulation of root hair density by phosphorus availability in Arabidopsis thaliana. Plant, Cell and Environment. 2001; 24:459-467. DOI: 10.1046/j.1365-3040.2001.00695.x

[88] Enstone DE, Peterson CA, Ma F. Root endodermis and exodermis: structure, function, and responses to the environment. Journal of Plant Growth Regulation. 2002; 21:335-351. DOI: 10.1007/s00344-003-0002-2

[89] Huang X, Liu Y, Li J, Xiong X, Chen Y, Yin X, Feng D. The response of mulberry trees after seedling hardening to summer drought in the hydro-fluctuation belt of Three Gorges Reservoir Areas. Environmental Science and Pollution Research. 2013; 20:7103-7111. DOI: 10.1007/s11356-012-1395-x

[90] Uga Y, Sugimoto K, Ogawa S, Rane J, Ishitani M, Hara N, Inoue H. Control of root system architecture by

deeper rooting 1. increases rice yield under drought conditions. Nature Genetics. 2013; 45:1097. DOI: 10.1038/ng.2725

[91] Steinemann S, Zeng Z, McKay A, Heuer S, Langridge P, Huang CY. Dynamic root responses to drought and rewatering in two wheat (Triticum aestivum) genotypes. Plant and Soil. 2015; 391:139-152. DOI: 10.1007/s11104-015-2413-9

[92] Morita S, Nemoto K. Morphology and anatomy of rice roots with special reference to coordination in organo- and histogenesis. In: Baluška F et al., editors. Structure and function of roots, Kluwer Academic Publishers, the Netherlands; 1995. p. 75-86. DOI: 10.1007/978-94-017-3101-0_9

[93] Ryser P. The mysterious root length. Plant and Soil. 2006; 286:1-6. DOI: 10.1007/s11104-006-9096-1

[94] Hodge A, Berta G, Doussan C, Merchan F, Crespi M. Plant root growth, architecture and function. Plant and Soil. 2009; 321:153-183. DOI: 10.1007/s11104-009-9929-9

[95] Wang H, Inukai Y, Yamauchi A. Root development and nutrient uptake. Critical Reviews in Plant Sciences. 2006; 25:279-301. DOI: 10.1080/07352680600709917

[96] Materechera SA, Alston AM, Kirby JM, Dexter AR. Influence of root diameter on the penetration of seminal roots into a compacted subsoil. Plant and Soil. 1992; 144:297-303. DOI: 10.1007/BF00012888

[97] Nguyen HT, Babu RC, Blum A. Breeding for drought resistance in rice: physiology and molecular genetics considerations. Crop Science. 1997; 37:1426-1434. DOI: 10.2135/cropsci1997.0011183X003700050002x

[98] Clark, LJ, Price AH, Steele KA, Whalley WR. Evidence from near-isogenic lines that root penetration increases with root diameter and bending stiffness in rice. Functional Plant Biology. 2008; 35:1163-1171. DOI: 10.1071/FP08132

[99] Ingram KT, Bueno FD, Namuco OS, Yambao EB, Beyrouty CA. Rice root traits for drought resistance and their genetic variation. In: Kirk GJD, editors. Rice Roots: Nutrient and Water Use. International Rice Research Institute, Manila, Philippines; 1994.

[100] Yambao EB, Ingram KT, Real JG. Root xylem influence on the water relations and drought resistance of rice. Journal of Experimental Botany. 1992; 43:925-932. DOI: 10.1093/jxb/43.7.925

[101] Hsiao TC. Plant responses to water stress. Annual Review of Plant Physiology. 1973; 24:519-570. DOI: 10.1146/annurev.pp.24.060173.002511

[102] Reddy AR, Chaitanya KV, Vivekananda M. Drought-induced responses of photosynthesis and antioxidant metabolism in higher plants. Journal of Plant Physiology. 2004; 161:1189-1202. DOI: 10.1016/j.jplph.2004.01.013

[103] Ji H, Liu L, Li K, Xie Q, Wang Z, Zhaom X, Li X. PEG-mediated osmotic stress induces premature differentiation of the root apical meristem and outgrowth of lateral roots in wheat. Journal of Experimental Botany. 2014; 65:4863-4872. DOI: 10.1093/jxb/eru255

[104] Ji H, Li X. ABA mediates PEG-mediated premature differentiation of root apical meristem in plants. Plant Signaling and Behavior. 2014; 9:e977720. DOI: 10.4161/15592324.2014.977720

[105] Westgate ME, Boyer JS. Osmotic adjustment and the inhibition if leaf, root, stem and silk growth at low water potentials in maize. Planta. 1985; 164:540-549. DOI: 10.1007/BF00395973

[106] Sharp RE, Davies WJ. Regulation of growth and development of plants growing with a restricted supply of water. In: Jones HG, Flowers TJ, Jones MB, editors. Plants under stress: biochemistry, physiology and ecology and their application to plant improvement. Cambridge: Cambridge press; 1989. p. 71-93.

[107] Spollen WG, Sharp RE, Saab IN, Wu Y. Regulation of cell expansion in roots and shoots at low water potentials. In: Smith JAC, Griffiths H, editors. Water Deficits: Plant Responses from Cell to Community. Oxford: BIOS Scientific Publishers; 1993. p. 37-52.

[108] Deak KI, Malamy J. Osmotic regulation of root system architecture. The Plant Journal. 2005; 43:17-28. DOI: 10.1111/j.1365-3040.2005.01306.x

[109] Russell EW. The Development of Plant Roots in Soil in Soil Conditions and Plant Growth. 9th ed. John Wiley and Sons, New York; 1961.

[110] van der Weele CM, Spollen WG, Sharo RE, Baskin TI. Growth of Arabidopsis thaliana seedlings under water deficit studied by control of water potential in nutrient-agar media. Journal of Experimental Botany. 2000; 51:1555-1562. DOI: 10.1093/jexbot/51.350.1555

[111] Anjum SA, Wang L, Farooq M, Khan I, Xue L. Methyl jasmonate-induced alteration in lipid peroxidation, antioxidative defence system and yield in soybean under drought. Journal of Agronomy and Crop Science. 2011; 197:296-301. DOI: 10.1111/j.1439-037X.2011.00468.x

[112] Bansal R, Pradheep K, Kumari J, Kumar S, Yadav MC, Gurung B, Kumari NK, Rana JC. Physiological and biochemical evaluation for drought tolerance in wheat germplasm collected from arid western plains of India. 2016.

[113] Turner NC. Drought resistance and adaptation to water deficits in crop plants. Stress physiology in crop plants. 1979; 343-372.

[114] Huang B, Fry JD. Root anatomical, physiological, and morphological responses to drought stress for tall fescue cultivars. Crop Science. 1998; 38:1017-1022. DOI: 10.2135/cropsci1998.0011183X003800040022x

[115] Price AH, Tomos AD, Virk DS. Genetic dissection of root growth in rice (Oryza sativa L.) I. A hydrophonic screen. Theoretical Applied Genetics. 1997; 95:132-142. DOI: 10.1007/s001220050542

[116] Narang RA, Bruene A, Altmann T. Analysis of phosphate acquisition efficiency in different Arabidopsis accessions. Plant Physiology. 2000; 124,1786-1799. DOI: 10.1104/pp.124.4.1786

[117] Price AH, Steele K, Gorham J, Bridges J, Moore B, Evans J, Richardson P, Jones RGW. Upland rice grown in soil-filled chambers and exposed to contrasting water-deficit regimes I. Root distribution, water use and plant water status. Field Crops Research. 2002; 76:11-24. DOI: 10.1016/S0378-4290(02)00010-2

[118] Ogawa A, Kawashima C, Yamauchi A. Sugar accumulation along the seminal root axis, as affected by osmotic stress in maize: A possible physiological basis for plastic lateral root development. Plant Production Science. 2005; 8:173-180. DOI: 10.1626/pps.8.173

[119] Bray RH. A nutrient mobility concept of soil-plant relationship. Soil Science. 1954;78:9-22. DOI: 10.1097/00010694-195407000-00002

[120] Kono Y, Yamauchi A, Nonoyama T, Tatsumi J. Comparison of growth responses to waterlogging of summer cereals with reference to

rooting ability. Japanese Journal of Crop Science. 1988; 57:321-331. DOI: 10.1626/jcs.57.321

[121] Yamauchi A, Kono Y, Tatsumi J, Inagaki N. Comparison of the capacities of water-logging and drought tolerances among winter cereals. Japanese Journal of Crop Science. 1988; 57:163-173. DOI: 10.1626/jcs.57.163

[122] Jackson MB, Drew MC. Effects of flooding on growth and metabolism of herbaceous plants. In: Kozlowski TT, editor. Flooding and plant growth. Academic Press, London, UK; 1984. p. 47-128.

[123] Pardales JRJr, Kono Y, Yamauchi A. Response of different root system components of sorghum to incidence of waterlogging. Environmental and Experimental Botany. 1991; 31:107-115. DOI: 10.1016/0098-8472(91)90013-E

[124] Lockhart JA. An analysis of irreversible plant cell elongation. Journal of Theoritical Biology. 1965; 8:264-276. DOI: 10.1016/0022-5193(65)90077-9

[125] Suralta RR, Inukai Y, Yamauchi A. Genotypic variations in responses of lateral root development to transient moisture stresses in rice cultivars. Plant Production Science. 2008; 11:324-335. DOI: 10.1626/pps.11.324

[126] An P, Inanaga S, Li X, Shimizu H, Tanimoto E. Root characteristics in salt tolerance. Root Research. 2003; 12:125-132. DOI: 10.3117/rootres.12.125

[127] Schleiff U, Muscolo A. Fresh look at plant salt tolerance as affected by dynamics at the soil/root-interface using Leek and Rape as model crops. The European Journal of Plant Science and Biotechnology. 2011; 5:27-32.

[128] Wang Y, Zhang W, Li K, Sun F, Han C, Wang Y, Li X. Salt-induced plasticity of root hair development is caused by ion disequilibrium in Arabidopsis thaliana. Journal of Plant Research. 2008; 121:87-96. DOI: 10.1007/s10265-007-0123-y

[129] Müller M, Schmidt W. Environmentally induced plasticity of root hair development in Arabidopsis. Plant Physiology. 2004; 134:409-419. DOI: 10.1104/pp.103.029066

[130] Ludlow MM, Muchow RC. A critical evaluation of traits for improving crop yields in water-limited environments. Advances in Agronomy. 1990; 43:107-153. DOI: 10.1016/S0065-2113(08)60477-0

[131] Crush J, Easton H, Waller J, Hume D, Faville M. Genotypic variation in patterns of root distribution, nitrate interception and response to moisture stress of a perennial ryegrass (Lolium perenne L.) mapping population. Grass and Forage Science. 2007; 62:265-273. DOI: 10.1111/j.1365-2494.2007.00583.x

[132] Whalley W, Dodd I, Watts C, Webster C, Phillips A, Andralojc J, White R, Davies W, Parry M. Genotypic variation in the ability of wheat roots to penetrate wax layers. Plant and Soil. 2013; 364:171-179. DOI: 10.1007/s11104-012-1342-0

[133] Hebbar KB, Rane J, Ramana S, Panwar N, Ajay S, Rao AS, Prasad P. Natural variation in the regulation of leaf senescence and relation to N and root traits in wheat. Plant and Soil. 2014; 378:99-112. DOI: 10.1007/s11104-013-2012-6

[134] Bouteille M, Rolland G, Balsera C, Loudet O, Muller B. Disentangling the intertwined genetic bases of root and shoot growth in Arabidopsis. PLoS One. 2012;7e32319. DOI: 10.1371/journal.pone.0032319

[135] De Deyn GB, Cornelissen JH, Bardgett RD. Plant functional traits and soil carbon sequestration

in contrasting biomes. Ecology Letters. 2008; 11:516-531. DOI: 10.1111/j.1461-0248.2008.01164.x

[136] Lynch JP. Roots of the second green revolution. Australian Journal of Botany. 2007; 55:493-512. DOI: 10.1071/BT06118

[137] Nupur JA, Hannan A, Islam MAU, Sagor GHM, Robin AHK. Root Development and Anti-Oxidative Response of Rice Genotypes under Polyethylene Glycol Induced Osmotic Stress. Plant Breeding and Biotechnology.2020; 8:151-162. DOI: 10.9787/PBB.2020.8.2.151

[138] Nemoto H, Suga R, Ishihara M, Okutsu Y. Deep rooted rice varieties detected through the observation of root characteristics using the trench method. Breeding Science. 1998; 48:321-324. DOI: 10.1270/jsbbs1951.48.321

[139] Hirayama M, Nemoto H, Hirasawa H. Relation between root system and drought resistance in Japanese upland rice (Oryza sativa L.) varieties with medium to late maturing under field conditions. Japanese Journal of Crop Science. 2007; 76:245-252. DOI: 10.1626/jcs.76.245

[140] Kato Y, Kamoshita A, Yamagishi J, Imoto H, Abe J. Growth of rice (Oryza sativa L.) cultivars under upland conditions with different levels of water supply. Plant Production Science, 2007; 10:3-13. DOI: 10.1626/pps.10.3

## Section 5

# Rice Economics, Valne Addition and Product Development

Chapter 10

# Does the Incidence of COVID-19 Pandemic Affect Rice Yield? Lessons from Southeast Nigeria

*Nnaemeka Success Esiobu*


## Abstract

Across Nigeria, while rice farmers are still battling the negative impact of climate change, the COVID-19 pandemic has brought a new risk that threatens not only farmers' livelihoods but also the most important global food security crop "rice". Every farming season, rice farmers face risks such as low rainfall, price volatility, and poor government policies. But the present risks from the COVID-19 pandemic are putting new challenges in front of rice value-chain that is already under serious threat. As a matter of urgency, farmers must respond to this new threat by choosing measures that increase their yield. Incidentally, empirical studies that documented the effect of COVID-19 pandemic on rice yield cannot be found as at the time of this study. These create emptiness in research. With this present threat, Nigeria is likely to experience a reversal in the development gains already achieved and will be unlikely to achieve the Agenda 2030 Goals.


**Keywords:** rice, perceived effect of COVID-19 pandemic, barriers, recommendations, Southeast Nigeria

## 1. Introduction

The first human cases of COVID-19, the disease caused by the novel coronavirus causing COVID-19, subsequently named SARS-CoV-2 were first reported by officials in Wuhan City, China, in December 2019 [1]. Retrospective investigations by Chinese authorities have identified human cases with onset of symptoms in early December 2019 [2]. While some of the earliest known cases had a link to a wholesale food market in Wuhan, some did not. Many of the initial patients were stall owners, market employees, or regular visitors to this market. Environmental samples taken from this market in December 2019 tested positive for SARS-CoV-2, further suggesting that the market in Wuhan City was the source of this outbreak or played a role in the initial amplification of the outbreak [3]. The market was closed on 1 January 2020. The World Health Organization declared COVID-19 a global pandemic on 11 March 2020 [4]. Since 31 December 2019 and as of 17 August 2020, 21,852,364 cases of COVID-19 (under the applied case definitions and testing strategies in the affected countries) have been reported, including 773,586 deaths globally and 1,123,493 cases across sub-Saharan African (SSA) [1].

The first confirmed case in Nigeria was announced on 27 February 2020, when an Italian citizen in Lagos tested positive for the virus [5]. Specifically, across sub-Saharan African (SSA) and as at 17th August 2020 the five countries reporting most cases are South Africa (587,345), Egypt (96,475), Nigeria (49,068), Ghana (42,653) and Algeria (38,583) [6]. The African continent is already grappling with food security challenges. Locust swarms in the Horn of Africa, regional insecurity and conflict, climate-change-related droughts and flooding are destroying crops and the livelihoods of millions of African smallholder farmers [7]. Furthermore, the COVID-19 pandemic potentially undermines the capacity of both short-term production and distribution. With its far-reaching geographical spread, the pandemic is projected to have devastating effects on the global economy, as attested by the projection of the International Monetary Fund (IMF) in April 2020 that the world economy would contract sharply by 3%, and that the economy of sub-Saharan Africa would contract by 1.60% in 2020 [8, 9]. Furthermore, Economic Commission for Africa (ECA) has projected that, in a worst-case scenario, economic activity for Africa as a whole would contract by 2.60%, with negative impacts on the employment rate, and that estimated that four out of five businesses in Africa would be significantly affected by the COVID-19 pandemic [10]. The African Development Bank (AfDB) predicted an economic contraction for the continent of 3.40% [11].

In Nigeria, the pandemic has already precipitated rice crises by disrupting rice value-chain system, thereby posing a great threat to actors of the rice value chain system livelihoods as well as national food and nutritional security. The Nigerian government had initiated an early coordinated response to minimize the impact by developing strategies to facilitate free movement of rice seedlings and agricultural inputs exempted from lockdown. Although agricultural value chain area exempted from lockdown, limits on the mobility of farmhands are contributing to labor shortages for rice sectors as it is characterized by periods of peak seasonal labor demand and labor-intensive production. Additionally, harvesting and distribution season is imminent in rice values-chain, and a shortage of labor is leading to production losses and shortages in the market. In Nigeria, this comes on one of the top existing difficulties in sourcing seasonal and harvesting labor for rice. Therefore, rice (*Oryza spp*.), which is the second-largest most consumed cereal (after wheat), shapes the lives of millions of households globally [12]. More than half of the worlds' population depends on rice for about 80% of its food calorie requirements [13]. It has become a staple food in Nigeria such that every household, both the rich and the poor, consumes a great quantity [14]. A combination of various factors seems to have triggered the structural increase in rice consumption over the years with consumption broadening across all socioeconomic classes, including the poor [15]. The rising demand could be as a result of increasing population growth and income level coupled with the ease of its preparation and storage [14]. In Nigeria, while rice farmers are still battling the negative impact of climate change and greenhouse gases (GHGs) emissions, the COVID-19 pandemic has brought a new risk that not only threatens farmers livelihoods but also the most important global food security crop "Rice". Every farming season in Nigeria, rice farmers face risks such as low rainfall, price volatility, methane emissions, rising debts and poor government policies. But the present risks from the COVID-19 pandemic are putting new challenges in front of rice value-chain that is already under serious threat. Additionally, the growing demand for rice across s sub-Saharan African (SSA) and particularly in Nigeria exceeds supply, resulting in a rice deficit. Currently, due to the present government objective on diversification of the economy, rice is grown in almost 36 states in Nigeria including Federal Capital Territory (FCT) under diverse production systems and agroclimatic conditions [14].

To the farmers across Southeast Nigeria, the effect of COVID-19 pandemic on agriculture, especially rice production, is very important. This is because the majority of the inhabitants depend on rice production for their livelihood. Rice is highly susceptible to seasonal shock, greenhouse gases and climate vagaries because of its sensitivity to changing climatic conditions [14]. The COVID-19 pandemic is already affecting the livelihoods of rural communities across Southeast Nigeria, which is predominately rice-based, and increasing their vulnerability as well. The increasing variability, greenhouse gases, intensity and erratic nature of rainfall, rising incidence of flood and soil erosion, and serious decline in agricultural yields are among the obvious climate change hazards ravaging communities across Southeast Nigeria [14, 16, 17]. To overcome this challenge, rice production/yield should be increased particularly now there is new threat for farmers. As a matter of urgency, farmers must respond to this new threat (COVID-19) by choosing measures that increase their yield.

Incidentally, empirical studies that documented the effect of COVID-19 pandemic on rice yield cannot be found as at the time of this study. These create emptiness in research and make it extremely difficult if not impossible for the government/interest group to know the method they can use in helping farmers mitigate the negative impact of COVID-19 pandemic in rice production. Southeast

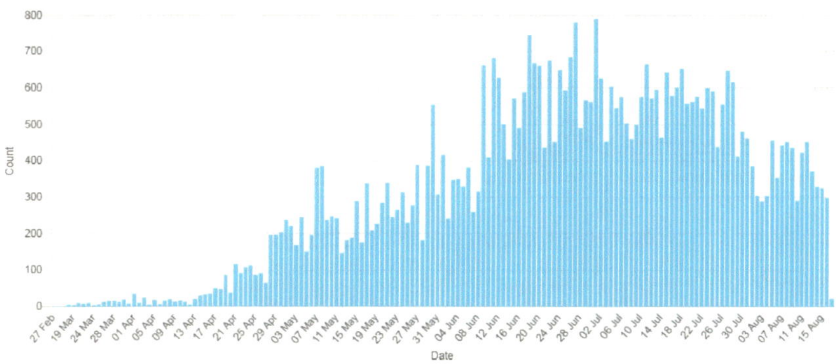


**Figure 1.**
*Epi-curve of confirmed COVID-19 cases in Nigeria as at 15th August 2020 [5].*

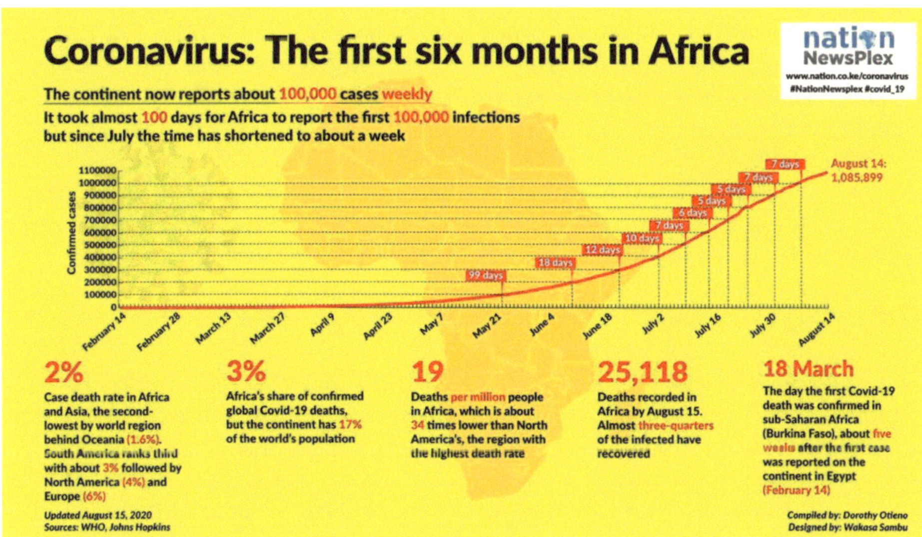


**Figure 2.**
*Confirmed COVID-19 cases in Africa overtime as at 14th August 2020 [6].*

Nigeria has an estimated land mass of 32,610 $km^2$ and a population of 22,583,076 persons [18, 19]. On average, the poverty level in South East is about 43.00% which means that about 43.00% of South East citizens are classified as very poor [20]. Also it is estimated that the number of people living below the poverty line in developing countries of which Nigeria is included will rise by 29 million [21]. Therefore, this demonstrates how vulnerable farmers in the States and Nigeria in general are to COVID-19 pandemic (**Figures 1** and **2**).

## 2. Specific objectives of the study

The study was logically and specifically guided by

i. describing the socio-economic characteristics of the rice farmers;

ii. developing and describing the conceptual framework linking the effect of COVID-19 pandemic in rice farming;

iii. determining the awareness of COVID-19 pandemic among rice farmers;

iv. identifying farmers sources of information on COVID-19 pandemic;

v. identifying the COVID-19 pandemic precautionary measures farmers are practicing;

vi. ascertaining the perceived effect of COVID-19 pandemic on rice yield and

vii. identifying the barriers in mitigating COVID-19 pandemic in rice farming.

## 3. Methodology

The study was carried-out in South-East agricultural zone of Nigeria. The zone is made up of five states, namely, Abia, Anambra, Ebonyi, Enugu and Imo. It has an estimated land mass of 32,610 $km^2$ and a population of 22,583,076 [18, 19]. The area lies between longitude $2^{o}61^{1}$ and $6^{0}32^{1}$ East and latitudes $6^{o}74^{1}$ and $8^{o}15^{1}$ North of Equator with the mean annual temperature ranges from 21.6 to 32.4°C while the annual rainfall ranges from 720 to 1440 mm in the rainforest region [22]. South-East Nigeria has two distinct ecological zones – the tropical rainforest to the South and derived guinea savanna to the North [23]. South-East Nigeria has fertile and well-drained soil and a good proportion of the population are essentially farmers. The sample for the study was drawn from rice farmers in the study area.

### 3.1 Sampling technique

A multi-stage random sampling procedure was adopted in the selection of respondents for the study. Secondly, three Local Government Areas (LGAs) were purposively selected from each of selected States giving a total of nine (9) local government areas for the study. Firstly three (3) States was randomly selected from the five (5) geopolitical zone of South-East, Zone of Nigeria. Secondly, three Local Government Areas (LGAs) was purposively selected from each of selected States giving a total of nine (9) local government areas for the study. Thirdly, three (3) communities were purposively selected from each of the nine (9) selected LGAs based on the concentration of rice farmers to give a total of twenty-seven (27) communities.

Finally, stratified random sampling techniques was used to select 14 crop rice farmers from each of the 27 selected communities to give a total sample size of 504 respondents for the study. The list of farmers in the communities, which forms the sample frame, was obtained from the zonal extension agents of each of the State Agricultural Development Programme in the selected States.

### 3.2 Data collection and data analysis

Primary data was used for the study. Primary data was collected through the use of structured questionnaire and it was supplemented with interview schedule in places where the respondents could neither read nor write. The author developed and used a COVID-19 risk and effectiveness questionnaire to elicit data from the respondents. The questionnaire was contained both open and closed ended form of questions. Similarly, the questionnaire was properly subjected to content validity with the help of the research supervisors before administrating it to the respondents. The questionnaire was divided into five sections. Section one covered the socio-economic characteristics of the farmers. Section two dealt identifying the awareness to COVID-19 and sources of information on COVID-19 pandemic; Section three focused on determining COVID-19 pandemic precautionary measures use among farmers; section four measure the perceived effect of COVID-19 pandemic on rice yield and while section five identified the constraints in mitigating COVID-19 pandemic in rice farming. The survey was conducted at the onset of discovering of the index case of COVID-19 in Nigeria (27 February 2020) that is, from March to July 2020 with the help of two-hundred (200) enumerators i.e. forty (40) for each State.

This study used descriptive statistical tools in analyzing the data. Descriptive statistics such as frequency distribution, percentages, mean, flow charts and 4-point likert scale type was used to analyze the data so as to realize objectives. The 4-point likert scale type was given as follows; 4 = Very High effectiveness; 3 = Highly effective; 2 = Moderately effective and 1 = Not effective.

$$4+3+2+1=10/4 \tag{1}$$

$$= 2.50 \text{ cut-off point} \tag{2}$$

## 4. Conceptual framework for the effect of COVID-19 in rice farming

The conceptual framework that guided this study was developed by the authors based on literature review and field work is presented in **Figure 4**, it provides a clear links between independent or explanatory variables and dependent variables as well as summarizes the mechanisms through which COVID-19 pandemic affects rice farmers livelihood, farm resources, labor, rice production, health status, access to market, food security. To examine the effects of COVID-19 on rice farming, the author developed a concept to link the important variables as shown in **Figure 1**. Microeconomic explanation of individual and household poverty revolves around the understanding that an individual's labor (health and energy), human capital (education and skills), and physical and social assets (such as land and access to a social network) determine the individual's ability to generate income both today and in the future [24–26].

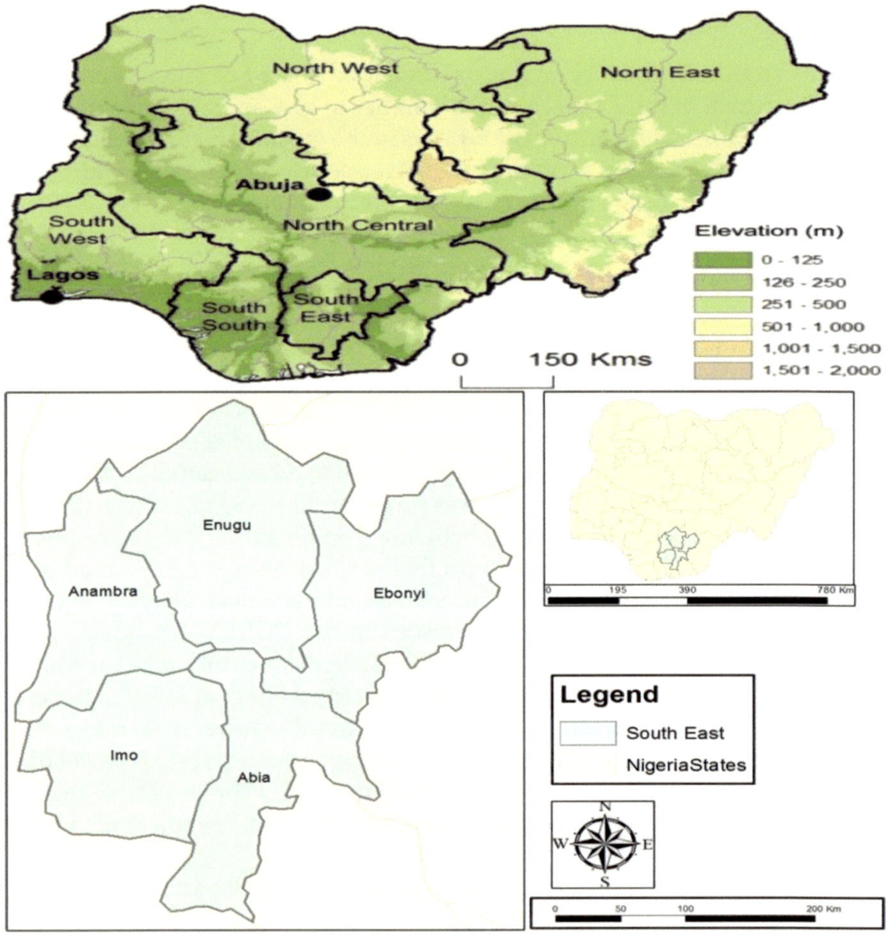


**Figure 3.**
*Map of Nigeria showing the Study Area [14].*

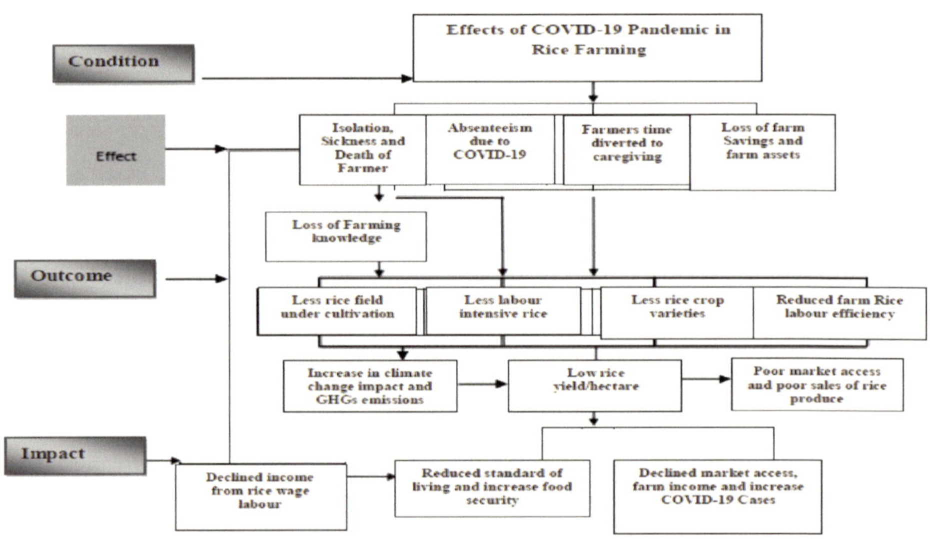


**Figure 4.**
*Conceptual framework for the effect of COVID-19 in rice farming (Researcher's own construct, 2020).*

From **Figure 3**, Illness results in morbidity and sometimes mortality. These reduce labor availability through absenteeism or death of agricultural producers/ worker. Through morbidity the quantity and quality of labor to the household is also affected because the sick may abstain completely or partially from work during the period of illness. Additionally, from **Figure 4**, labor availability is also indirectly affected when labor of productive members of the household is diverted from the farm to caring for the sick. When sick persons die, the knowledge that they acquired through learning and experience is no longer available for others to use and considerable rice agricultural knowledge is lost. However, it must be pointed out that although chronic illness (COVID-19) results in a net decrease in household labor (the ill and the caretaker), the death of a chronically ill person who had a full-time caretaker can result in increased household labor when the caretaker returns to the workforce. This is true as production, adaptation to climate change and mitigation GHGs emissions in rice field are costly [14, 16, 27].

Moving forward, from **Figure 3**, the cost of healthcare for sick persons from COVID-19, and its protocols and of funerals drive many households into debt, and they resort to using their savings and remittances they receive or even sell household and farm assets to defray the costs. For instance, due to costs of treatment, cost of COVID-19 protocol and other expenses as well as lower incomes from loss of labor, COVID-19 affected households usually draw on assets from the farm. After the onset of COVID-19, savings and financial assets are usually the first to be depleted. Then, non-productive assets, such as furniture, cooking utensils and clothes follow. Ultimately, households may have to sell off productive assets such as equipment, draught animals, and land.

The outcomes of these effects include reduction of farm sizes, cultivation of less-intensive improved rice varieties, reduction in the variety of rice field planted, and the increase in climate change impact and greenhouse gases emission. The study of [14, 28] found that rice field is a significant anthropogenic source of methane ($CH_4$) and nitrous oxide ($N_2O$), two important greenhouse gases (GHGs). These outcomes affect livelihoods in terms of reduction in rice farm outputs and income, a decline in income from wage labor, rice yield, market access, off-farm activities, and food insecurity.

## 5. Results and discussion

### 5.1 Socio-economic characteristics of rice farmers

**Table 1** revealed that majority (58.53%) fell within the age range of 41–50 years. The mean age was 45.00 years. This shows that farmers in the area are vibrant, young and still within the active age. Rice farming is so strenuous. The implication is that younger farmers are more likely to practice more and modern farming method in reducing the effect of COVID-19 faster than the older ones. Young farmers are more likely to know about new strategies to avert COVID-19 effect with the willingness to bear risk. **Table 1** also reveals that majority (61.30%) of the farmers were males. The finding implies that both sexes are involved in rice farming but male are more in number in the area. This is true as male farmers has been found to be more relatively efficient that women farmers [29].

Entries in **Table 1** also show that greater proportion (57.73%) had secondary school education. The mean educational level is 12 years which is equivalent to secondary school education. The finding implies that approximately 96.83% of the farmers had formal education which is expected to increase their level of understanding on the effect of COVID-19 in rice farms and various strategies to practice in thwarting these new threats.

| Age (years) | Frequency | Percentage (%) | Mean (X) |
|---|---|---|---|
| 21–30 | 8 | 1.58 | |
| 31–40 | 49 | 9.72 | |
| 41–50 | 295 | 58.53 | |
| 51–60 | 108 | 21.42 | |
| 61–70 | 32 | 6.34 | |
| 71–80 | 12 | 2.38 | |
| **Total** | **504** | **100.0** | **45.00 years** |
| **Sex** | **Frequency** | **Percentage (%)** | |
| Male | 309 | 61.30 | |
| Female | 195 | 38.69 | |
| **Total** | **504** | **100.0** | |
| **Educational level** | **Frequency** | **Percentage (%)** | |
| No formal education | 16 | 3.17 | |
| Primary | 161 | 31.94 | |
| Secondary | 291 | 57.73 | |
| Tertiary | 36 | 7.14 | |
| **Total** | **504** | **100.0** | **12 years equivalent to secondary education** |
| **Marital status** | **Frequency** | **Percentage (%)** | |
| Married | 270 | 53.57 | |
| Single | 157 | 31.15 | |
| Widowed | 54 | 10.71 | |
| Divorced | 23 | 4.56 | |
| **Total** | **504** | **100.0** | |
| **Farming experience (years)** | **Frequency** | **Percentage (%)** | |
| 01–10 | 15 | 2.97 | |
| 10–19 | 34 | 6.74 | |
| 20–30 | 284 | 56.34 | |
| 31–40 | 101 | 20.03 | |
| 41–50 | 59 | 11.70 | |
| 51–60 | 11 | 2.18 | |
| **Total** | **504** | **100.0** | **21 years** |
| **Household size (number of persons)** | **Frequency** | **Percentage (%)** | |
| 1–2 | 3 | 0.59 | |
| 3–4 | 14 | 2.78 | |
| 5–6 | 29 | 5.78 | |
| 7–8 | 124 | 24.60 | |
| 9–10 | 222 | 44.04 | |
| 11–12 | 101 | 20.03 | |

| | | | |
|---|---|---|---|
| 13–14 | 11 | 2.18 | **9 persons** |
| **Total** | **504** | **100.00** | |
| **Extension contact** | **Frequency** | **Percentage (%)** | |
| Not at all | 430 | 85.32 | |
| Once in a fortnight | 47 | 9.32 | |
| Once in a month | 23 | 4.56 | |
| Once in a year | 4 | 0.79 | |
| **Total** | **504** | **100.0** | |
| **Access to credit** | **Frequency** | **Percentage (%)** | |
| Access | 401 | 79.56 | |
| No-access | 103 | 20.44 | |
| Total | 504 | 100.0 | |
| **Total** | **504** | **100.0** | |
| **Farm size (Ha)** | **Frequency** | **Percentage (%)** | |
| 0.1–1.0 | 52 | 10.32 | |
| 1.1–2.0 | 89 | 17.66 | |
| 2.1–3.0 | 263 | 52.18 | |
| 3.1–4.0 | 100 | 19.84 | **3.10 ha** |
| **Total** | **504** | **100.0** | |
| **Annual farm income (N)** | **Frequency** | **Percentage (%)** | |
| 100,000–200,000 | 13 | 2.58 | |
| 200,001–300,000 | 36 | 7.14 | |
| 300,001–400,000 | 28 | 5.56 | |
| 400,001–500,000 | 23 | 4.56 | |
| 600,001–700,000 | 74 | 14.70 | |
| 700,001–800,000 | 96 | 19.05 | |
| 800,001–900,000 | 223 | 44.25 | |
| 900,001–1,000,000 | 11 | 2.18 | |
| **Total** | **504** | **100.0** | **N880,200.00 (2267.62USD)** |

*Source: Field Survey Data, 2020.*

**Table 1.**
*Socio-economic characteristics of rice farmers.*

Result in **Table 1** shows that majority (53.57%) were married. The finding implies that rice farming is an enterprise of married individual who are expected to be responsible according to societal standard [14]. Married farmers have likelihood of adapting to the effect of COVID-19 pandemic easily than their unmarried counterpart since they have access to labor. During this period of COVID-19 pandemic, non-availability of labor is interrupting some harvesting and distribution activities of rice.

There are disruptions in supply chains for purchase of inputs because of poor transportation system and partial ease of lockdown. Therefore, married farmers have increased chance of adapting easily to COVID-19 than their counterpart since they have access to household size. Result of farming experience is shown in **Table 1** and it shows that about 56.34% of the farmers had a farming experience ranging from 20 to 30 years. The mean year of experience in farming was 21.00 years. This shows that the farmers were quite experienced in rice farming and have started been practicing several COVID-19 pandemic strategies to increase their farm yield and income in the area. It is expected that farmers with more experience are more likely to accept innovations and new COVID-19 pandemic strategies to increase their farm yield and income than inexperienced farmer. The number of years of farming helps to cushion the effects of COVID-19 pandemic, since the incidence of diseases and infection is a recurring decimal globally such as the Spanish flu, 1918; AIDS pandemic and epidemic, 1981; H1N1 Swine Flu pandemic, 2009–2010; West African Ebola epidemic, 2014 and Zika Virus epidemic, 2015 among others.

Result in **Table 1** also show that majority (85.32%) of the farmers had no contact with extension agents. The COVID-19 procotol such as the restriction of movement have made it very difficult for extension agents to visit various farm families. Although, the use of video conferences have been identified as a way to bridge the gap. Regrettably, rural farmers lack access, fund and technical-know-how to use it. The implication is that majority of the farmers may not have the opportunity of learning new COVID-19 farming strategies and consequently exposing their rice farming to incidence COVID-19 pandemic impact in the area. It becomes clear that there is need for the government to strengthen the Agricultural Development programme (ADP) of the various Southeast State with modern ICTs gadgets and skills for personnel to facilitate timely extension contacts with farmers in the area. Entries in **Table 1** revealed that about 44.04% had a household size ranging from 9 to 10. The mean household size was found to be 9.00 persons. The result shows that farmers had large households.

The lockdown created both a shortage of farm labor. The implication is that they could draw farm labor from their households during this period of COVID-19 to increase their farm yield and income. **Table 1** revealed that majority (52.18%) of the farmers had farm size of between 2.10 and 3.00 hectares. The average farm size is 3.10 ha. The finding implies that the farmers in the area are mainly smallholder farmers operating on less than or equal to 3.50 hectares of farmland. This could be as a result of land tenure system or increasing population prevalent in the area. Additionally, the small farm size is not even contiguous plot but rather small plots scattered in different areas of the community. It is expected that farmers with large farm size will practice more and better COVID-19 strategies than those with lesser farmland in the area.

Finally, **Table 1** indicates that majority (54.14%) had an average annual farm income of between N800,001 and N900,000. The mean annual farm income was N880,200.00 (2267.62USD) while monthly farm income was estimated to be N73,350.00 (188.97USD). The finding implies that the farmers have a relatively low farm income despite the larger household size which they recorded. The implication of the findings is that farmers may not have the much needed financial capacity to mitigate to the negative effect of COVID-19 pandemic in their rice farming. This is true as some mitigation strategies to COVID-19 pandemic strategies are costly. Hence, farmers may have several COVID-19 pandemic strategies they want to practice but limited fund may continue to hinder them.

## 6. Rice farmers' awareness of COVID-19 pandemic

The result in **Figure 5**, shows farmers' awareness of COVID-19 Pandemic in the area. It shows that larger percentages (97.92%) of the farmers are aware of COVID-19

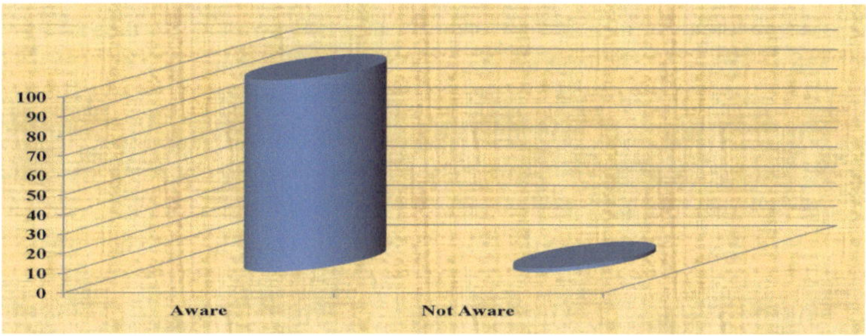


**Figure 5.**
*Rice farmers' awareness of COVID-19 pandemic.*

Pandemic while approximately 2.08% are not aware of COVID-19 Pandemic in the area. The finding shows greater proportion of that farmers are aware of COVID-19 Pandemic and may have been witnessing the negative effect. Additionally, farmers may have also developed sustainable COVID-19 pandemic strategies to thwart the negative effect of COVID-19 Pandemic to improve their farm yield, income and standard of living. The high awareness could be attributed to the massive publicity about the virus by various Nigeria news agencies, and agricultural extension agents. The extension agents have been called to help bridge gap in information for farmers.

The COVID-19 pandemic is not the first time extension agents have been called to action in an unfolding disaster. As an institution with trained technical staff who are trusted by communities, and with local reach and communication skills, extension agents has supported efforts and educated communities during crises such as HIV/AIDS, Ebola, avian influenza, natural disasters, and pest infestations. In the same way, even though, awareness does not really signify development and use/practice of modern COVID-19 pandemic strategies, however, it could serve as an essential determinate. Ultimately, quick action from the governments coupled with credible, regular information is critical in dealing with emergencies such as COVID-19. As a critical actor in providing such information to rural areas, extension agents can do several things globally to help mitigate the economic and health impacts of COVID-19.

## 7. Farmers' sources of information of COVID-19 pandemic

The result in **Figure 6**, shows farmers' sources of information of COVID-19 pandemic in the area. It reveals that about 99.50 and 93.02% of the farmers source of information was radio and mobile phone (Bulk Text messages) respectively. The merging and use of technology particularly during this period of COVID-19

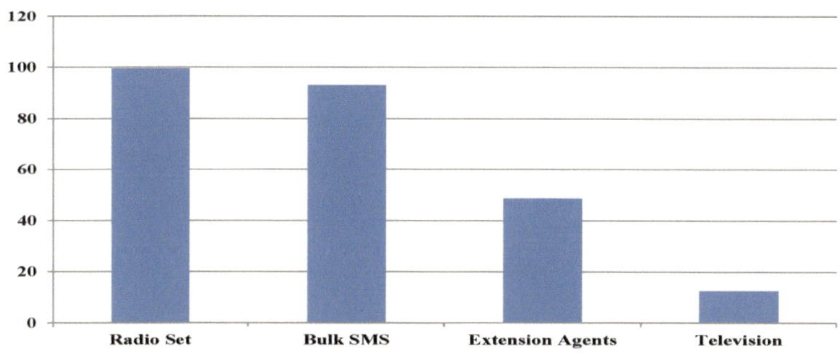


**Figure 6.**
*Farmers' sources of information of COVID-19 pandemic.*

pandemic, that is, telephone (Bulk SMS) and radio has made it more accessible to many. Farmers are able to go everywhere with their radios, including to their farms. Every home had a radio while others had more than one radio. Radio is one of powerful communication tool and has proved to be the most effective media in promoting agriculture and development in rural areas, particularly as a tool for the delivery of quick information.

The result is strengthened by the finding of [30] who found that radio leads in overall media consumption followed by the mobile phone (Bulk Text messages) with the wake of many vernacular radios which are able to effectively communicate to the targeted groups. The result also shows that simple majority (48.73 and 12.62%) of the farmers identified extension agents and television as one of the sources of information on COVID-19 pandemic. Therefore, it should be a call for serious concern. For poor extension service delivery, it could be attributed to poor extension contact observed earlier from the socio-economic variable of the farmers in the study. Agricultural extension agents can offer great support to farmers during uncertainty and sudden changes that come with the pandemic, and strategies to bounce back from shocks and enhance resilience. Television is one of the possible mechanisms to activate already existing mental skills in the farmers. Television combines multiple symbol systems, such as visual images, sounds, music, spoken and written language, and presents them simultaneously. Regrettably, poor power supply and high cost of television-set have made it difficult for farmers to assess COVID-19 information thorough it.

## 8. Farmers' health precautionary measures in face of COVID-19 pandemic

The result in **Figure 7**, shows farmers health precautionary measures in face of COVID-19 pandemic in the area. It reveals that about 99.99, 99.78, 98.00 and 95.72% of the farmers have started hand washing, wearing a face mask, practising no-shake and physical/social distancing, respectively. The [1] reported that masks can help prevent the spread of the virus from the person wearing the mask to others. The report when further to state that masks alone do not protect against COVID-19, and should be combined with physical distancing and hand hygiene. It is important to note that farmers who say to have started using a face mask were because of the local force abuse of the process. Most farmers said they wear a mask to avoid Nigerian police, taskforce and other health agency harassment and not for the fear of contracting the virus. In addition, [3] washing of hands is one of the most important steps one can take to avoid getting sick and spreading germs to others.

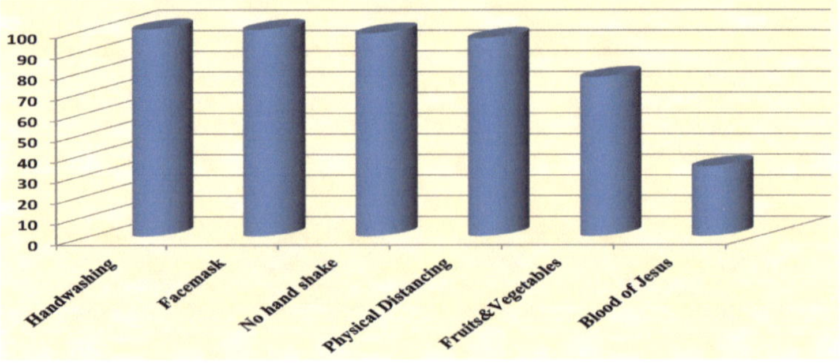


**Figure 7.**
*Farmers' health measures in COVID-19 pandemic.*

Many diseases and conditions are spread by not washing hands with soap and clean, running water. The [1] recommends cleaning hands in a specific way to avoid getting sick and spreading germs to others. The guidance for effective hand-washing and use of hand sanitizer was developed based on data from a number of studies [8, 9]. About 76.83% have started heavy consumption of fruits and vegetables while a smaller proportion (33.72%) said that Blood of Jesus Christ protects them and they do not and will not practice any measure. The report of [8, 9] has advised people to Use fresh ingredients and those that have a shorter shelf life first.

The study of [1, 8, 9] further states that if fresh products, especially fruits, vegetables and reduced-fat dairy products continue to be available, prioritize these over non-perishables. Frozen fruits and vegetables can also conveniently be used over longer periods of time and often have a similar nutrient profile to fresh foods. The author founds out that most farmers do not trust the government and think that COVID-19 is a hoax and means the Nigerian Government and other world leaders want to use and embezzle public fund. This situation, therefore, calls for urgent concern for the global and Nigerian Government to show great sincere and accountable leadership to its citizens at all time.

## 9. Farmers' perceived effect of COVID-19 pandemic in rice farming

The result of the farmers' distribution based on perceived effect of COVID-19 pandemic in rice farming is compiled in **Table 2**. The various attributes of were rated in a 4- point likert scale type of questions of Strongly Agreed (4); Agreed (3); Disagreed (2) and Strongly Disagreed (1). Using the method of mean score analysis, a discriminatory mean of 2.50 was produced which divided the distribution into four scale was produced. The mean value of each attribute equal to or above ($X = \geq 2.50$) was regarded as being an accepted decision while attributes with mean value less than ($X = \leq 2.50$) was regarded as a rejected decision. The values of standard deviation (SD) denote the degree of variation in the responses of the farmers. Standard deviation with values equal or above one indicates high variances in farmers a response regarding level of knowledge of organic farming practices. The standard deviation value which ranged from 0.10 to 0.40 indicated that farmers were in agreement in their perceived effective of organic farming practices on poverty alleviation in the area. Additionally, all the items were rated high and had an acceptable overall discriminatory score ($\bar{x} = \geq 2.50$). This result strongly confirmed that COVID-19 pandemic is seriously affecting rice farmers in the area. A very salient finding was that COVID-19 pandemic is disrupting activities in rice value-chains ($\bar{x} = \geq 3.11$).

The result also showed that the COVID-19 pandemic has led to the partial restriction of all agricultural institutions in the country since March 2020 and disrupted rice activities, like harvesting of 2019/2020 season's rice produce that was ongoing and land preparations and planting for 2020/2021 season. Result show that the non-availability of labor is interrupting planting, harvesting and distribution activities of rice ($\bar{x} = \geq 3.32$). There are disruptions in supply chains for purchase of inputs because of poor transportation system and partial lockdown ($\bar{x} = \geq 3.12$). Rice farmers are also paying more on inputs ($\bar{x} = \geq 3.43$). Result also shows that extension visits to farmers have been seriously and market access limited ($\bar{x} = \geq 3.44$) due to the COVID-19 pandemic. With its far-reaching geographical spread, the findings confirmed the study of [8, 9] which reported that the pandemic is projected to have devastating effects on the global economy, as attested by the projection of the International Monetary Fund (IMF) in April 2020 that the world economy would contract sharply by 3.00%, and that the economy of sub-Saharan Africa would contract by 1.60% in 2020.

| S/No | Items | Strongly agreed | Agreed | Disagreed | Strongly disagreed | Mean ($\bar{x}$) (≥2.50) | SD | Decision |
|---|---|---|---|---|---|---|---|---|
| 1 | COVID-19 pandemic has brought a new threat to our rice farming | 298 (59.12) | 139 (27.58) | 65 (12.90) | 2 (0.40) | 3.22 | 0.85 | Accepted |
| 2 | COVID-19 pandemic is disrupting planting of 2020/2021 season seedling, harvesting of 2019/2020 season produce and distribution activities in rice farming. | 261 (51.79) | 155 (30.75) | 83 (16.47) | 5 (0.10) | 3.11 | 0.72 | Accepted |
| 3 | COVID-19 pandemic has led to a shortage of farm labor | 327 (64.88) | 132 (26.19) | 42 (8.33) | 3 (0.60) | 3.32 | 0.56 | Accepted |
| 4 | COVID-19 pandemic has decreased supply chains for purchase of inputs because of the poor transportation system and partial lockdown | 281 (55.75) | 128 (25.40) | 84 (16.67) | 11 (2.18) | 3.12 | 0.52 | Accepted |
| 5 | COVID-19 pandemic has led to a decrease in the provision of basics food for our children | 301 (59.72) | 103 (20.45) | 98 (19.44) | 2 (0.40) | 3.17 | 0.81 | Accepted |
| 6 | COVID-19 pandemic has led to an increase in the cost of input and machinery | 375 (74.40) | 97 (19.25) | 31 (6.15) | 1 (0.20) | 3.43 | 0.87 | Accepted |
| 7 | COVID-19 pandemic has decreased my rice production/yield and aggregate farm income | 293 (58.13) | 144 (28.57) | 58 (11.51) | 9 (1.79) | 3.20 | 0.83 | Accepted |
| 8 | COVID-19 pandemic has made us unable to provide the recommended health precautionary measures for our family | 382 (75.79) | 112 (22.22) | 7 (1.39) | 3 (0.60) | 3.48 | 0.68 | Accepted |
| 9 | COVID-19 pandemic have negatively affected our different rice value chain | 317 (62.90) | 146 (28.97) | 33 (6.55) | 8 (1.59) | 3.30 | 0.75 | Accepted |
| 10 | COVID-19 pandemic have greatly increased our farm production, harvesting and distribution cost | 303 (60.12) | 101 (20.04) | 88 (17.46) | 12 (2.38) | 3.15 | 0.78 | Accepted |
| 11 | COVID-19 pandemic have negatively drew | 297 (58.93) | 120 (23.81) | 61 (12.10) | 26 (5.16) | 3.14 | 0.80 | Accepted |

| S/No | Items | Strongly agreed | Agreed | Disagreed | Strongly disagreed | Mean ($\bar{x}$) (≥2.50) | SD | Decision |
|---|---|---|---|---|---|---|---|---|
| | us back in adapting and mitigating to climate change | | | | | | | |
| 12 | COVID-19 pandemic have negatively drew us back in mitigating to greenhouse gases emission in our rice field | 308 (61.11) | 105 (20.83) | 87 (17.26) | 4 (0.79) | 3.20 | 0.79 | Accepted |
| 13 | COVID-19 pandemic have negatively affected our access to market and agricultural extension services | 134 (53.17) | 101 (40.08) | 12 (4.76) | 5 (1.98) | 3.44 | 0.81 | Accepted |

*SD; standard deviation; discriminatory index: cut off point x̄ ≥ 2.50 accepted;*
**Figures in parenthesis are percentage; Field Survey Data, 2020.*

**Table 2.**
*Farmers' perceived effect of COVID-19 pandemic in rice farming.*

Furthermore, Economic Commission for Africa (ECA) has projected that, in a worst-case scenario, economic activity for Africa as a whole would contract by 2.60%, with negative impacts on the employment rate, agriculture and that estimated that four out of five businesses in Africa would be significantly affected by the COVID-19 pandemic [10]. Going forward, the African Development Bank (AfDB) predicted an economic contraction for the continent of 3.40% [11]. The study of [31] asserted that in order to mitigate the shock of the pandemic and its related effects on smallholder farmers building capacities and providing financial and marketing support for farmers would be essential. Similarly, food security is fragile under normal circumstances for smallholder farm families and other vulnerable populations across Southeast, Nigeria. As noted in an impact analysis from International Food Policy Research Institute (IFPRI), projections for increased poverty levels and food insecurity are dire as a result of the pandemic [32].

With this present threat, African countries and particularly Nigeria are likely to experience a reversal in the development gains already achieved and will be unlikely to achieve the United Nations 2030 Sustainable Development Goals and the aspirations of African Agenda 2063. The study therefore identified the need for governments at all levels in the country to design a sustainable policy in such a way that farmers should have access to affordable farm credit as well as subsidized agricultural inputs in order to increase their ability and flexibility to change in production strategies in response to COVID-19 pandemic.

## 10. Farmers' barriers in mitigating the effect of COVID-19 pandemic in rice farming

The finding in **Figure 8** shows farmers' barriers in mitigating the effect of COVID-19 pandemic in rice farming in the area. The finding reveals that about 99.82% of the farmers identified inadequate information. This could be attributed to the dearth in research on COVID-19 pandemic mitigation strategies for rice

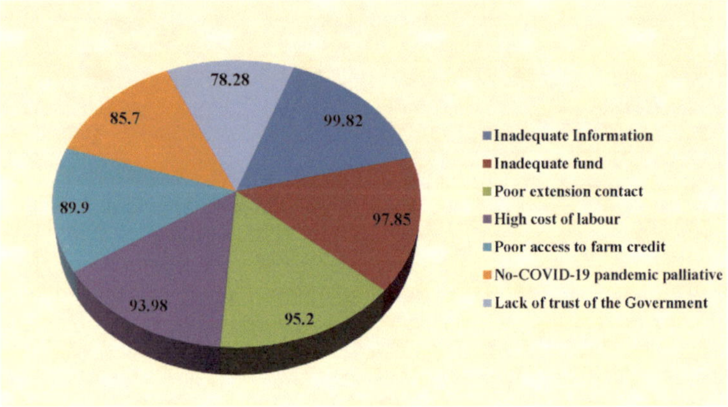


**Figure 8.**
*Farmers' barriers in mitigating the effect of COVID-19 pandemic in rice farming.*

farmers as well as lack of information on clarity on the actionable guidance and precautionary measures by World Health Organization (WHO), Centre for Disease Control and Prevention (CDC), Nigeria Centre for Disease Control (NCDC) and local health authority. This constraint left the farmers unable to get the much-needed information on COVID-19 pandemic and sustainable rice farming. In the present information age, inadequate information could pose serious challenges to the farmers' coping strategies as they may not be aware of recent developments regarding COVID-19 pandemic, mitigation strategies and the necessary re-adjustments needed. Meanwhile, crop farmers have been badly hit due to misinformation, particularly on social media, that crops and animals are the carriers of COVID-19. Also, poor/inadequate information on COVID-19 pandemic and sustainable strategies for rice farming may rise the food insecurity situation and lead to unsustainable rice production overtime.

Approximately 97.85% complained of the inadequate fund. The inadequate fund left most of the rice farmers unable to get necessary productive resources in mitigating the effect of COVID-19 pandemic in the area. This could be attributed to the sudden rise of rice productive resources due to lockdown and food shortage. Inadequate fund hinders farmers from getting the necessary resources and technologies which assist to mitigate efficiently to the effect of COVID-19 pandemic. The COVID-19 pandemic mitigation options are costly because of restrictions and input factory closure hence, farmers need adequate fund to adapt. Going forward, poor extension contact, high cost of labor, poor access to farm credit and no-COVID-19 pandemic palliative by the government were complained by 95.20, 93.98, 89.90 and 85.70% respectively of the rice farmers. The high cost of farm inputs could also be attributed to inadequate fund. With the limited fund, the acquisition of necessary production equipment will be difficult. They may not only be costly but may also appear scarce for poor farmers particularly ducting the period of the pandemic. Besides, the farmers may not also have the necessary facilities for information like radio and television to obtain COVID-19 report.

Poor access of credit could be linked to lack of information or awareness of the presence of loan facilities, high collateral and location of banks in urban which is far from the rural areas where farmers live. In the same way, the COVID-19 pandemic related transport restrictions make it particularly difficult for suppliers to get inputs like seeds, fertilizers, crop protection products and equipment to rice farmers in time for planting season, disrupting production of rice. Restrictions are also triggering additional higher labor cost, post-harvest losses, as unsold and rotting food accumulates on farms. Similarly, planting and harvesting season is imminent for rice farming and a shortage of labor and poor access to farm credit is leading to production losses and shortages in the market. In Nigeria, this comes on top of as one of the existing difficulties in sourcing rice production labor and finance. Most Nigerian were left unhappy with government selective provision of palliative. Some reports claimed that South-East Nigeria was the study was conducted never receive any form of palliative from the government. This situation left most of the farmers unable to mitigate the effect of COVID-19 pandemic in their rice farms.

About 78.28% identified lack of trust. Trust is the foundation upon which the legitimacy of public institutions is built and is crucial for maintaining social cohesion [33]. Most farmers do not trust the government and this leads to the farmers believing that the COVID-19 is real. Some believed that the government wants to use the media to embezzle public fund. Similarly, many farmers gloat that COVID-19 is mainly targeting the politicians, despite warnings that the life-threatening respiratory illness could hit the poor farmers as well. Some farmers also hoped that the "selectiveness" of the virus might be God's way of bringing about changes leadership of Nigerian government. During this current COVID-19 pandemic, including containment, mitigation and recovery, trust in public institutions is vital for governments' ability to respond rapidly and to secure citizen support. Government's ability to harness public trust as a force out of the crisis will be crucial for planning and implementing an inclusive recovery from the COVID-19 emergency. It is important for the government to show great transparent and accountable leadership for the people to trust. Curbing this barrier will be vital in promoting not just local COVID-19 mitigation strategies in increasing rice yield, farmers' livelihood, farmers' income and standard of living but global strategies in area and perhaps beyond.

## 11. Conclusion

The COVID-19 pandemic is having a great effect on the entire rice value chain, confirming in the most terrible way that we are all part of a rice value-chain system that is interconnected and fragile and that solutions must be developed together. Rice farming has been recognized almost worldwide as an essential activity, and farmers were asked to keep on working, even harder, to cope up with this new challenge. Nevertheless, each component of the rice value chain system has an impact on farming activities, so each bottleneck occurring somewhere on the rice value-chain system is having an impact on farmers livelihood, farm yield, and farm income. Countries lock-down and borders' closures are strongly impacting rice farmers' access to input like seeds, fertilizers, and agrochemicals. In Nigeria, while rice farmers are still battling the negative impact of climate change and greenhouse gases (GHGs) emissions, the COVID-19 pandemic has brought a new risk that not only threatens farmers livelihoods but also the most important global food security crop "Rice". Every farming season in Nigeria, rice farmers face risks such as low rainfall, price volatility, methane emissions, rising debts, and poor government policies. But the present risks from the COVID-19 pandemic are putting new challenges in front of rice value-chain that is already under serious threat.

Additionally, the growing demand for rice across SSA and particularly in Nigeria exceeds supply, resulting in a rice deficit. To overcome this challenge, rice yield should be increased particularly now there is new threat for farmers. As a matter of urgency, farmers must respond to this new threat (COVID-19) by choosing measures that increase their yield. Incidentally, empirical studies that documented the effect of COVID-19 pandemic on rice yield cannot be found as at the time of this study. These create emptiness in research and make it extremely difficult if not impossible for the government/interest group to know the method they can use in helping farmers mitigate the negative impact of COVID-19 pandemic in rice production. The study was logically guided by describing the socio-economic characteristics of the rice farmers; determining the awareness of COVID-19 pandemic among rice farmers; identifying farmers sources of information on COVID-19 pandemic; identifying the COVID-19 pandemic precautionary measures farmers are using; ascertaining the perceived effect of COVID-19 pandemic on rice yield and constraints in mitigating COVID-19 pandemic in rice farming.

The survey was conducted at the onset of discovering of the index case of COVID-19 in Nigeria (27 February 2020) that is, from March to July 2020 with the help of two-hundred (200) enumerators, that is, forty (40) for each State. A cross-sectional data was elicited from 504 rice farmers selected from Abia, Anambra, Ebonyi, Enugu, and Imo State. Data collected were analyzed using descriptive statistical tools such as means, flow-chart, percentage and Likert scale-type. Result show that the mean age was 45.00 years. Greater proportions (61.30%) were male. Majority (53.57%) were married with an average household size of 8 persons. Average educational level, farming experience and annual farm income were 12.00 years (equivalent to secondary education), 21.00 years and N880,200.00 (2267.62USD), respectively. Additionally, farmers may have also developed sustainable COVID-19 pandemic strategies to thwart the negative effect of COVID-19 Pandemic to improve their farm yield, income, and standard of living. The high awareness could be attributed to the massive publicity about the virus by various Nigeria news agencies, and agricultural extension agents. The extension agents have been called to help bridge gap in information for farmers. The finding reveals that about 99.82% of the farmers identified inadequate information. This could be attributed to the dearth in research on COVID-19 pandemic mitigation strategies for rice farmers as well as lack of information and clarity on the actionable guidance and precautionary measures by World Health Organization (WHO), Centre for Disease Control and Prevention (CDC), Nigeria Centre for Disease Control (NCDC) and local health authority. This constraint left the farmers unable to get the much-needed information on COVID-19 pandemic and sustainable rice farming. In the present information age, inadequate information could pose serious challenges to the farmers' coping strategies as they may not be aware of recent developments regarding COVID-19 pandemic, mitigation strategies, and the necessary re-adjustments needed.

## 12. Recommendations

The following recommendations were made based on the major research observations and findings of the study.

i. The study identified inadequate and misinformation as one of the barriers farmers complained about, therefore, it is important that the government should intensify mass mobilization campaign against COVID-19 pandemic alongside the community leaders and extension agents to reduce, if not eliminate the spread of the deadly virus.

ii. The study authenticated the importance of linking education to the mitigation of COVID-19 pandemic in improving understanding, farm yield, farm income, the standard of living of the rice farmers. Therefore, it is necessary for the rice farmers on their owe should enroll for adult education to increase their ability and flexibility to change in production strategies in response to COVID-19 pandemic.

iii. Rice production-related constraints such as labor and other inputs mainly hindered farmers from mitigating the effect of COVID-19 effectively. Labor and input are in high demand and cost now due to restrictions. Therefore, the government must also design policy in such a way that farmers' should have access to affordable credit, farm labor as well as subsidized input to increase their rice production, the standard of living as well as encourage more farmers to go into rice farming.

iv. Effective agricultural policies and programmes should focus on how to intensify awareness on COVID-19 pandemic in rice farm as well as its mitigation strategies. This should be done through strengthened agricultural extension service delivery.

v. Agricultural extension agents can offer support during uncertainty and sudden changes that come with the pandemic, and strategies to bounce back from shocks and enhance resilience. Therefore, capacity strengthening, the right tools and channels are necessary to provide tailored effective agricultural extension agents messages. The government should strengthen agricultural extension agents' service delivery.

vi. It is important to protect the community where these rice farmers are domiciled from exposure to the infection, all residents in the potential risk areas should be encouraged to stay at home, which is one of the most effective ways of blocking the transmission routes. Local community health workers and volunteers, after the specific training, should be encouraged to proactively participate in screening the suspicious infections and help in implementing proper quarantine measures by providing support services, such as driving patients to the mobile hospitals. All those activities should logistically be managed at the community level.

vii. The finding observed that there is distrust between the government and farmers. Farmers do not trust the government on the issue of COVID-19 pandemic. Some farmers still do not believe that there is even COVID-19. Therefore, it is important that the government should show transparent and accountable leadership so as to get full farmers to support in implementing strategies.

## Drawbacks and areas for further study

Constraints on restrictions of movement due to COVID-19 pandemic, the poor transportation network in rural areas, time and limited resources of the researcher forced to select only 504 rural-based rice farmers for the study across the Southeast States. Hence results are largely applicable to other areas of Nigeria not selected. Additionally, the questionnaire and personal interview method of data collection require the respondents to recall from their memories about their socio-economic variables, income, government support, COVID-19 precautionary measures and perceived effect of COVID-19 pandemic in rice farming among others. Most rural farmers do not keep farm records hence, the findings may be subject to memory lapses of the respondents.

Therefore, the author and further studies may attempt to rigorously and systematically link farmers perceived effect of COVID-19 pandemic in rice farming with the quantitative measurement of farm harvest, farm output per hectare, farm sales, farm income and other associated production variables. This is to evaluate the actual impact (before and after) of the pandemic in rice farming with a view of checking if farmers' perception were right. For instance, farmers may perceive there is a significant decrease in yield per hectare due to COVID-19 pandemic, but the actual measurement may say otherwise and vice versa. This further study (impact) may require monitoring one or more rice farming seasons for systematic and logical evaluation.

## Acknowledgements

The author is very grateful to the field volunteers of *Centre for Building Farmers Capacity to Climate Change (CBFCCC)*; a rural-based Not-for-Profit Organization committed to the attainment of United Nations Agenda-2030 goals no: 2, 12, 13 and 15 and Africa Agenda 2063 through partnership for goals (17). The *CBFCCC* filed volunteers helped in visiting the sampled farmers in their remote rice farms for evidence-based data collection even with restriction of movement in the country. Additionally, special thanks to the local rice farmers in the study area who provided the data for the study even amid hunger and loss of farm yield as a result of the COVID-19 pandemic. Thanks to all those involved in data entry, data cleaning, data coding, and analysis. We cannot thank you all enough.

## Conflicts of interest

The author declares no conflict of interest.

## Funding

No external funding was received for this research.

## Author details

Nnaemeka Success Esiobu
Department of Agricultural Economics, Extension and Rural Development,
Imo State University, Owerri, Nigeria

*Address all correspondence to: esiobunnaemekasuccess@gmail.com

## References

[1] World Health Organization (WHO). Infection Prevention and Control During Health Care When COVID-19 Is Suspected. 2020a. Available from: https://www.who.int/publications-detail/infection-prevention-and-control-during-health-care-when-novel-coronavirus-(ncov)-infection-is-suspected-20200125

[2] Wang H, Wang Zb, Dong Y, Chang R, Xu C, Yu X, et al. Phaseadjusted estimation of the number of coronavirus disease 2019 cases in Wuhan, China. Cell Discovery. 2020; **6**(1):2-8

[3] Yang J, Zheng Y, Gou X, Ke P, Zhaofeng C, Qinghong G, et al. Prevalence of comorbidities in the novel Wuhan coronavirus (COVID-19) infection: A systematic review and meta-analysis. International Journal of Infectious Diseases. 2020;**94**:91-95. DOI: 10.1016/j.ijid.2020.03.017

[4] World Health Organization (WHO). WHO Director General's Opening Remarks at the Media Briefing on COVID-19. 2020b. Available from: https://www.who.int/dg/speeches/detail/who-director-general-s-opening-remarks-at-the-media-briefing-on-covid19 [Accessed: 11 March 2020]

[5] Nigeria Centre for Disease Control (NCDC). Covid-19 Page. Nigeria Centre for Disease Control; 2020. Available from: https://covid19.ncdc.gov.ng/

[6] African Centre for Disease Control (ACDC). COVID-19 Scientific and Public Health Policy Update 1. 2020. Available from: https://africacdc.org/download/covid-19-scientific-and-public-health-policy-update-31-july-2020/#

[7] PricewaterhouseCoopers (Pwc). Responding to the Impact of Agriculture and Food Security. 2020. Available from: https://www.pwc.com/ng/en/assets/pdf/impact-covid19-food-security-nigeria.pdf [Accessed: June 2020]

[8] Centre for Disease Control and Prevention. Take Steps to Care for yourself and Help Protect Others in your Home and Community. 2020. Available from: https://www.cdc.gov/handwashing/when-how-handwashing.html [Accessed: 02 August 2020]

[9] International Monetary Fund (IMF) (2020). World Economic Outlook, April 2020: The Great Lockdown. 2020. Available from: https://www.imf.org/en/Publications/WEO/Issues/2020/04/14/weo-april-2020 [Accessed: 21 August 2020]

[10] Economic Commission for Africa (ECA). COVID-19 in Africa: Protecting Lives and Economies. 2020. Available at www.uneca.org/publications/covid-19-africa-protecting-lives-and-economies

[11] African Development Bank (AfDB). African Economic Outlook 2020 amid COVID-19: Supplement. 2020

[12] Akande T. An Overview of the Nigerian Rice Economy; Agriculture and Rural Development Department the Nigerian Institute of Social and Economic Research (NISER). 2020. Available from: https://unep.ch/etb/etp/events/Agriculture/nigeria.pdf

[13] Food and Agricultural Organisation (FAO). The State of Food and Agriculture; 2010–2019. FAO: Rome; 2020. p. 2020

[14] Esiobu NS, Onubuogu GC, Njoku SM, Nwachukwu BC. Sustainability and Determinate of Farmers' Mitigation Strategies to Greenhouse Gases (GHGs) Emission in Rice Agric-Food System in Nigeria. Rijeka, Croatia: IntechOpen; 2020. pp. 1-16. DOI: 10.5772/intechopen.93188

[15] Ojo OT, Ogundeji AA, Babu SC, Alimi T. Estimating financing gaps in rice production in southwestern Nigeria. Journal of Economic Structures. 2020;**9**: 12. DOI: 10.1186/s40008-020-0190-y

[16] Esiobu NS, Onubuogu GC. Trend, perceptions and adaptation options of arable crop farmers to climate change in Imo State, Nigeria; a multinomial logit approach. World Science Journal. 2014; **5**(9):12-24

[17] Kushal KB. Analysis of greenhouse gas (methane and nitrous oxide) emission and global warming potential from rice fields: With reference to biological mitigation of climate change. Journal of Earth Sciences & Environmental Studies. 2018;**3**(2): 395-407

[18] National Bureau of Statistics (NBS). National Bureau of Statistics Official Gazette (FGP 71/52007/2, 500 (OL24)): Legal Notice on Publication of the Details of the Breakdown of the National and State Provisional Totals, 2006 Census. 2007. Available from: www.nigerianstat.gov.ng [Accessed: 28 February 2016]

[19] Nigeria Population Commission (NPC). Nigeria Federal Government Initiative of individual head count by gender. Spread, state by state. MOFINEWS. 2006;**6**(3):01-32

[20] Central Bank of Nigeria/Business day Nigeria (CBN/BD). Insecurity and Poverty Crisis in South East: Evidence of Leadership Failure. 2020. Available from: https://businessday.ng/opinion/article/insecurity-and-poverty-crisis-in-south-east-evidence-of-leadership-failure/#:~:text=Following%20Enugu%20is%20Abia%20with,are%20classified%20as%20very%20poor

[21] United Nations (UN). Policy Brief: Impact of COVID-19 in Africa. 2020. Available from: https://www.uneca.org/sites/default/files/PublicationFiles/sg_policy_brief_on_covid19_impact_on_africa_may_2020.pdf

[22] Nigerian Meteorological Agency (NiMET). Drought and Flood Monitoring in South East Bulletin. 2020. Available from: www.nimet.gov.ng [Accessed: 31 July 2020]

[23] National Bureau of Statistics and Federal Ministry of Agriculture and Rural Development (NBS/FMARD). Collaborative Survey on National Agricultural Sample Survey 2014/2015 Draft Report. Abuja, Nigeria: NBS Press; 2016

[24] Kwadwo A, Catherine A, Paul T, Kwaw A, Mekonnen A. HIV and AIDS and farm labor productivity: A review of recent evidence in Africa. Journal of Development and Agricultural Economics. 2010;**2**(12):406-415

[25] Von Braun J, Hill RV, Pandya-Lorch R. The poorest and hungry: A synthesis of analyses and actions. In: Von Braun J, Hill RV, Pandya Lorch R, editors. The Poorest and Hungry: Assessments, Analyses and Actions, Washington, D. C.: International Food Policy Research Institute. 2009. pp. 421–438

[26] Esiobu NS. Relative efficiencies of resource use among Cassava Farmers in Imo State, Nigeria [M. Sc thesis]. Owerri, Nigeria: Department of Agricultural Economics, Extension and Rural Development, Faculty of Agriculture and Veterinary Medicine, Imo State University; 2018

[27] Esiobu NS, Onubuogu GC. Determinant of risk-smart options among farming households in agricultural risk management in Imo State, Nigeria; (a multinomial logit model approach). Journal of Environment Protection and Sustainable Development. 2018;**3**(10):20-39

[28] Ngonidzashe C, Laura A, Maria K, Sandra L, Fernando C, Manabu I, et al.

Sustainable and low greenhouse gas emitting rice production in Latin America and the Caribbean: A review on the transition from ideality to reality. Sustainability. 2018;**2018**(10):671

[29] Onubuogu GC, Esiobu NS. Determinants of allocative (pricing) efficiency of cassava farms in Imo State, Nigeria. Journal of Agriculture and Food Sciences. 2020;**17**(2):86-99. Available from: https://www.ajol.info/index.php/jafs/article/view/194543

[30] George A, Stylianou A. Evaluation of the radio as an agricultural information source in rural areas. Journal of Agricultural & Food Information. 2018;**19**(4):362-376. DOI: 10.1080/10496505.2017.1401480

[31] International Crops Research Institute for the Semi-Arid Tropics (ICRISAT). What African Farmers and Processors Say about the COVID-19 Pandemic and Lockdowns. 2020. Available from: https://reliefweb.int/report/world/what-african-farmers-and-processors-say-about-covid-19-pandemic-and-lockdowns [Accessed: 03 August 2020]

[32] International Food Policy Research Institute (IFPRI). Poverty and Food Insecurity Could Grow Dramatically as COVID-19 Spreads. 2020. Available from: https://www.ifpri.org/blog/poverty-and-food-insecurity-could-grow-dramatically-covid-19-spreads [Accessed: 03 August 2020]

[33] Organisation for Economic Co-operation and Development (OECD). Trust in Government. 2020. Available from: https://www.oecd.org/gov/trust-in-government.htm [Accessed: 05 August 2020]

Chapter 11

# Advances in Rice Postharvest Loss Reduction Strategies in Africa through Low Grade Broken Rice Fractions and Husk Value Addition

*Danbaba Nahemiah, Iro Nkama, Idakwo Paul Yahaya, Mamudu Halidu Badau and Aliyu Umar*

## Abstract

Paddy production in African is increasing at a significantly impressive rate due to increased public and private sector investment, the introduction of high yielding varieties and improved production practices. But about 40% or more of this quantity does not reach the table of consumers largely due to post-harvest losses. These losses are subdivided into physical grain loss (PGL) and grain quality loss (GQL). Efforts towards reducing these losses through valorisation of low quality rice and processing by-products has received attention over the last few years. Innovative development and out scaling of simple, cost effective, adoptable and well-defined practical technology to convert low grade milled rice to nutrient dense value-added products that could be used for family meals or weaning purposes and utilization of rice husk for energy is the new way to go. This paper reviews major advance made especially by the Africa-Wide Taskforce on rice processing and value addition and its partners in developing strategies for minimizing postharvest loss in Africa through the development of technologies for utilization of broken rice fractions and rice husk to reduce postharvest losses. Major challenges mitigating the adoption of this technologies and possible opportunities in the rice postharvest value chain that can attract investment for the improvement of rice production and reduction in rice postharvest losses are also outlined. This synthesis we believe will help in providing future direction for research and support for sustainable rice postharvest system in Africa.

**Keywords:** rice, Africa, postharvest losses, broken rice value addition, husk gasification, innovation

## 1. Introduction

Rice (*Oryza sativa* L.) is one of the most important crop in the world in terms of total developing world production (480 x $10^6$ tonnes of rough rice in 2012) and the number of consumers (3.5 billion) dependent on it as their staple food and is cultivated in over 100 countries in every continent (except Antarctica), from 53°N to 40°S and from the sea level to an altitude of 3 kilometres high [1]. In 2019, the total world rice production amounted to approximately 738.75 million metric tons

(MMT) from total harvested area of approximately 162.71 million ha, making rice the world's third most-produced cereal crop after maize (1.12 billion metric tons) and wheat (731.45 MMT) [2]. On the African continent, especially in sub-Saharan Africa (SSA), rice has become a staple food crop and constitutes major part of the human diet [3]. Over the last three decades, African countries has experienced a consistent increase in rice production and consumption demand making rice the fastest growing staple food especially among low income earners [4]. In countries such Tanzania, Niger and Nigeria transformational changes in the production practices and shift of consumer preference from other coarse grain such as corn, sorghum and millet towards rice is particularly glaring and fuelling increased local production and consumption demand. Available statistics indicated that Africa produce an estimated 20.5 million tonnes of paddy rice annually [5], and West Africa is the continent's rice powerhouse, producing about 66% of the total paddy in Africa, mostly by smallholder farmers [4].

The growth in rice production, processing and consumption in many Africa countries has been shown to have direct correlation with growing income, rapid urbanization, population growth, and change in the occupational structure of African families. It is believed that as more and more women and young girls in Africa join the workforce, and more men live and work in urban area, there is a shift toward food that is more convenient and cooks fast such as rice. Although the per capita consumption of rice is declining in many parts of Asia, in Africa, especially the SSA region, the demand for rice is increasing and at a faster rate than in any part of the world [6]. However, rice production in Africa has not kept pace with the increasing demand, resulting in huge volume of rice imported to fill the gap at a significantly high cost to Africa external reserves. Rice farmers in Africa, especially in Nigeria, Niger and Tanzania, have responded to the increasing demand for rice, as reflected in upward trends in total production in recent years [7]. But, when compared with population increases, the rice production trends are much less impressive and many of the countries are becoming increasingly dependent on rice imports, fuelled by growing production-to-consumption gaps [8].

Geographically, according to International Rice Research Institute (IRRI), Africa has the highest reserves of untapped natural resources for food production globally, especially water and land (130 million ha of inland valley) which are essential for rice production [5]. In spite of these sizeable land and favourable agro-ecological conditions, the Food and Agriculture Organization [7] and The World Bank, [9] states that significant number of population are undernourished while poverty and unemployment levels in country such as Nigeria is significantly high (69%). Added to the high level of unemployment, food insecurity and under nutrition, there is huge food losses and waste along the entire food value chain. It has therefore become imperative to make concerted efforts to reduce losses especially postharvest losses to improve food and nutrition security in Africa [10]. Huge volume of rice produced in Africa for instant like in most developing countries does not reach the table of the final consumers due to significant post-harvest losses in terms of physical grain loss (PGL) and grain quality loss (GQL) [11].

Research for development (R4D) in Africa have developed technologies and innovations and made recommendations for increasing rice productivity through the use of high yielding varieties, expansion of area under cultivation and reducing postharvest losses through good production practices and adoption of improved technologies [10, 11]. However, in most African countries, where tropical weather and poorly developed infrastructure contribute to the problem of food loss, wastage can regularly be as high as 40–50% and has been one of the key encumbrances to farmers' income and sustainable food security in this region [12, 13]. Postharvest losses have therefore contributed significantly to African's inability to attain self-sufficiency in local food production and also a huge drain to local production and food security, as colossal quantities of food, including rice are lost, year after year [13]. Globally, Gustavsson *et al.*, [14] noted that about 1.3 billion tons of food are wasted or lost annually, while in the local context such as Nigeria, the country's agricultural productivity has been generally low, mostly due to post harvest losses of farm produce (20% for grains such as rice and over 40% for fruits and vegetables), and attributed these to poor post-harvest handling, inadequate agro-processing development among other critical factors.

The adoption of good agronomic practices, favourable government policies and shift in consumer preference from other staple coarse grains toward rice have fuelled increased production and yield per hectare of rice across Africa. However, postharvest losses that have been relatively small in absolute terms have increased proportionally with increased yield per ha. Therefore, integrated management of postharvest operations such as threshing, cleaning, drying, parboiling, milling, grading and branding and storage have now been adopted in many rice producing clusters to reduce losses at each stage of the chain [13].

Ndindeng *et al*, [11] observed that resolving the critical issues along the rice value chain in many SSA countries is also impeded by the lack of a simple, adoptable and well- defined practical methodology on how to estimate PGL and GQL after harvest. This makes it impossible to have credible data during the various operations along the rice value-chain. Secondly, there is also wide quality gap between imported milled rice and domestically processed rice. The locally processed rice in Africa including Nigeria tend to be of poor quality due to high level of impurities (stones, weed seeds, sand and insect residues), high level of broken fractions, variability in grain size and colour and off-flavour perceived when cooked. However, many cost effective and efficient postharvest handling machines and practices developed and recommended by R4D organizations are not available for farmers, probably due to poor extension and funding challenges. In postharvest operation such as parboiling, the use of rudimentary technologies has resulted in high losses estimated at 15–20% with high energy and water demand which contributes to the final cost of the final product and environmentally unsustainable practices because of dependent of wood fuel [10]. They recommended the valorisation of rice processing by products to enhance income for the rice value chain actors and also improve food security and sustainable environment.

Broken rice fractions, bran and husk are major by-products of rice processing operations. They account for about 25–50% by weight of milled rice depending on variety and technology of milling. In many rice producing communities in Africa, rice processing by-products such as husk and bran are generally dispose and dumped as hips of wastes in many rice processing sites with little or no environmentally friendly ways of disposal. This has resulted in dusk related health challenges for people living nearby and methane emission during its natural decomposition [15]. But research in many parts of the world including Africa has indicated that rice husk if properly harnessed can serve as good raw materials for fuel [16, 17] and low grade broken fractions could be used for the production of other value added products [10] that may increase farmer's income, safe guide the environment and improve food and nutrition security. Broken rice fractions can be converted to high quality flour and used for the production of value added products that can enhance nutrition and food security and livelihood of smallholder farmers and profitability of small-scale food processing industries [18]. It can also employ huge number of youths and women and serve as sources of employment and reduce restiveness.

This chapter will cover selected innovative techniques and technology advancement made especially by the Africa-Wide Taskforce on Rice Processing and Value Addition and its partners in developing strategies for minimizing postharvest loss in Africa through the development of technologies for utilization of broken rice fractions and rice husk to reduce rice postharvest losses in Africa. Major challenges mitigating the adoption of this technologies and possible opportunities in the rice postharvest value chain that can attract investment for the improvement of rice production and reduction in rice postharvest losses are also outlined. This synthesis we believe will help in providing future direction for research and support for sustainable rice postharvest system in Africa.

## 2. Understanding the rice postharvest value-chain in Africa

Rice postharvest value chain is a set of unit operations in which well matured harvested paddy rice pass through from the point of harvest to consumption. Efficient and sustainable rice postharvest value chain therefore, aimed at minimizing losses and maximizes quality of the harvested grains until it reaches the consumer [10]. At each level of the value chain, several actors are involved and different values of losses are recorded. In Africa, especially in West Africa, several actors using diverse kinds of equipment and techniques are involved in primary, secondary and tertiary postharvest operations of the rice value chain (**Figure 1**).

Losses particularly along the value chain [1–18] has been highlighted as a major source of lost in revenue and productivity among value chain actors as both quantitative and qualitative losses occur during any of the stages [19]. This is an indication that critical attention need to be given to the postharvest value chain to reduce loss in productivity and make rice production a sustainable venture. Technically, when paddy is harvested, it passes through the first routes (A), before storage, but may also be traded directly by farmers to middle men or collected together by farmers' cooperative groups where this exists before marketing at a favourable period. Currently in Africa, especially SSA, little or no value addition is carried out at the primary postharvest level. At the second level (B), some levels of value addition are made where the paddy is either milled after parboiling or directly after winnowing to produce white rice which is traded as milled rice and used for the preparation of traditional whole kernel rice-based foods [20]. At this point where appropriate technologies are used, grain quality is improved which translate into improved economic value and competitiveness of milled rice.

Over the last few years, in Nigeria and other African countries, several large scale integrated mills have been installed which combined parboiling and milling operations and coupled with grading and packaging system. In these mills, parboiling and drying energy are generated by combusting the husks, while milling uses electricity from national grid or private generators. Recently, a third level have been added to the chain, where low grade broken fractions, a by-product of rice milling is converted to rice flour and used for the production of diverse rice-based products (C) or other by-products such as husk are used for energy for artisanal rice parboiling and household cooking [16, 17, 21]. The tertiary postharvest level is built on broken rice, bran

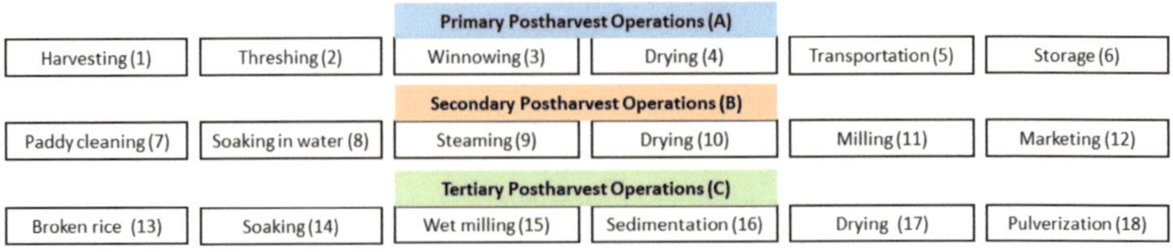


**Figure 1.**
*Unit operations at different levels of rice postharvest system in Africa.*

and husk utilization where low quality rice is converted into flour and used for the production of flour-based products, while bran is used in combination of legumes for the production of animal feeds and sold to animal husbandry firms and husk for energy sources. It is important to note that rice postharvest operations in SSA consist mainly of manual operations resulting in high crop losses and contamination.

## 3. Postharvest losses situation in Africa

Postharvest losses in food production including rice not only have effects on social and economic scales, but also represent a waste of resources used in production such as land, water, energy and other inputs. Report by Africa Postharvest Loss Information System [22] indicated that losses occurs hugely at all levels of the rice postharvest operations. Harvesting operations including harvesting, threshing, winnowing and drying resulted in an average of 11.2% loss due to grain spillage and poor threshing where grains are left on panicles. Transportation resulted in 2.3% (to farm and market) and storage 3.4% indicating an approximately 15.91% average postharvest loss across the continent. Report by Sallah, [23] on the postharvest losses of rice and its implication on livelihood and food security in Africa taking a case of Cameroon and The Gambia indicated that losses at threshing operation were 19 and 17%, drying 9.3 and 7.0%, storage 4.2 and 6.0%, milling 1.3 and 1.0% and transportation 1.33 and 0.8% respectively for Cameroon and The Gambia. This results in reduced income and employability of the people in the study area. Loss was aggravated by lack of or poor processing equipment, poor storage facilities, poor knowledge and skills on postharvest reduction strategies.

It has been estimated in Nigeria by Oguntade *et al.*, [24] that rice post-harvest losses may be as high as 20 to 40%, implying conservatively between 10 and 40% of rice that grown in the country never reaches the market or consumers table or are traded at a discounted price due to loss of quality resulting from poor postharvest management. The high postharvest losses slowdown the marginal increase in rice production recorded over the last few years in many African countries and also threatened food and nutrition security. Because of the adoption of improved technology in rice production in developed countries, postharvest losses occur primarily at the consumer level, with minimal losses at the field or after harvesting or at the other stages of the value chain [10, 25]. In contrast, postharvest losses in Africa occur mainly during harvesting through to market stages, with slightest share of losses occurring at the consumption level [25, 26].

According to Oguntade *et al.* [24], huge losses totalling about 11.39% is recorded during rice postharvest level in Nigeria, with harvesting accounting for 4.43%, threshing and cleaning (4.97%), transporting paddy from field to homes (0.34%), paddy drying and storage (1.53%) and transporting of paddy to local markets (0.12%). At secondary postharvest levels (**Figure 1**), rice parboiling process, an essential pre-treatment given to paddy rice before milling accounted for 5.19% paddy loss, while milling at the village level and milled rice transportation, marketing and storage results in 4.40% and 7.54% losses respectively. Danbaba *et al.*, [10] correlated the data with rice production statistics of 17.5 MMT of paddy produced in Nigeria in 2016 [27], considering postharvest losses of 11.39% paddy from harvest to market and 135 Naira per Kg market price of paddy (as at November, 2018), Nigeria losses about 1.99 MMT of paddy representing 269.09 billion naira annually. These losses are huge and unsustainable if added up to the estimated 123 billion naira losses during the parboiling and milling processes. Situations from the three African countries classically indicates the unfavourable postharvest loss situation on the continent which calls for urgent action and intervention.

## 4. Constraints and need for innovative loss reduction strategies

The continues increase in rice consumption together with minimal increase in domestic production coupled with high postharvest losses, high rice import cost and glaring impacts of climate change and conflicts in Africa, research and development organizations are working together under a coordinated strategy lead by Africa Rice Centre (Africa*R*ice) to provide innovative approach for improving productivity and food and nutrition security through postharvest loss reduction. The rapid advances in small and intermediate technology development, formulation and production of new value added products from low grade broken rice fractions and other rice processing by-products demonstrated the ability to improve food and nutrition security in Africa through novel postharvest loss reduction strategies [10, 16, 21, 28]. Until recently, rice research for development has focused on yield improvement without much emphasis on postharvest practices especially as it relates to loss reduction, quality improvement and marketability. But Nguyen and Ferrero [29] opined that in near future, the possibility of expanding rice production area will remain limited in SSA due to high cost of developing new land suitable for rice production combined with water scarcity for rice production and urban and industrial expansion, implying that loss at any point of the value chain need to minimized to save food and nutrition security in SSA.

In 2008, the SSA countries were faced with significant hike in food price [30]. Milled rice in the international market grow by almost 400% and combined with about 40% rice deficit in SSA, it become highly vulnerable to global rice prize shock and probably was the major cause of 'food riot' in 2008 in countries such as Burkina Faso, Cameroon, Cote d'Ivoire, Mauritania and Senegal [31, 32]. The riot of 2007–2008 [32] triggered renewed focus and investments in rice production together with postharvest operations in many African countries. Nigeria, Ghana, Togo, Cote d'Ivoire, The Gambia, Senegal and Burkina Faso developed a national strategic plan to attain rice self-sufficiency in medium and long time by increasing public and private sector investment into rice sub-sector of their economy, but quality and postharvest losses are least emphasised [33]. In 2011, Africa*R*ice lead a consortium of research organizations in major rice producing countries of Africa to implement and innovative postharvest loss reduction model 'enhancing food security in Africa through the improvement of rice postharvest handling, marketing and development of new rice-based products'. The project emphases the utilization of flour from low grade broken rice fractions to prepare value added food products such as snacks, biscuits, and porridges. This innovative uses of rice can catalyse rural enterprises and raise income, especially for women farmers and processors in Africa [33]. The project also developed innovative technology to utilize rice husk for energy as a strategy to add value to rice husks which are hitherto stockpiled and dumped near mills where they rot and produces methane (a potential greenhouse gas) or burned in the open fields, thus causing pollution.

By improving harvest and postharvest system of rice value chain in Africa, small holder farmer's income will be enhanced through time saving on processing, reduction in qualitative and quantitative postharvest losses which will translate to higher income and better quality of locally milled rice which may compete favourably with imported brands and fetch better price, thereby enhancing the incomes of various actors along the value chain. New rice products containing high nutrients will improve nutrition security and provide employment for women and youths and the overall industrial development of rural communities. The utilization of rice husks for energy will certainly reduce deforestation which is currently threatening significant number of countries of Africa, especially the Sahel region.

## 5. Innovative strategies for rice postharvest loss reduction in Africa

Innovative production is a concept that describes an on-going re-engineering process with the major aims of evolving products and production engineering from prevalent trends based on advances in research for development [34]. Innovative rice postharvest loss reduction trends in Africa is being re-engineered by evolving new value added products based on prevalent research trends. Since production innovation strengthens the productivity and resource use efficiency of production system, recent trends in Africa in the field of rice postharvest system development is the innovative approach to the utilization of rice processing by-products as a strategy to strengthen the productivity of rice and resource use efficiency. The following sections describes the innovative strategies currently used in Africa to reduce postharvest through efficient postharvest system management.

### 5.1 Utilization of broken rice fractions for rice flour production

Fissuring cause by poor postharvest handling of paddy results in broken kernels upon milling, and consequently lost in quality and economic values of milled rice [35]. However, recent increase in the use of rice flour has promoted interest in broken rice fractions utilization as raw materials in many foods especially snacks, porridges and others [36]. Rice flour has been used traditionally for the production of traditional stiff dough (*tuwo*) in Nigeria and many West African countries [20]. Its application in the production of high quality flour that could be used in baking has been hampered by lack of improved rice flour production process that produces flour of particles sizes that could be considered suitable as baking flour and improved functionality [21].

Chiang and Yeh [37] proposed wet milling of rice kernels to produce flour of desirable functionality. As a strategy to valorised broken rice fractions resulting from poor milling processes and rice of low grain quality characteristics, broken rice fractions are processed through wet milling process to produce high quality rice flour that has appreciably acceptable baking quality [21]. The innovative technique which is being commercialized in Africa, involves repeated wet grinding of soaked broken rice fractions and sieving through a fine cloth mesh until virtually all the slurries are made to pass through the sieve. The filtrate is allowed to stand for 3–4 hours depending on the variety and water temperature and decanted to obtain smooth sediment at the bottom. The solid sediment is broken into pieces and dried in an oven before pulverizing and sieving (200 μm) to obtain rice flour (**Figure 2**). The United States Code of Federal Regulation (CFR) state that for a product of milling of grains to be considered as flour, not less than 98% of the particles of the milling process must pass through a sieve having opening not larger than 212 μm [38]. Flour of this particle size characteristics has been demonstrated to impact positively on the end-use application [21, 38–40] studied the physicochemical and functional properties of flours from some common Nigerian rice varieties and concluded that these properties are promising for their application in food systems.

Production of flour from broken rice fraction has been shown to improve the economic value of broken rice kernels by 38% and significant consumer preference for snacks and other baked products. This has significantly reduced qualitative losses incurred during rice processing and improved income of smallholder food processors. The high quality rice flour is also blended with legume based flour (**Figure 3**) to improve protein content and quality to enhance nutrition and product specifications [21] which is an innovative production system.

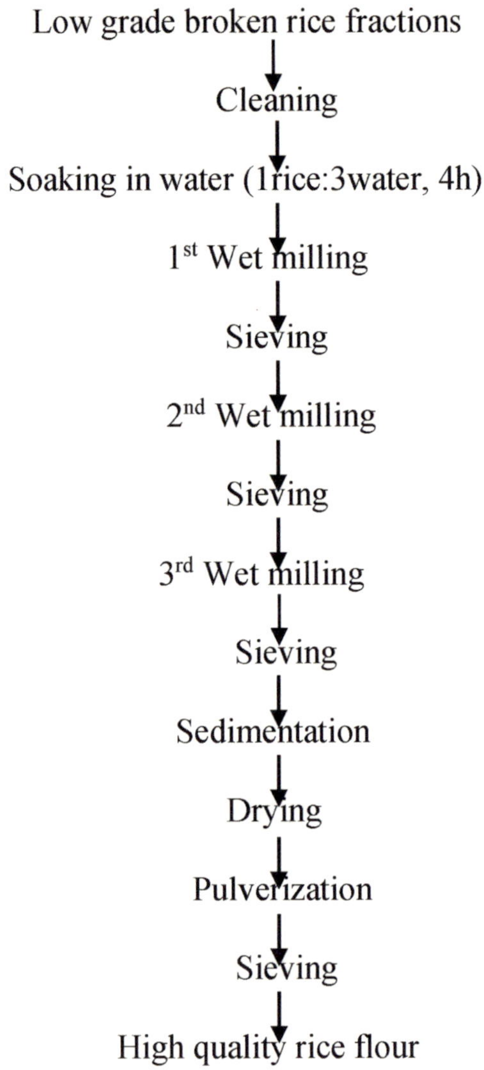


**Figure 2.**
*Flow chart for the production of high quality rice flour from broken rice fractions. Danbaba et al. [21].*

**Figure 3.**
*High quality rice flour from broken rice fractions (left), branded rice flour (centre) and rice flour blended with cowpea flour for the production of high protein baked products [21].*

### 5.2 Development of ready-to-eat (RTE) high protein extruded snacks and porridges from broken rice fractions

Recent changes in social life of many population across the world and the development of middle class worker in developing countries of Africa has resulted in high population of people who are inclined to eat 'ready-to-eat' food, because of its convenience, easy to consume, low to moderate price with minimal need for further processing. Extruded snacks are example of such products and their consumption is growing by day. Extrusion cooking technology is a continuous mixing, cooking and shaping process carried out at high temperatures over short times [41]. It is a very versatile, low-cost and highly energy efficient technology for snack or expanded foods production. Extrusion of cereal-based flours or other starchy raw materials is widely used in the food industry in developed countries to produce snack foods [42]. Little of extrusion cooking is being practiced in Africa especially as it relates to value added rice processing, but recent advances in rice postharvest science has introduce the use of low grade broken rice fractions as raw material for the production of extruded snack foods [10, 43].

However, when starchy raw materials such as rice are subjected to extrusion cooking, there is a chemical and structural transformation such as starch gelatinization, protein denaturation, complex formation between amylose, lipids and/or proteins, and degradation of pigments and vitamins [44]. Under the Africa-Wide Taskforce on Rice Processing and Value Addition of Africa Rice Centre and its national partners, low grade broken rice fractions from different milling operations have been tested and validated for the production of snacks that are high in protein and acceptable to consumers [21, 43]. Through process modelling and optimization, optimum moisture content, barrel temperature and level of legume flour for extrusion have been established for the blends of broken rice fractions with cowpea, bambara groundnut and soybean, keeping other extrusion parameters within range [21, 28, 43]. This optimized process conditions produces extruded snacks with smooth outer-surface (**Figure 4**) and uniform air spaces with regular shape, this according Ryu et al., [45] are features of good quality extrudates.

Because extrusion cooking process allows for the production of low-fat snacks and induces the formation of resistant starch, which makes no caloric contribution

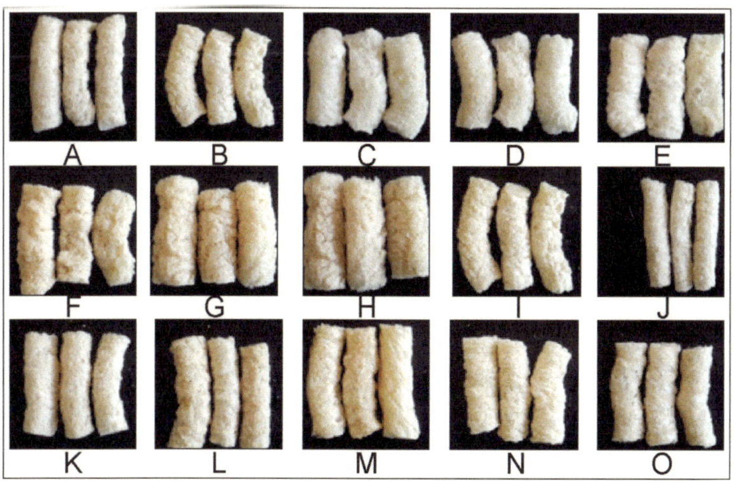


**Figure 4.**
*Photographic images (longitudinal section) of the physical state of rice-cowpea blend extruded snacks.*

and behaves physiologically like dietary fibre [46], rice-based extruded snacks in Africa have received satisfactory acceptability among consumers that are concerned with nutritional quality of food they eat. As a result, therefore, the application of extrusion cooking is increasingly becoming popular for snack production in Africa using raw materials such as rice [28, 43], sorghum [47], and millet [48] containing protein, starch and dietary fibre in an effort to create novel food products such as snacks with a more adequate nutritional value. This new product is expected to improve rice postharvest system through qualitative loss reduction and improve overall food and nutrition security of the populace.

In some instance, it has been demonstrated that when crushed and pulverized, extruded broken rice fractions could be used as porridge or weaning foods. Danbaba *et al* [21, 28] introduced extruded ready-to-eat rice porridge (**Figure 5**) as part of valorisation of low quality broken rice fractions after blending with appropriate amount of legume flour. Protein-energy malnutrition (PEM) and micronutrient deficiency is a severe problem facing developing countries and particularly children

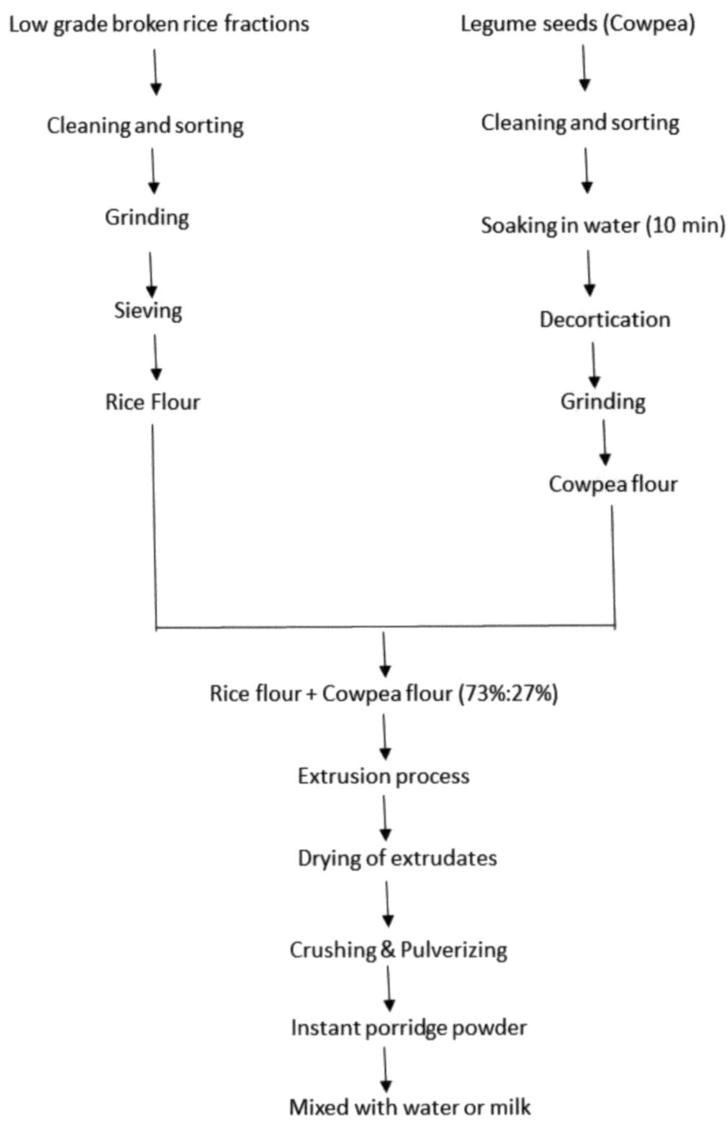


**Figure 5.**
*Production of extruded high protein-energy weaning porridge from blends of broken rice and cowpea.*

under the age of 5 years. This has resulted in more than 50% of childhood death in developing countries including Africa [49, 50]. Blending cereals with legumes in the production of complementary foods has been shown to improve childhood nutrition and significantly reduce mortality [21, 28, 43]. Several authors including Stojceska *et al.*, [51]; Obradović *et al.*, [52]; Panak Balentić *et al.*, [53, 54] have also shown in other parts of the world that it is possible to enrich extruded cereal-based snacks with nutritionally valuable ingredients such as protein from ingredients like legumes. The utilization and application of extrusion cooking in Africa provides an alternative for producing high protein-energy weaning porridges from the blends of low grade broken rice and legumes. This process according to Pathania, *et al.*, [55] credible alternative from the traditional practices for the manufacturing of re-constitutable foods for blended flours (**Figure 5**). Extrusion cooking therefore is expected to impact positively on the rice postharvest system in Africa in the years to come.

### 5.3 Development of third-generation snacks

The increased demand by more consumers for gluten-free products has over the few decades necessitated the quest for suitable alternative raw materials to wheat for the production of third-generation snacks, and the use of rice flour is gaining greater interest because of its favourable attributes of negligible gluten content, good expansion during extrusion and bland taste [56]. Third-generation snacks (3G), also called semi or half products, during production undergo cooking after extrusion and are dried to a stable moisture content (approximately 12%) and then expanded by frying in hot oil, puffing in hot air or microwaving and infrared heating as new variants [57]. In developed world or where extrusion cooking technology has gained popularity, 3G snacks are common. After expansion products are spiced with various types of spices and then packaged and sold as ready-to-eat (RTE) snacks [57]. The products can also be flavoured before expansion and sold as pellets, for preparation at home [58]. In Nigeria, under a strategy to improve postharvest quality of rice, especially poor quality rice varieties having poor parboiling characteristics, kernels are converted to high quality flour of specific particle size and used innovatively for the production of 3G snacks (**Figure 6**) that are current popular among snack producers in many African countries [21].

**Figure 6.**
*Some rice-based 3G snacks produced from low grade broken rice flour.*

Cold forming extrusion (40–70°C, 60–90 bar) of pre-gelatinized rice flour blended legume flour is used for the production of rice-based 3G snacks. Adjusting extrusion temperature, residence time and initial ingredient moisture facilitate complete gelatinization of starch component of the ingredients before frying [57, 59, 60]. Extruded snacks from rice will significantly take some market share as more and more countries in Africa are increasingly improving their rice production and more consumers are becoming more interested in non-gluten baked snacks. Badau *et al.* [61] state that the addition of 30% cowpea to rice flour for the production of traditional Nigeria snack (*Garabia*) significantly improves protein content, metabolizable energy and vitamin $B_2$, while consumer rating based on 9-point hedonic scales was above 6.0 indicating that with the addition of cowpea, the snacks are well-liked by consumers.

## 6. Utilization of rice husk for energy

In 2014, it was estimated that Sub-Saharan Africa produces about 22.1 million tonnes of paddy, which represent about 4.6% of the total global production [5]. Structurally, paddy consists of about 72% kernel, 5–8% bran and 20–22% husk [62]. Therefore, when 22.1 million tonnes of paddy are subjected to milling, it produces about 4.8 million tonnes of husk [11]. With the increased production of paddy in Africa over the last 2 decades, the annual production of rice husk has also proportionally increased. The utilization of rice husk for economic purposes hitherto in Africa especially SSA is very low even though by-products such as rice husk is suitable raw material for energy generation and bran is a nutritive ingredient for food formulation [11, 63]. The high amount of silica in rice husk even when mixed with bran as obtained from village mills (Engelberg type mill) is not suitable for animal feeding purposes. In SSA, significant proportion of rice husk produced is disposed of by burning in open fields or abandoned around rice milling facilities [11]. These practices have resulted in the pollution of air, land and water through the generation of greenhouses gases and particles in water and air [64]. This situation calls for urgent and innovative technique to economically utilize the husk and improve rice postharvest handling for sustainable environment.

Rice husk, a by-product of rice milling is about 20% by weight of paddy and chemically contains about 20% $SiO_2$. Gasification technique for rice husk as energy for rice parboiling and household cooking has been recently developed and is being commercialized across the continent of Africa [65]. Five different rice husk top-lit updraft (TLUD) gasifier household cooking stoves for use in rice processing clusters of Africa has been evaluated under a study to select technically feasible rice husk stove for rural and semi urban household cooking and artisanal rice processing in Africa. Ndindeng, *et al.* [65] study demonstrated that fan-assisted cook stoves especially PO150 recorded better thermal and emission indices and are safer to use than the natural draft gasifiers stove and is therefore recommended for household cooking in rice processing communities of Africa.

Gasification is the process of converting biomass such as rice husk into a combustible gas through thermo-chemical reaction of oxygen in the air and carbon available in the biomass during combustion. In other to gasify rice husk therefore, about 4.7 kg of air per kg of rice is needed [66, 67] and has resulted in the development of several models of fan-assisted rice husk gasifier [65]. The energy obtained are environmentally friendly and the technology easy to use by rural households. Using biomass such as rice husk in Africa for energy generation offers several advantages, including the mitigation of gaseous emissions such as $CO_2$, $SO_x$, and $NO_X$ [68]. This is probably due to low amount of sulphur and nitrogen present in agricultural residues as well as minimal chlorine content [69]. But the question arises as to whether some components of emitted gasses by the stove during burning can contaminate the food being processed and exert toxic effects on consumers. Germaine *et al.* [70] evaluated in vivo toxicity of rice husk used as fuel for household cooking and indicated significantly non toxicity of water boiled with rice husk gasifier. The results obtained by Germaine *et al.* [70] suggested that rice husk used as fuel in household cooking using a fan-assisted rice husk stove is not toxic at 0.5, 1.0 and 2 ml/100100 g body weight and did not produce any evident symptoms in the acute and sub-chronic oral toxicity studies. Even though no evident symptom of toxicity was observed, Quispe *et al.*, [69] suggested that the use of agricultural residues such as rice husk for energy purpose require the performance of integral assessment considering all stage of its life cycle and comparing same with the use of fossil fuels as a means of identifying the conditions and scenarios for a lower environmental impact. Ndindeng, *et al.*, [65], McKendry, et al., [71, 72] illustrated the following as the main advantages of the innovative rice husk gasification cooking stove introduced in Africa:

1. Newly introduced rice husk stove had better performance metrics than that of existing brands in the region.

2. Rice husk mixed with palm kernel shell or other biomass significantly increase burning time but not flame temperature.

3. Data from end-user evaluation were in conformation with stove performance metrics determined instrumentally.

4. If the rice husks are completely burned, the amount of $CO_2$ produced is equal to the amount taken from the environment during the growing stage, making it husk gasification and environmentally sustainable practice.

5. Another advantage is the diversification of energy supply avoiding non-renewable resources depletion which is challenging African forest and farming lands.

## 7. Conclusion

Significant improvement has been made in Africa in terms of rice production mainly as a results of the development of new improved varieties, expansion of area under rice cultivation and huge public and private sector investments. This increased production has resulted in increased by-products such as broken rice fractions and husk. Poor utilization of the broken fractions resulted in reduction of productivity of rice and the husks have become of huge environmental and health changes. The high postharvest losses recorded in Africa has become of great concern to research and development experts, and new innovative methodologies were developed to use broken rice fractions for the production of high quality rice flour that could be used to produce high nutrients and consumer acceptable value added products that improve income and food security of smallholder rice value chain actors. The utilization of rice husk for energy generation has also become a fast moving technology where fan-assisted cooking stoves are developed and provide efficient alternative to fossil fuel. Both qualitative and quantitative postharvest losses in rice are being aggressively managed as a strategy to improve food and nutrition security, environmental sustainability and overall productivity of rice production system. Stakeholder including policy-makers, environmental experts, among others, should as a matter of urgency priority consider the use of biomass as sources of energy for home cooking to reduce over dependence on forest woods and popularize the fan-assisted cooking stove among rural dwellers especially among populations in the Sahel region of Africa where desert is moving fast. Utilization of broken rice fraction as raw materials for flour, snacks, porridges and other foods should be encouraged as means of improving food and nutrition security as well as the socioeconomic development of rural areas.

## Conflict of interest

Authors declare no conflict of interests.

## Author details

Danbaba Nahemiah[1*], Iro Nkama[2], Idakwo Paul Yahaya[3], Mamudu Halidu Badau[3] and Aliyu Umar[1]

1 Rice Research Program, National Cereals Research Institute (NCRI), Badeggi, Niger State, Nigeria

2 Department of Food Science and Technology, University of Nigeria, Nsukka, Enugu State, Nigeria

3 Department of Food Science and Technology, University of Maiduguri, Borno State, Nigeria

*Address all correspondence to: dnahemiah@gmail.com

## References

[1] Juliano B.O. (2016) Rice: Overview. In: Wrigley, C., Corke, H., and Seetharaman, K., Faubion, J., (eds.) Encyclopaedia of Food Grains, Volume. 1, pp. 125-129. Oxford: Academic Press.

[2] Shahbandeh, M (2020). Total global supply of rice 2000/2001-2018/2019. www.statista.com access on 2nd June, 2020.

[3] Oteng, J. W and Sant'Anna, R. (1999). *Rice production in Africa: current situation and issues*. International rice commission newsletter, 48.

[4] GIZ. Competitive African Rice Initiative (CARI) Overview – Rice in Africa. German Ministry for Economic Cooperation and Development: Deutsche Gesellschart Fur Internationale Zusammenarbeit, GmbH; 2018 https://www.cari-project.org

[5] International Rice Research Institute. (IRRI) (2015). World rice statistics online query facility. http://ricestat.irri.org:8080/wrs2/entrypoint.htm. Assessed 16 April 2020.

[6] Mohanty, S. (2013). Trends in Global Rice Consumption. Rice Today, 12, 44-45. https://www.irri.org

[7] Food and Agricultural Organization IFAO. FAOSTAT Database. In: Food and Agriculture Organization. Italy: Rome; 2014 http://faostat.fao.org

[8] Nasrin, S., Lodin, J. B., Jirström, M., Holmquist, B., Agnes Djurfeldt, A. A., and Djurfeldt, G. (2015). Drivers of rice production: evidence from five Sub-Saharan African countries. Agriculture and Food Security, 4:12. http://www.doi.org.10.1186/s40066-015-0032-6

[9] The World Bank (2017). Nigeria Bi-Annual Economic Update: Fragile Recovery. The World Bank, Washington DC, 49. https://doi.org/10.1596/26658

[10] Danbaba, N., Idakwo, P.Y., Kassum, A.L., Bristone, C., Bakare, S.O., Aliyu, U., Kolo, I.N., Abo, M.E., Mohammed, A., Abdulkadir, A.N., Nkama, I., Badau, M.H., Kabaraini, M.A., Shehu, H., Abosede, A.O. and Danbaba, M.K. (2019a) Rice Postharvest Technology in Nigeria: An Overview of Current Status, Constraints and Potentials for Sustainable Development. Open Access Library Journal, 6: e5509. https://doi.org/10.4236/oalib.1105509

[11] Ndindeng, S. A., Manful, J., Futakuchi, K., Moreira, J., Graham-Acquaah, S., Houssou, P., Mapiemfu, D. L., Johnson, P. T., Jarju, O. M., Coulibaly, S. S., Danbaba, N., Sarr, F., Jean G. Nemlin J. G., and Candia, A (2015a). Proceedings of The First International Congress on Postharvest Loss Prevention, Developing Measurement Approaches and Intervention Strategies for Smallholders, ADM Institute for the Prevention of Postharvest Loss, Rome, Italy, October 4-7.

[12] SPORE (2011). Post-harvest management. Adding value to crops. The magazine for agricultural and rural development in ACP countries. N° 152. http://spore.cta.int

[13] Coker, A.A and Ninalowo, S.O. (2016). Effect of post-harvest losses on rice farmers' income in Sub-Saharan Africa: a case of Niger State, Nigeria. Journal of Agricultural Science and Food Technology, 2 (3): 27-34.

[14] Gustavsson J, Cederberg C, Sonesson U, Van Otterdijk R, Meybeck A. (2011). Global Food Losses and Food Waste: Extent Causes and Prevention. United Nations, Food and Agriculture Organization. Rome, Italy.

[15] Contreras, L. M., Schelle, H., Sebrango, C. R and Pereda, I. (2012). Methane potential and biodegradability of rice straw, rice husk and rice residues from drying process. Water Science and Technology, 65(6):1142-1149.

[16] Ndindeng, S.A., Mbassi, J.E.G., Mbacham, W.F., Manful, J., Graham-Acquaah, S., Moriera, J., Dossou, J., and Futakuchi, K (2015b). Quality optimization in briquettes made from rice milling by-products. Energy for Sustainable Development, 29: 24-31. http://dx.doi.org/10.1016/j.esd.2015.09.003

[17] Ndindeng, S.A., Marco W., Sidi, S., Koichi, F. (2019). Evaluation of fan-assisted rice husk fuelled gasifier cooking stoves for application in Sub-Saharan Africa. Renewable Energy, 925-937.

[18] Prasad, K., Singh, Y and Anil, A. (2012). Effects of grinding methods on the characteristics of Pusa 1121 rice flour. Journal of Tropical Agriculture and Food Science, 40(2): 193-201.

[19] Babatunde, R., Omoniwa, A., and Aliyu J. (2019). Postharvest losses along the rice value chain in Kwara State, Nigeria: An assessment of magnitude and determinants. *Cercetari Agronomice in Molldova*, LII (2)178: 141-150. http://doi.org/10.2478/cerce-2019-0014

[20] Danbaba, N., Alabi, M.O., Nkama, I. and Abo, M.E (2004). *Utilization of Rice and Rice by-Products in Nigeria*. In: Abo, M.E. and Abdullahi, A.S., Eds., Nigerian Rice Memorabilia, Project Synergy Limited, Abuja, Nigeria, 343-349.

[21] Danbaba, N., Idinoba, P., Nwelene, F., Umekwue, J. and Maji, A.T. (2017). Entrepreneurship in Rice-Based Food Products for Women and Youths in Nigeria. An Africa Rice Centre/National Cereals Research Institute Training Manual on Rice Product Diversification for Entrepreneurial Development, 32 pp.

[22] APHLIS. Losses at different postharvest stages of rice in sub-Saharan Africa. APHLIS: Africa Postharvest Loss Information System; 2011 www.aphlis.net

[23] Sallah A (2017). Postharvest losses of rice and its implication on livelihood and food security in Africa: the case of Cameroon and The Gambia. MSc. thesis submitted to the Department of Development Studies of the Pan African Institute for Development – West Africa (PAID-WA), Buea. www.paidafrica.org

[24] Oguntade, A.E., Thyimann, D. and Deimling, S. (2014). *Postharvest Losses of Rice in Nigeria and Their Ecological Foot Print.* Federal Ministry of Economic Cooperation and Development. Deutsche Gesellschaft fur Internatinale Zusammenarbeit (GIZ), German Food Partnership/Competitive African Rice Initiative (CARI), 51.

[25] Aulakh, J., Regmi, A., Fulton, J. and Alexander, C. (2013) Estimating Post-Harvest Food Losses: Developing a Consistent Global Estimation Framework. The Agricultural and Applied Economics Association 2013 AAEA and CAES Joint Annual Report.

[26] Olukunle, O.T. (2017). Determinants of postharvest grain Losses in Nigeria. International Journal of Current Research, 9, 63533-63540.

[27] GEMS4 (2017). Mapping of rice production clusters in Nigeria. A project report on growth and employment in states by the United Kingdom Department of International Development (DFID) in Nigeria, 66.

[28] Danbaba N., Nkama I., Badau, M.H and Idakwo, P. Y (2019b). Statistical modelling and optimization of processing conditions of twin-screw extruded rice-legume instant breakfast gruel. *Arid Zone Journal of Engineering, Technology and* Environment, 14(4): 693-712. www.azojete.com.ng

[29] Nguyen, N. V and Ferrero, A. (2006). Meeting the challenges of global rice production. Paddy Water Environ, 4: 1-9. http://dx.doi.10.1007/s10333-005-003-5

[30] Smith T. G (2013). Food price spikes and social unrest in Africa. Climate Change and Political Stability (CCAPS) Research brief No. 11, www.file.ethz.ch

[31] Sneyd LQ, Legwegoh A, Fraser ED. Food riots: media perspectives on the causes of food protest in Africa. Food Security. In: Springer Science + Business Media Dordrecht and International Society for Plant Pathology. 2013 http://doi.10.1007/s12571-013-0272-x

[32] Berazneva J and Lee D. R. (2013). Explaining the African food riots of 2007-2008: An empirical analysis, Food Policy, 39: 28-39. http://dx.doi.10.1016/j.foodpol.2012.12.007

[33] Savitri Mohapatra (2014). Adding value to Africas rice. *Rice Today*, www.ricetoday.irri.org/adding-value-to-africas-rice/

[34] Romero D, Larsson L, Ronnback AO, Stahre J. Strategies for production innovation. In: A conference paper. 2017 http://doi.10.1007/978-3-319-66923-6_1

[35] Owusu, R. K., Sugri, I., Dogbe, W., Adogoba, D. S. and Kugbe, J. (2020). Cracking and breaking response in four rice varieties as influenced by fertilization regime and storage duration, African Journal of Agricultural Research, 16(7): 1044-1049, http://doi.10.5897/AJAR2020.14762

[36] Mukhopadhyay S and Siebenmorgen, T. J (2017). Physical and functional characteristics of broken rice kernels caused by moisture-adsoprtion fissuring. Cereal Chemistry, 94(3): http://doi.org/10.1094/CCHEM-08-16-0214-R

[37] Chiang, P. Y and Yeh, A. I. (2002). Effect of soaking on wet-milling of rice. Journal of Cereal Science, 35:85-94. http://doi.10.1006/jcrs.2001.0419

[38] CFR (2013). CFR 137.105: Requirements for specific standardized cereal flours and related products. Washington, D.C: Code of Federal Regulations; 2013.

[39] Chinma, C.E., Azeez, S.O., Olaitan, O.H., Danbaba, N., Anuonye, J.C and Oloyede, O.O (2018a). physicochemical properties of flours from five Nigerian rice cultivars. Nigerian Food Journal, 36(1):135-140.

[40] Chinma, C.E., Azeez, S.O., Ezekiel, M.Y., Adetutu, M.A., Ocheme, O.B and Danbaba N. (2018b). Functional properties of flours from five Nigerian rice cultivars. Nigerian Food Journal, 36(1): 141-149.

[41] Kebede, L., Worku, S., Bultosa, G. and Yetneberek, S. (2010). Effect of extrusion operating conditions on the physical and sensory properties of tef (Eragrostis tef [Zucc.] Trotter) flour extrudates, EJAST 1(1): 27-39.

[42] Hernandez-Dianz, J. R., Quintero-Ramos, A., Barnard, J and Balandran-Quintana, R. R. (2007). Functional properties of extrudates prepared with blends of wheat flour/pinto bean meal with added wheat bran. Food Science and Technology International, 13(4): 301-308, http://doi.10.1177/1082013207082463

[43] Danbaba N., Nkama I., Badau, M.H and Idakwo, P. Y (2019c). Influence of extrusion conditions on nutritional composition of rice-bamabara groundnut complementary foods. *Arid Zone Journal of Engineering, Technology and* Environment, 14(4): 559-582. www.azojete.com.ng

[44] Ding, Q. B., Ainsworth, P., Plunkett, A., Tucker, G. and Marson, H. (2006).

The effect of extrusion conditions on the functional and physical properties of wheat-based expanded snacks. Journal of Food Engineering, 73: 142-148

[45] Ryu, G. H.; Neumann, P. E. and Walker, C. E. (1993). Effects of some baking ingredients on physical and structural properties of wheat flour extrudates. Cereal Chemistry, 70(3): 291-297.

[46] Larrea, M. A., Chang, Y. K. and Martinez-Bustos, F. C. (2005). Some functional properties of extruded orange pulp and its effect on the quality of cookies. LWT-Food Science and Technology, 38(3): 213-220. http://dx.doi.10.1016/j.lwt.2004.05.014

[47] Jiddere, L and Filli, K. B. (2016). Physicochemical properties of Sorghum malt and Bambara Groundnut Based extrudates. *Journal of Food Science and Technology* Nepal, 9 (55-65).

[48] Filli, K. B., Nkama, I., Jideani, V. A and Abubakar, U.M (2013). Application of response surface methodology for the study of composition of extruded millet-cowpea mixtures for the manufacture of *fura*: A Nigerian food. African Journal of Food Science, 5(17): 884-896. http://www.doi.10.5897/AJFS11.169

[49] Thaoge, M. L., Adams, M. R., Sibara, M. M., Watson T. G., Taylor, J. R., and Goyvaerts, G. E. (2003). Production of improved infant porridges from peal millet using a lactic acid fermentation step and addition of sorghum malt to reduce viscosity of porridge with high protein, energy and solids (30%). World Journal of Microbiology and Biotechnology, 19, 305-310.

[50] Akande O. A., Nakimbugwe, D. and Mukisa, I. M (2017). Optimization of extrusion conditions for the production of instant grain amaranth-based porridge flour. Food Science and Nutrition, 2017; 5:1205-1214. http://www.doi.10.1002/fsn3.513

[51] Stojceska, V., Ainsworth, P., Plunkett, A. and Ibanoglu, S. (2009). The effect of extrusion cooking using different water feed rate on the quality of ready-to-eat snacks from food by-products. Food Chemistry, 114(1): 226-232. http://doi.10.1016/j.foodchem.2008.09.043

[52] Obradovic, V., Babic J., Subaric, D., Ackar, D. and Jozinovic, A. (2014). Improvement of nutritional and functional properties of extruded food products. Journal of Food and Nutrition Research, 53(3): 189-206.

[53] Panak Balentić, J., Ačkar, D., Jozinović, A., Babić, J., Miličević, B., Jokić, S., Pajin, B., and Šubarić, D (2017). Application of supercritical carbon dioxide extrusion in food processing technology, Hemijska industrija 71(00):24-24. http://doi.10.2298/HEMIND150629024P

[54] Panak Balenti, J., Ackar, D., Jokic, C., Jozinovic, A., Babic, J., Borislav Milicevic, B., Šubaric, D and Pavlovic N. (2018). Cocoa Shell: A By-Product with Great Potential for Wide Application. Molecules, 23(6):1404. http://doi.10.3390/molecules23061404

[55] Pathania, S., Singh, B. and Sharma, S. (2013). Development of lowcost multipurpose instant mix by extrusion process using response surface methodology. International Journal of Scientific Research, 2(6): 317-320.

[56] Brunatti, A. C. S., Garcia, E. L., Mischan, M. M. and Leonel, M (2018). Gluten-free puffed snacks of rice and cassava. Australian Journal of Crop Science, 12(2): 185-192. http://doi.10.21475/ajcs.18.12.02.pne477

[57] Panak Balentic, J., Babic J., Jozinovic, A., Ackar, D., Milicevic, B., Muhamedbegovic, B. and Subaric, D.

(2018). Production of third generation snacks. Croatian Journal of Food Science and Technology, 10(1): 98-105. http://doi.10.17508/CJFST.2018.10.1.04

[58] Sevatson, E., and Huber, G. R. (2000). Extruders in Food Industry. In: Extruders in Food Applications, Riaz, M. N. (ed.), Boca Raton, USA: CRC Press, pp. 193-204

[59] Schaaf, H.-J. (1992): Method of making expanded foodstuffs. Germany. US 5153017 A http://www.google.com/patents/US5153017 Accessed July 26, 2017.

[60] Huber, G. (2001): *Snack foods from cooking extruders*. In: Snack Foods Processing, Lusas, E. W., Rooney, L. W. (eds.), Boca Raton, USA: CRC Press, pp. 356-359.

[61] Badau, M. H., Ngozi, C and Danbaba, N. (2013). Quality of *Garabia* (A Nigerian Traditional Snack) From Four Varieties of rice as affected by the addition of cowpea. Advance Journal of Food Science and Technology 5(3): 249-254

[62] Juliano, B.O and Tuano, A. P. P. (2019). Gross structure and composition of rice grain. In Rice: Elsevier and AACC International, 31-53 p, http://doi.10.1016/B978-0-12-811508-4.00002-2

[63] Belewu, M. A and Akinladenu, H. A. (1998). A note on the apparent digestibility of rice husk and hatchery by-product meal based diets fed to West African dwarf goats, Journal of Applied Animal Research, 13(1-2): 197-200, https://doi.org/10.1080/09712119.1998.9706687

[64] Thi Mai, T. P., Kurisu, K. and Hanaki, K (2011). Greenhouse gas emission potential of rice husk for An Giang Province, Vietnam. Biomass and Bioenergy, 35(8): 3656-3666. http://dx.doi.1016/j.biombioe.2011.05.023

[65] Ndindeng, S.A., Marco W., Sidi, S., Koichi, F. (2019). Evaluation of fan-assisted rice husk fuelled gasifier cooking stoves for application in Sub-Saharan Africa. Renewable Energy, 925-937. http://doi.10.1016/j.renene.2019.02.132

[66] Kaupp, A. (1984). *Gasification of rice hull*. Theory and practice. 303

[67] Jameel, H., Kashweni, D., Carter, S., and Treasure T. (2010). *Thermochemical conversion of biomass to power and fuels*. In: Biomass to Renewable Energy Process. Cheng. J. (ed), CRS Press, Florida, 447-487. http://doi.org/10.1201/9781315152868

[68] Saidur, R., Atabani, A. E., Demirbas, A., Hossain, M. S., and Makhilef, S. (2011). A review on biomass as fuel for boilers. Renewable and Sustainable Energy Review, 15(5): 2262-2289.

[69] Quispe, I., Navia R and Kahhat R. (2017). Energy potentials from rice husk through direct combustion and fast pyrolysis: A review. Waste Management, 59: 200-210, http://dx.doi.org/10.1016/j.wasman.2016.10.001

[70] Germaine, M.J.E., Ndindeng, S.A., Loh, A.M.B., Theophile, D and Wilfred, M.F (2020). Evaluation of in vivo toxicity of rice husk used as fuel for cooking in households. African Journal of Biochemistry Research, 14(2): 46-56. http://doi.org/10.5897/AJBR20.1078

[71] McKendry P., (2002a). Energy production from biomass (Part 1). Overview of biomass, *Bioresources Technology*, 83: 37-46. http://doi.10.1016/S0960-8524(10)00118-3

[72] McKendry P., (2002b). Energy production from biomass (Part 2). Conversion te chnologies, *Bioresources Technology*, 83: 47-54. http://doi.10.1016/S0960-8524(01)00119-5

Chapter 12

# Improving the Efficacy of Climate Policy in the Indonesian Rice Sector: The Potential Use of Perceived-Impact Measures in Targeting Policy Beneficiaries

*Rokhani, Mohammad Rondhi, Anik Suwandari, Ahmad Asrofi, Ahmad Fatikhul Khasan, Yasuhiro Mori and Takumi Kondo*

## Abstract

Climate change (CC) increases the frequency of flood and drought and is a significant threat to smallholder rice farming in Indonesia. Adapting to these changes is crucial to minimize the damages to the Indonesian food system. Accordingly, the Indonesian government has formulated National Adaptation Plans (NAPs) to mitigate the effect of climate change on priority sectors, including rice farming. To this end, the Indonesian government included climate change adaptation into the National Development Plan (2019–2024). Selecting the appropriate beneficiaries of this program is crucial to improve the efficacy of Indonesian climate policy. In the case of rice farming, farmers with a high probability to adapt are the appropriate beneficiaries of this program. Thus, this chapter aimed to identify the characteristics of Indonesian smallholder rice farmers with a high probability to adapt to climate change. To this end, this chapter used the findings of the study on 87,330 rice farmers in Indonesia. Education, gender, land tenure security, presence of irrigation infrastructure, application of chemical fertilizer, cropping system, access to extension services, and participation in farmer group are significant determinants of adaptation practices. The finding suggests that prioritizing farmers based on these characteristics are crucial to improve the efficacy of climate policy.

**Keywords:** climate policy, smallholder rice farming, climate change perceived-impact, national adaptation plans, the efficacy of public policy

## 1. Introduction

It is estimated that the Indonesian economy will suffer a loss of at least IDR 100 trillion (~USD 6.7 billion) annually in the period of 2019–2024 due to climate change (CC). The estimation comes only from four economic sectors that are heavily impacted by CC: agriculture, marine and fisheries, water resources, and health. Among others, agriculture (rice sector) is the hardest hit with a total loss amounted to IDR 30 trillion per year, and more than 10 million farmers suffer a production loss of 30–60%. Furthermore, the production loss in the rice sector will impact national food security since rice is the staple food in Indonesia. Adaptation to climate change both at the macro and micro level are required to minimize and even reverse the negative impact of climate change.

Adaptation strategy, both at the national and farm level, are needed to moderate the negative impact of climate change [1]. Adaptation to climate change is grouped into two categories: autonomous and planned adaptation [2]. Autonomous adaptation is adaptation practices taken by farmers using their resources. In contrast, the planned adaptation requires government roles to conceptualize, formulate, and implement the adaptation practices using government resources. The former implemented at the micro/farm-level while the latter conducted at the macro/national level. Several studies have reported that farmers in developing countries have adjusted their farming practices in response to CC and found that the adaptation has a positive effect on crop yield [3, 4]. However, several barriers limit adaptation practices, such as financial barriers (lack of financial resources and lack of supporting institutions, whether public or private, to finance adaptation), social and cultural barriers (individuals and group perspectives, values, and beliefs toward CC), and informational and cognitive barriers (individual perceptions, values, and opinions about the risk of CC) [5]. This study focuses on the third barrier and specifically individual climate risk perception.

A farmer's perception of climate risk is essential because it represents the degree of perceived impact (P-I)—a measure of how a farmer personally feels about the impact of a particular occurrence [6, 7]. Past exposure to climate-related disaster increases the degree of P-I, which in turn drives farmers to undertake adaptation actions [8, 9]. Although some studies stressed the benefits of autonomous adaptation, other studies reported that it ultimately results in unintended maladaptive outcomes, such as increasing the farmer's vulnerability to CC, shifting the vulnerability to other stakeholders or sectors, and decreasing the quality of common pooled resources [10, 11]. Thus, assessing a farmer's P-I toward CC is essential in two aspects: first, it provides valuable information about the efforts to encourage autonomous adaptation; second, it provides crucial insight into the effort to avoid maladaptation practices. As most developing countries have a national adaptation policy [12], this study contributes to addressing the question of which farmers should be prioritized and through what channel the content of a policy should be delivered. **Figure 1** shows how climate risk perception is related to autonomous adaptation and adaptation outcome.

The role of climate risk perception in CC adaptation has received considerable attention. A study in Bangladesh showed that farmer perceptions of CC are mostly aligned with observed meteorological data and are correlated positively with the rate of adopted adaptation practices [9]. Similarly, a study in French coastal populations showed that they perceive the local changes in climate, weather, coral, and beaches. Still, they only regard it as a problem instead of a danger [13]. In contrast to the result from France, a study on the peri-urban community in Mexico that experiences a risk of drought indicated that the community perceives CC and treats it as a threat because their livelihood as brick producers is severely impacted by climate change [14]. A study on Canadian bivalve aquaculture indicated the importance of stakeholder perceptions of CC in adapting to these changes and further expanding the industry [15]. A cross-country analysis in Europe indicated that the perception of CC is affected by individual-level factors such as gender, age, political orientation, and education. Still, the size of the effects of each variable varies across countries [16].

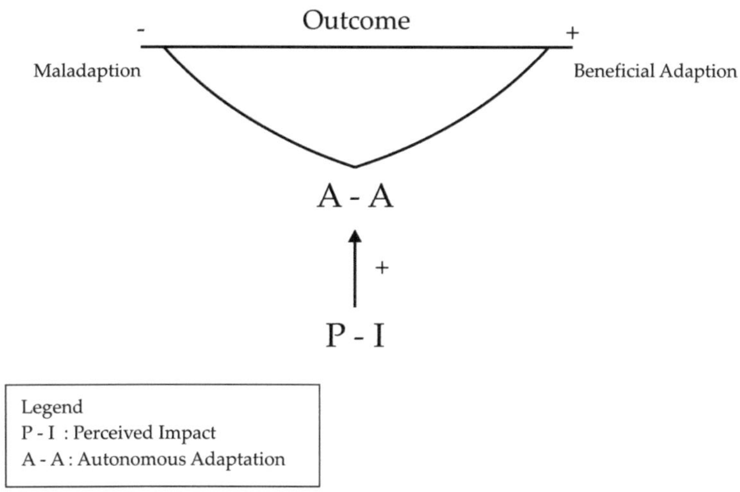


**Figure 1.**
*Perceived-impact of climate change increase autonomous adaptation and affect adaptation outcome (source: Authors work).*

This chapter focuses on identifying Indonesian rice farmer's characteristics that affect the perceived-impact of climate change. Understanding these characteristics is crucial to improve the effectiveness of climate policy since the probability of a farmer to adopt adaptation practices increase with the degree of their perceived-impact of climate change. Furthermore, understanding these characteristics assists the policy implementation-body to locate policy beneficiaries. This chapter is structured as follows; the next section provides an overview of Indonesian smallholder rice farming. The third section discusses Indonesia national framework of climate change adaptation, the RAN-API. The fourth section reviews the findings of our study on factors affecting the degree of perceived-impact of climate change on Indonesian smallholder rice farming. The fifth section explains the policy instruments to improve the effectiveness of Indonesian climate action. And the last section concludes the chapter.

## 2. Overview of Indonesian rice farming

Rice is the most important food crop in Indonesia. The harvested area for Indonesian rice in 2019 reached 10.68 million hectares, with the highest harvest occurring in March of 1.72 million hectares. Indonesia's rice production in 2019 reached 54.60 million tons, with the highest amount of rice production occurring in March at 9.17 million tons. Rice is cultivated throughout Indonesia, but the majority of farmers are on the island of Java. The percentage of rice farmers in Java Island is 55%, followed by Sumatra Island at 9%, Kalimantan Island at 7%, and Sulawesi Island at 6%, while the remaining 20% is spread across other islands. 46% of paddy land is on the island of Java, 14% on the island of Sulawesi, 12% on the island of Sumatra, 8% on the island of Kalimantan, and the remaining 20% is spread over other islands. In 2019, rice production in Indonesia decreased by 4.60 million tons compared to the previous year. Climate change is the leading cause of this decline in production [17].

Climate change is the main factor causing a decrease in rice production in Indonesia. Changes in the intensity and frequency of rainfall, as well as an increase in air temperature, have a significant effect on decreasing rice production in Indonesia. For example, during El Nino, rainfed lowland rice production in Maluku decreased by 2.9% [18]. Then, an increase in temperature of 10°C increases the rainfall by 5% and decreases rice production in North Sulawesi from 6.86 tons/ha to 6.33 tons/ha [19]. An increase in air temperature of 0.4–0.6°C and a decrease in rainfall of 0–197 mm reduce rice productivity in South Sumatra by 0.59% annually. Similarly, Rice productivity in Malang Raya also decreased by 1.59% per year due to an increase in air temperature of 0.7–0.8°C and a decrease in rainfall of 0–550 mm [20].

At the farm level, climate change increases the risk of farming and decreases farmers' income. For example, climate change increases the risk of pest and disease attacks in Subak Penebel District, Tabanan Regency [21]. Then, the flood that occurred in Rawang Panca Arga District, Asahan Regency reduced rice production by 0.60% and caused a total farm loss of IDR 1,256,036 [22]. Furthermore, the floods that occurred during 2006–2010 reduced rice production in West and Central Java by 2.5 tons/hectare and 3.0 tons/hectare; as a result, farmers suffered losses of Rp. 6.5–7.0 million/hectare [23]. These studies show that climate change harms the agricultural sector, especially rice farming, in Indonesia. Government policies to mitigate the impacts of climate change on the agricultural sector are required.

Government policies to mitigate the impacts of climate change are crucial to minimizing losses due to climate change [24]. To overcome the risk of crop failure and decreased productivity due to climate change and pest outbreaks, the government issued a rice farming insurance program (AUTP). The AUTP has a premium price of IDR 36,000/hectare/season and provides benefits for farmers of IDR 6,000,000/ha/season if damage to farming reaches ≥75% [25]. Farmers well receive this program because it helps them provide farming capital for the next planting season when they experience failure in farming [26]. However, farmer participation in this program is low, even though the government subsidized the premium by 80% [27]. It shows that a more comprehensive policy framework and a careful implementation of that policy are required to mitigate the impact of climate change on rice farming in Indonesia.

## 3. Indonesian National Action Plan for climate change adaptation

Indonesia is one of the agricultural countries most vulnerable to the impacts of climate change. Data from the study conducted by the Indonesian Ministry of Environment shows a trend of increasing temperatures of 1°C during the 20th century [28]. Climate change has a significant negative impact on various development sectors in Indonesia, such as food security, health, infrastructure and settlements, and ecosystems. Efforts and strategies to mitigate the impacts of climate change in the short and long term are essential to reduce higher economic losses in the future. For this purpose, the Government of Indonesia has formulated a strategy to deal with the negative impacts of climate change in the form of a national policy framework to address the impacts of climate change, namely the National Adaptation Plan and the National Action Plan for Adaptation to Climate Change Indonesia [29].

Indonesia's National Action Plan for Adaptation to Climate Change (RAN-API) is a national policy framework for dealing with and adapting to climate change. The RAN-API program is an integrated policy concept and involves all relevant parties from the government, society, community organizations, and industry. The purpose of RAN-API is to ensure the achievement of the national development plan and to increase the physical, economic, social, and environmental resiliency of the community against the impacts of climate change. The inclusion of climate change adaptation in national development plans was aimed to reduce the risk of climate change to national development. The targets and policy directions of the RAN-API are in the form of adaptation strategies, policy adjustments, management, technology, and attitudes to minimize the negative impacts of climate change. The strategic targets of RAN-API cover several areas, namely economic resiliency, livelihood resiliency, ecosystem resiliency, and special region resiliency. **Figure 2** shows the schematic representation of the strategic goals and targets of RAN-API.

One of the strategic targets of the national action plan for climate change adaptation (RAN-API) is economic resilience. Climate change harms economic stability and economic development efforts. The emphasis of the RAN-API strategy in the field of economic resilience is divided into two aspects, namely food and energy security. The RAN-API has three priorities to achieve food security. First, reducing the climate-change-related loss rate of food and fisheries production. Second, establish new sources for food production and inland fisheries in areas with low climate risk and minimum environmental impact. Third, develop a food security system for farmers, fishers, and the community with a healthy, nutritionally balanced diet and food diversification.

The primary targets of RAN-API in the field of food security are realized by several means. These are adapting and developing farming systems that are resistant to climate change, developing and applying technologies that are adaptive to climate change, as well as developing and optimizing land, water, and environmental ecosystem resources. The existence of seven main programs supports these strategies—these programs are (1) Adapting the food production system to climate change; (2) Expanding the area of food production; (3) Improvement and development of climate-proof agricultural infrastructure; (4) Food diversification; (5) Development of innovative and adaptive technologies; (6) Development of information and communication system (for climate and technology); and (7) Establishing supporting programs.

To achieve these targets, the Indonesian government established a coordinating body specialized in managing the impact of climate change. This coordinating body

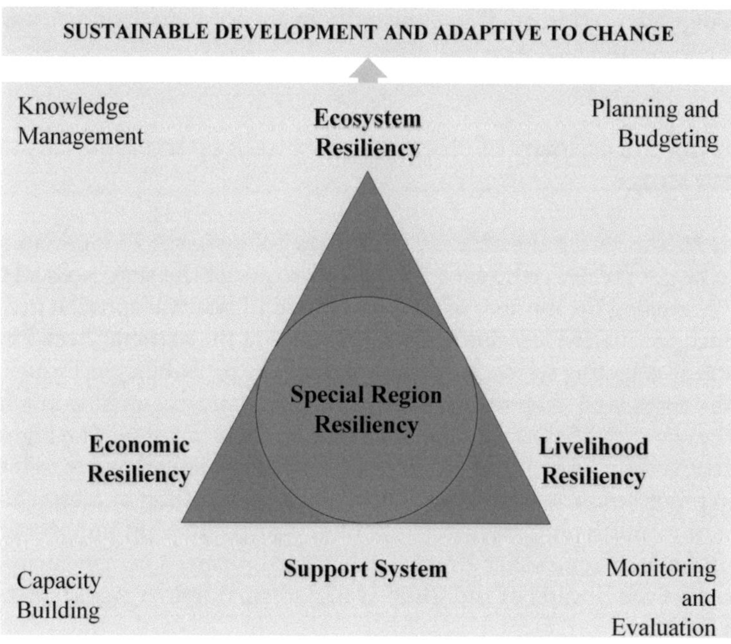


**Figure 2.**
*The strategic goals and targets of RAN-API (source: BAPPENAS [29]).*

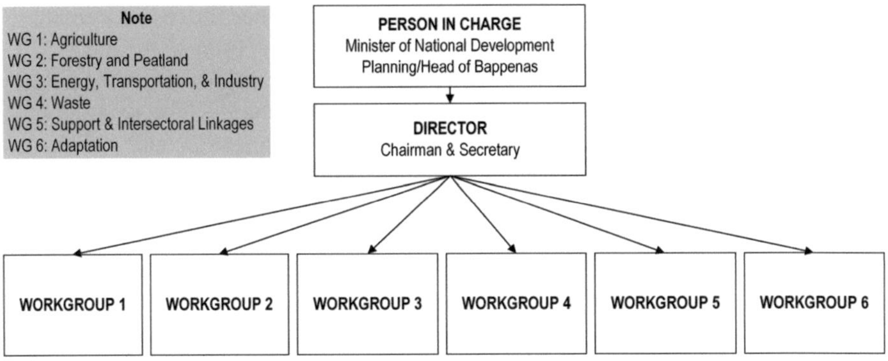


**Figure 3.**
*The organizational structure of the CCCT (source: Authors works).*

is under the Ministry of National Development Planning/National Development Planning Agency and was named *Tim Koordinasi Perubahan Iklim* (Climate Change Coordination Team/CCCT). The primary task of CCCT is to coordinate the national government bodies (ministries and institutions) and regional government (province and regency) in the preparation and implementation of climate policy both at the national and regional levels. Structurally, the CCCT is directed by a chairman and secretary who are under the responsibility of the Minister of National Development Planning/Head of Bappenas. The CCCT has six workgroups which specialized in agriculture; forestry and peatland; Energy, Transportation, and Industry; Waste; Support and Intersectoral Linkages; and Adaptation. **Figure 3** shows the organizational structure of the CCCT.

Currently, the CCCT's primary task is to coordinate the adaptation efforts conducted by both national and regional governments. However, the government has no specific budget for the adaptation program. The current budget for the climate change adaptation program comes from the national and regional government budgets as a part of a general development budget. Besides, the fund for financing climate change adaptation programs comes from government debt, private investment, and corporate social responsibility.

## 4. Improving the efficacy of climate policy using perceived-impact of climate change

The key to improving the efficacy of climate policy in the Indonesian rice sector is to target farmers who are highly likely to adopt the suggested adaptation practices. Increasing the number of farmers who implement adaptation practices would reduce production loss due to climate change at the national level. The problem remains on selecting which farmers to target. The probability of farmers implementing the suggested adaptation practices depends mostly on how much farmers perceive the severity of climate change impact on their farming. The higher farmer perceived the severity of CC impact, the more likely they are to implement adaptation practices. And vice versa. Thus, an understanding of factors that affect the degree of farmer's perceived-impact of CC is of paramount importance. This section will review factors that affect the degree of farmer's perceived-impact of CC based on the finding of the study of 87,330 smallholder rice farmers in Indonesia [30].

This section grouped these factors into the driving and inhibiting factors to adaptation practices. In doing so, the perceived-impact measure was used to determine the category of each factor. A factor that increases the degree of perceived-impact of climate change was categorized into the driving factors of adaptation practice, and those that decrease the degree of perceived-impact were categorized into the inhibiting factors of adaptation. **Figure 4** shows the distribution of driving and inhibiting factors of climate adaptation. The driving factors are those with a positive probability to adapt. In contrast, the inhibiting factors are those with a negative probability of adopting the suggested adaptation practices.

This study suggests that targeting farmers with a high probability of adopting the suggested adaptation practices will improve the efficacy of Indonesian climate policy. Below we describe each variable and its spatial distribution across Indonesian territory.

### 4.1 Education

The first variable is the farmer's education. Education decreases the degree of perceived-impact of climate change. Thus, farmers with a high formal education reported a low degree of climate change impact while those with a low formal education perceived a high degree of climate change impact. Consequently, the latter group has a higher probability of adopting the suggested adaptation practices. Farmer education has received considerable attention in the previous studies. A study on farmers in Pakistan shows that the higher the education of the farmers, the better the adaptation will be. The form of adaptation taken is planting drought-tolerant crops, crop diversification, and these adaptations have a positive impact on food security [31]. Other similar studies on farmers in Ethiopia show that education has a positive impact on climate change adaptation through land conservation and changes in planting time [32]. In general, education plays a crucial role in farm decisions making in Indonesia. Farmer's education significantly affects participation in contract farming for tobacco [33], sugarcane [34], and broiler farmers [35, 36]. Furthermore, education also significantly affects the adoption of certified seed plants for Indonesian sugarcane farmers [37].

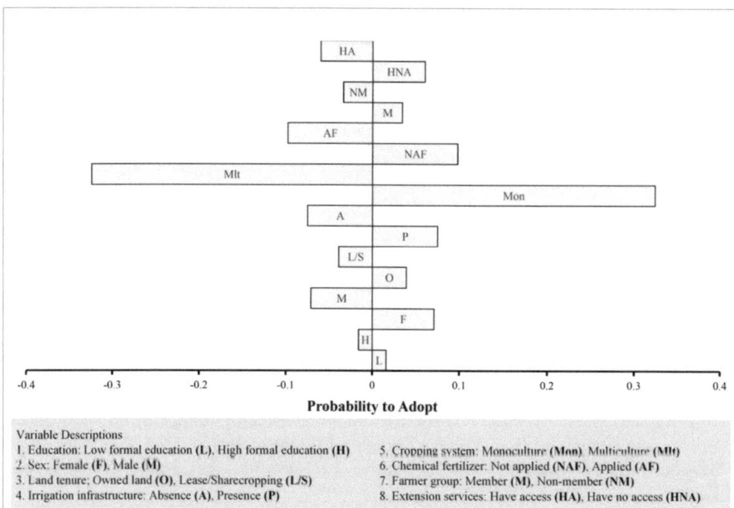


**Figure 4.**
*The driving and inhibiting factors of farmer adoption of the suggested adaptation practices (source: Authors works).*

The majority of Indonesian rice farmers have low-level formal education. 70% of Indonesian rice farmers fall in the category of low education farmers having attended only elementary or never attended formal education. Then, 27.6% of Indonesian rice farmers have a middle-level formal education, both junior (SMP) and senior high school (SMA). Meanwhile, only 2.4% of farmers completed higher education. The majority of low education farmers are located in Java and Sumatera (65.78%), followed by Sulawesi, Bali and Nusa Tenggara, Maluku, and Papua. Even though the majority of farmers have low-level education, each region has a farmer who attended higher education. This figure is crucial to plan a farmer-to-farmer extension program, where high education farmers acted as the key farmer. Establishing a farmer-to-farmer extension is promising in improving the adaptation program and is crucial in the general agricultural development program. **Table 1** shows the distribution of Indonesian rice farmers based on their level of formal education.

## 4.2 Gender

The second variable is the farmer's gender. The study shows that female farmers perceived a lower impact of climate change on their farming. The finding implies that female rice farmers in Indonesia are better at adapting to climate change than male farmers. The literature found a varied effect of farmer's gender on climate adaptation. Studies conducted in Ethiopia [32] and Kenya [38] show that male farmers are more able to adapt to climate change than female farmers. However, research on farmers in Pakistan shows that female farmers are better able to implement climate change adaptation strategies and can maintain food security [31]. Also, research on farmers in Europe suggests that female farmers are more sensitive to climate change [16]. This finding suggests that involving female farmers and policymakers in climate action and climate-related decision making are crucial to improve the effectiveness of the adaptation program. Furthermore, female farmers are present in each region in Indonesia (**Table 2**).

| **Region** | **Low** | **Middle** | **High** |
|---|---|---|---|
| Sumatera | 14,442 (23.65) | 8486 (35.14) | 537 (25.50) |
| Java | 25,729 (42.13) | 6749 (27.95) | 708 (33.62) |
| Kalimantan | 5962 (9.76) | 2466 (10.21) | 348 (16.52) |
| Sulawesi | 6997 (11.46) | 2490 (10.31) | 175 (8.31) |
| Bali & Nusa Tenggara | 6846 (11.41) | 3442 (14.25) | 308 (14.62) |
| Maluku | 579 (0.95) | 282 (1.17) | 16 (0.76) |
| Papua | 520 (0.85) | 234 (0.97) | 14 (0.66) |
| **Indonesia** | **61,075 (70)** | **24,149 (27.6)** | **2106 (2.4)** |

*Notes:*
1. *The region of* ***Sumatera*** *consists of ten provinces (Aceh, North Sumatera, Riau, Jambi, Riau Islands, Bengkulu, Bangka Belitung Islands, West Sumatera, South Sumatera, and Lampung);* ***Java*** *consists of six province (Jakarta, Banten, West Java, Central Java, Yogyakarta, and East Java);* ***Kalimantan*** *has five provinces (West, Central, South, East, and North Kalimantan);* ***Sulawesi*** *has six provinces (North, Central, West, South, and Southeast Sulawesi, and Gorontalo)* ***Bali & Nusa Tenggara*** *has three provinces (Bali, East Nusa Tenggara, and West Nusa Tenggara);* ***Maluku*** *has two provinces (Maluku and North Maluku);* ***Papua*** *has two provinces (Papua and Papua Barat).*
2. *The value in brackets indicates the percentage within each education level.*
3. *The value in brackets on the lowest row indicates the percentage within the total number of the farmer.*

**Table 1.**
*The distribution of Indonesian rice farmers based on their level of education.*

| Region | Male | Female | Density |
|---|---|---|---|
| Sumatera | 19,762 (25.53%) | 3793 (37.06%) | 6 |
| Java | 29,701 (38.53%) | 3485 (34.05%) | 9 |
| Kalimantan | 7909 (10.26%) | 867 (8.47%) | 10 |
| Sulawesi | 8255 (10.71%) | 1407 (13.75%) | 6 |
| Bali & Nusa Tenggara | 10,004 (12.98%) | 592 (5.78%) | 17 |
| Maluku | 844 (1.09%) | 33 (0.32%) | 26 |
| Papua | 709 (0.92%) | 59 (0.58%) | 13 |
| **Indonesia** | **77,094 (88.28%)** | **10,236 (11.72%)** | 8 |

*Notes:*
*1. The region classification is similar to those presented in* ***Table 1.***
*2. The value in brackets indicates the within-group percentage (male/female).*
*3. The density is the male to female ratio. For example, in Maluku, for every 26 male farmers, there is only one female farmer.*
*4. The value in brackets on the lowest row indicates the percentage of male/female farmers to total farmers.*

**Table 2.**
*The distribution of male and female Indonesian rice farmers.*

**Table 2** shows that the majority of Indonesian rice farmers are male (88.28%), and only 11.72% are female. Similar to the previous variables, both male and female farmers were concentrated in Sumatera and Java. However, the data indicate that female farmer density varies across the region. Sumatera and Sulawesi have the highest female farmer's density of 6, which means that there is one female farmer for every six male farmers in the region. Java and Kalimantan have a somewhat similar density of nine and ten. The region of Papua, Bali & Nusa Tenggara and Maluku has the lowest density of 13, 17, and 26 respectively. These figures suggest that involving female farmers in climate action and climate-related decision making in Sumatera, Java, and Sulawesi is a good option. 75.9% of Indonesian rice farmers were located in Sumatera, Java, and Sulawesi. Also, 84.85% of female farmers were located in these regions. Thus, focusing on the gender-related program in these regions is crucial to improve the efficacy of climate policy.

### 4.3 Land tenure

The third variable is land tenure. Land tenure represents the security of farmland ownership. A higher land tenure security encourages the farmer to provide farm investment in the form of adaptation practices. This study categorized land tenure into three groups: owned land, leased land, and sharecropping land. As expected, a farmer who cultivates their land perceives a lower impact of climate change. In contrast, those who cultivated on leased or sharecropped land perceive a higher impact of climate change. This finding suggests that targeting farmers with insecure land tenure is crucial to mitigate the negative impact of climate change.

The importance of land tenure security on smallholder farming has been studied extensively. A study on farmers in Ghana shows that land tenure increases a farmer's willingness to adapt to climate change [39]. The results of this study also show that landowners have a high level of willingness to adapt to climate change compared to tenants and sharecropping farmers. Similarly, farmers in Rome, Italy, with the status of landowners, have a high level of initiative to adapt to climate change [40]. Land tenure security also plays a crucial role in Indonesian agriculture. Land tenure security increases the productivity and profitability of rice farming in Indonesia [41], the probability of participating in contract farming [34], and the adoption of certified seed plants [37]. A weak land tenure security lessens farm investment and makes the farmer more vulnerable to shocks, including climate change. Thus, focusing the climate action on farmers with a weak land tenure security is crucial in moderating the impact of climate change. **Table 3** shows the distribution of Indonesian rice farmers based on land tenure security.

**Table 3** shows that about 30% of Indonesian rice farmers have weak land tenure security and perform farming either on leased or sharecropped land. The majority of these farmers were located in Java and Sumatera. Furthermore, the majority ratio of own farmer to lease/sharecropping farmer is lowest in Java and Sumatra. Both Java and Sumatera have 3 and 2 own to lease/sharecropping farmer ratio, indicating that there is one lease/sharecropping farmer for every three own farmers in Java, and two in Sumatera. Focusing climate action on lease/sharecropping in Java and Sumatera is crucial to improve the effectiveness of the adaptation program.

## 4.4 Irrigation

The fourth variable is the irrigation infrastructure. Historically, irrigation infrastructure has played a central role in Indonesian rice farming. The development of the irrigation system in Indonesia started in the colonial era and continued in each government regime. In the independence period, irrigation development was characterized by the construction of massive irrigation infrastructure [42]. The finding of this study shows that the presence of irrigation infrastructure reduces the perceived-impact of climate change. It indicates that irrigation infrastructure moderates the negative impact of climate change, especially during drought. It also implies that farmers with no access to irrigation are vulnerable to climate change and is appropriate as the beneficiary of climate adaptation program.

Several studies have found the benefit of irrigation infrastructure in mitigating the impact of climate change. The development and improvement of irrigation infrastructure are crucial in mitigating the negative impact of drought due to climate changes in the Vietnamese rice sector [43]. Irrigation development in Vietnam is also able to overcome the excess of water during the rainy season and provide water during the

| Region | Own | Lease | Sharecropping | Ratio |
|---|---|---|---|---|
| Sumatera | 14,452 (23.39%) | 5826 (36.90%) | 3101 (33.16%) | 2 |
| Java | 24,708 (39.99%) | 4976 (31.52%) | 3319 (35.49%) | 3 |
| Kalimantan | 6646 (10.76%) | 1243 (7.87%) | 829 (8.87%) | 4 |
| Sulawesi | 7535 (12.20%) | 1414 (8.96%) | 676 (7.23%) | 4 |
| Bali & Nusa Tenggara | 7064 (11.43%) | 2222 (14.07%) | 1270 (13.58%) | 3 |
| Maluku | 747 (1.21%) | 72 (0.46%) | 57 (0.61%) | 6 |
| Papua | 632 (1.02%) | 35 (0.22%) | 99 (1.06%) | 5 |
| **Indonesia** | **61,784 (70.75%)** | **15,788 (18.08%)** | **9351 (10.71%)** | **3** |

*Notes:*
1. *The region classification is similar to those presented in* ***Table 1****.*
2. *The value in brackets indicates the within-group percentage (own/lease/sharecropping).*
3. *The data contains 407 observations with missing value, thus excluded from data presented in* ***Table 3****.*
4. *The data reported three categories of land tenure owned, leased, and lease-free. However, we categorized 'lease-free' as sharecropping since there are a resource and result-sharing agreement between the farmer and the landowner.*
5. *The ratio is own to the farmer lease+sharecropping ratio. The value indicates the number of lease/sharecropping farmers for every own-farmer.*

**Table 3.**
*The distribution of Indonesian rice farmers based on land tenure.*

dry season [44]. Research conducted on rice farmers in Indonesia shows that irrigation infrastructure development can increase yields of crops, planting season, and planting intensity [45]. Furthermore, the presence of irrigation infrastructure increases agricultural land value in the rural and peri-urban areas in Indonesia [46] and drives the way farmers govern water-user associations [47]. However, the current irrigation infrastructure only covers less than half of Indonesian rice farmers (**Table 4**).

**Table 4** shows that irrigation infrastructure covers only 45.27% of Indonesian rice farmers. Furthermore, the majority of farmers who have access to irrigation infrastructure were located in Java (42.30%), followed by Sumatera, Bali & Nusa Tenggara, Kalimantan, Sulawesi, Maluku, and Papua. However, the percentage of farmers with having no access to irrigation infrastructure was also located in Java (34.45%). But, looking at the ratio, Sulawesi has the highest ratio of 96. It means that for every ten farmers in Sulawesi who have access to irrigation infrastructure, 96 farmers do not. Java, Kalimantan, Bali & Nusa Tenggara, and Maluku have a ratio of less than ten. It means that the number of farmers that have access to irrigation infrastructure is higher than those who do not. This figure indicates that irrigation development was concentrated on the island of Java.

### 4.5 Farming system

The fifth and six variables are cropping system and chemical fertilizer application and belong to the farming system category. Multicultural rice farming system increases farmer's perceived impact of climate change. Thus, farmers applying multi-culture rice farming perceived a high degree of climate change impact. Fortunately, the majority of Indonesian rice farmers applied monoculture rice farming (96.1%). 65.8% of farmers applied monoculture rice farming were located in Sumatera and Java. In contrast, the application of chemical fertilizer decreases the degree of perceived impact of climate change. Chemical fertilizer increases rice yield. Thus, the use of chemical fertilizer is associated with higher rice production. Consequently, it is associated with the low degree of perceived impact of climate change. The majority of Indonesian rice farmers applied chemical fertilizer (91.4%), and 67.8% of them were located in Java and Sumatera. **Table 5** shows the distribution of farmers based on the cropping system and chemical fertilizer application.

| Region | Non-irrigated land | Irrigated land | Ratio |
|---|---|---|---|
| Sumatera | 13,674 (28.61%) | 9.791 (24.77%) | 13 |
| Java | 16,466 (34.45%) | 16,720 (42.30%) | 9 |
| Kalimantan | 3212 (6.72%) | 8776 (14.08%) | 5 |
| Sulawesi | 8757 (18.32%) | 905 (2.29%) | 96 |
| Bali & Nusa Tenggara | 4988 (10.44%) | 5608 (14.19%) | 8 |
| Maluku | 226 (0.47%) | 651 (1.65%) | 3 |
| Papua | 477 (1%) | 291 (.74%) | 16 |
| **Indonesia** | **47,800 (54.73%)** | **39,530 (45.27%)** | **12** |

*Notes:*
*1. The region classification is similar to those presented in* ***Table 1****.*
*2. The value in brackets indicates the within-group percentage (irrigated/non-irrigated land).*
*3. The data contains 407 observations with a missing value.*
*4. The ratio represents the number of the farmer with no access to irrigation infrastructure for every ten farmers with access to irrigation infrastructure.*

**Table 4.**
*The distribution of Indonesian rice farmers based on access to irrigation infrastructure.*

| Region | Cropping system | | Chemical Fertilizer | |
|---|---|---|---|---|
| | Multiculture | Monoculture | Not applied | Applied |
| Sumatera | 881 (26.2) | 22,584 (27.1) | 2633 (35) | 20,832 (26.3) |
| Java | 880 (26.2) | 32,306 (38.7) | 225 (3) | 32,961 (41.5) |
| Kalimantan | 829 (24.7) | 8833 (10.6) | 2947 (39.2) | 6715 (8.5) |
| Sulawesi | 183 (5.4) | 10,413 (12.5) | 820 (10.9) | 9776 (12.3) |
| Bali and Nusa Tenggara | 551 (16.4) | 8225 (9.9) | 805 (10.7) | 7971 (10) |
| Maluku and Papua | 39 (1.2) | 1120 (1.3) | 84 (1.1) | 1075 (1.4) |
| **Indonesia** | **3363 (3.9)** | **83,481 (96.1)** | **7514 (8.6)** | **79,330 (91.4)** |

*Notes:*
*1. The value in brackets indicates the within-group percentage (multi-culture/monoculture and not applied/applied chemical fertilizer).*
*2. The data contains 486 observations with a missing value.*

**Table 5.**
*The distribution of Indonesian rice farmers based on the cropping system and chemical fertilizer application.*

| Region | Extension services | | Farmer group | |
|---|---|---|---|---|
| | Have no access | Have access | Non-member | Member |
| Sumatera | 18,748 (28.8) | 4717 (21.8) | 11,299 (27.2) | 12,166 (26.8) |
| Java | 24,022 (36.9) | 9164 (42.3) | 17,794 (42.9) | 15,392 (34) |
| Kalimantan | 7331 (11.3) | 2331 (10.8) | 4465 (10.8) | 5197 (11.5) |
| Sulawesi | 7442 (11.4) | 3154 (14.5) | 4176 (10.1) | 6420 (14.2) |
| Bali and Nusa Tenggara | 6937 (10.6) | 1839 (8.5) | 3703 (8.9) | 5073 (11.2) |
| Maluku and Papua | 684 (1) | 475 (2.2) | 84 (0.2) | 1075 (2.4) |
| **Indonesia** | **65,614 (75.2)** | **21,680 (24.8)** | **41,521 (47.8)** | **45,323 (52.2)** |

*Notes:*
*1. The region classification is similar to those presented in **Table 5**.*
*2. The value in brackets indicates the within-group percentage (have no access/have access and non-member/member).*
*3. The data contains 39 observations with a missing value for Extension services and 486 missing values for the farmer group.*

**Table 6.**
*The distribution of farmers based on access to extension services and farmer group membership.*

## 4.6 Farm institutions

Farm institutions consist of two variables: agricultural extension services and farmer groups. The farmer group plays a crucial role in Indonesian rice farming. The government used farmer groups to deliver various programs related to rice farming, such as the distribution of subsidized fertilizer, seed, agricultural machinery, and program related to the improvement of farming practices. Thus, participation in farmer group improves farmer ability to adapt to climate change and decreases their perceived impact of climate change. Unfortunately, only 52.2% of Indonesian farmers registered in the farmer group. It suggests that there is still a large proportion of farmers with no access to farmer groups and their subsequent facility.

Agricultural extension service is the second farm institution. Agricultural extension plays a crucial in the dissemination of agricultural technology, farming practices, and current agricultural knowledge and information to farmers. Agricultural extension plays an even more crucial role in climate change adaptation efforts. Farmers require accurate information to adapt to climate change. Furthermore, the existence of agricultural extension officers is crucial in informing the appropriate adaptation practices to farmers and improves farmer's awareness of the threat of climate change on their farming. Unfortunately, the coverage of agricultural extension in Indonesia is relatively low. The current agricultural extension service covers only 24.8% of Indonesian rice farmers. Increasing the coverage of agricultural extension service should be the long-term priority of the Indonesian government in the rice sector policy. Currently, the government should optimize the channel of agricultural extension to identify and target the most vulnerable farmers to climate change (**Table 6**).

## 5. Policy implications

The results of this study indicate that several factors influence the impact of climate change on farmers. Female farmers are more able to adapt to climate change than male farmers. Socialization theory suggests that women are sensitive to respond to change. Thus, female farmers have relevant characters for implementing climate change adaptation and mitigation than male farmers. Therefore, increasing the participation of women farmers in climate change adaptation activities and making adaptation decisions is a crucial factor in increasing farmers' resilience to climate change. Besides, focusing on the region with a high number of vulnerable farmers is crucial to improve the effectiveness of the policy. **Figure 5** shows the distribution of farmers with characteristics of those who perceived a high climate change impact.

The level of education and access to extension reduce the negative impacts of climate change. The information provided by extension agencies can encourage farmers to adapt to face the risks of climate change. Studies on farmers in Mali and Senegal [48] and West Africa [49] show that information is a crucial factor for determining the type of adaptation and increasing farmers' resilience to the risks of climate change. Therefore, it is essential to increase the ability to provide information to minimize the impact of climate change. Strengthening information on climate change can be done in two ways, namely increasing the reach of extension agencies, as well as increasing the role of individual extension workers or extension carried out from farmer to farmer.

The government can strengthen the role of individual farmer extension agents by selecting key farmers or farmers with high levels of education and knowledge. Conduct intensive training for these farmers, and provide supporting programs through these farmers. This strategy is feasible to implement, considering that rice farmers with a high level of education are at the age of 20–40 years.

Land ownership has a positive effect on minimizing the risk of climate change impacts. This study shows that farmers with owned land types have a greater incentive to adapt than farmers with production sharing or rental models. The implementation of climate change policies based on land ownership types can be focused on farmers with weak ownership status. So that farmers with a land lease and sharecropping status will apply adaptation strategies. This strategy is crucial to be implemented, considering that 30% of rice farmers in Indonesia are classified as lease and sharecropping farmers. If these farmers do not adapt to climate change, rice production will decline substantially and threaten food security in Indonesia.

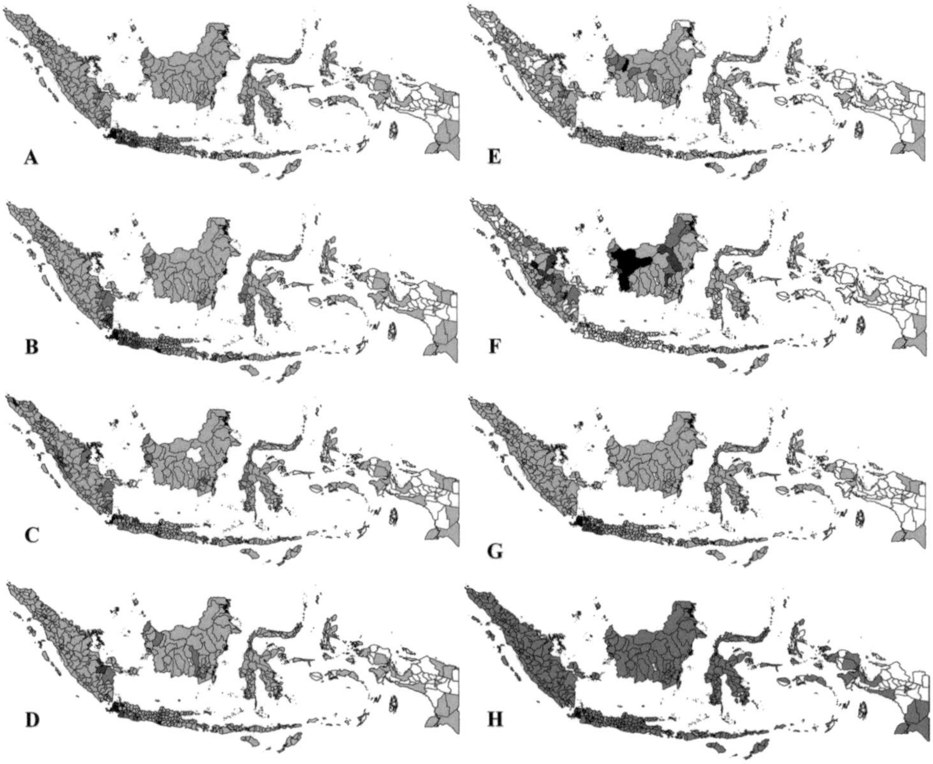


**Figure 5.**
*The distribution of farmers with a high perceived-impact of climate change (A) low education farmers, (B) male farmers, (C) lease/sharecropping farmers, (D) farmers with no irrigation, (E) multiculture farmers, (F) farmers not applied fertilizer, (G) non-member in farmer group, and (H) No access to extension services (the color indicate the farmer density, the darker the color, the higher the density) (source: Authors work).*

## 6. Conclusions

The purpose of this chapter is to review the Indonesian framework for climate change adaptation and identify factors affecting the perceived-impact of climate change on smallholder rice farmers in Indonesia. Eight factors significantly affect how a farmer perceived the impact of climate change. The study indicates that Indonesian rice farmers are concentrated in the Java and Sumatera island. However, a substantial number of farmers outside these areas are heavily impacted by climate change. Thus, targeting program beneficiaries based on the characteristics of vulnerable farmers are crucial in improving the efficacy of Indonesian climate policy.

## Acknowledgements

We wish to acknowledge the Editor and the Author Services Manager of IntechOpen for proving meaningful support during the preparation of the manuscript.

## Conflict of interest

The authors declare no conflict of interest.

## Author details

Rokhani[1], Mohammad Rondhi[2]*, Anik Suwandari[2], Ahmad Asrofi[2], Ahmad Fatikhul Khasan[2], Yasuhiro Mori[3] and Takumi Kondo[4]

1 Department of Agricultural Extension, Faculty of Agriculture, University of Jember, Jember, Indonesia

2 Department of Agribusiness, Faculty of Agriculture, University of Jember, Jember, Indonesia

3 Department of Sustainable Agriculture, Rakuno Gakuen University, Ebetsu, Japan

4 Research Faculty of Agriculture, Hokkaido University, Sapporo, Japan

*Address all correspondence to: rondhi.faperta@unej.ac.id

## References

[1] Seo SN. An analysis of public adaptation to climate change using agricultural water schemes in South America. Ecol Econ [Internet]. 2011;70(4):825-34. Available from: http://dx.doi.org/10.1016/j.ecolecon.2010.12.004

[2] Stage J. Economic valuation of climate change adaptation. Ann N Y Acad Sci. 2010;1185:150-63.

[3] Khanal U, Wilson C, Hoang VN, Lee B. Farmers' Adaptation to Climate Change, Its Determinants and Impacts on Rice Yield in Nepal. Ecol Econ [Internet]. 2018;144(February 2017):139-47. Available from: http://dx.doi.org/10.1016/j.ecolecon.2017.08.006

[4] Waha K, Müller C, Bondeau A, Dietrich JP, Kurukulasuriya P, Heinke J, et al. Adaptation to climate change through the choice of cropping system and sowing date in sub-Saharan Africa. Glob Environ Chang. 2013;23(1):130-43.

[5] Adger WN, Agrawala S, Mirza MMQ, Conde C, O'Brien K, Pulhin J, et al. Assessment of adaptation practices, options, constraints and capacity. Climate Change 2007: Impacts, Adaptation and Vulnerability. Contribution of Working Group II to the Fourth Assessment Report of the Intergovernmental Panel on Climate Change. 2007.

[6] Hossain K, Quaik S, Ismail N, Rafatullah M, Ali I. Climate Change-Perceived Impacts on Agriculture , Vulnerability and Response Strategies for Improving Adaptation Practice in Developing Countries ( South Asian Region ). Int J Agric Res [Internet]. 2016;11(1):1-12. Available from: http://dx.doi.org/10.3923/ijar.2016.1.12

[7] Makuvaro V, Walker S, Masere TP, Dimes J. Smallholder farmer perceived effects of climate change on agricultural productivity and adaptation strategies. J Arid Environ [Internet]. 2018;152(January):75-82. Available from: https://doi.org/10.1016/j.jaridenv.2018.01.016

[8] Gawith D, Hodge I. Moving beyond description to explore the empirics of adaptation constraints. Ecol Indic [Internet]. 2018;95(May):907-16. Available from: https://doi.org/10.1016/j.ecolind.2018.08.022

[9] Hasan MK, Kumar L. Comparison between meteorological data and farmer perceptions of climate change and vulnerability in relation to adaptation. J Environ Manage [Internet]. 2019;237(February):54-62. Available from: https://linkinghub.elsevier.com/retrieve/pii/S0301479719301793

[10] Neset TS, Wiréhn L, Opach T, Glaas E, Linnér BO. Evaluation of indicators for agricultural vulnerability to climate change: The case of Swedish agriculture. Ecol Indic [Internet]. 2019;105(May 2017):571-80. Available from: http://dx.doi.org/10.1016/j.ecolind.2018.05.042

[11] Juhola S, Glaas E, Linnér BO, Neset TS. Redefining maladaptation. Environ Sci Policy. 2016;55:135-40.

[12] Parry J-E, Terton A. How Are Vulnerable Countries Adapting to Climate Change – Frequently Asked Questions (FAQ's) [Internet]. International Institute for Sustainable Development. 2016 [cited 2019 Feb 20]. Available from: https://www.iisd.org/faq/adapting-to-climate-change/

[13] Goeldner-Gianella L, Grancher D, Magnan AK, de Belizal E, Duvat VKE. The perception of climate-related coastal risks and environmental changes on the Rangiroa and Tikehau atolls, French Polynesia: The role of

sensitive and intellectual drivers. Ocean Coast Manag [Internet]. 2019;172(January):14-29. Available from: https://doi.org/10.1016/j.ocecoaman.2019.01.018

[14] Soto-Montes-de-Oca G, Alfie-Cohen M. Impact of climate change in Mexican peri-urban areas with risk of drought. J Arid Environ. 2019;162(December 2017):74-88.

[15] Steeves L, Filgueira R. Stakeholder perceptions of climate change in the context of bivalve aquaculture. Mar Policy [Internet]. 2019;103(January):121-9. Available from: https://linkinghub.elsevier.com/retrieve/pii/S0308597X18301118

[16] Poortinga W, Whitmarsh L, Steg L, Böhm G, Fisher S. Climate change perceptions and their individual-level determinants: A cross-European analysis. Glob Environ Chang [Internet]. 2019;55(May 2018):25-35. Available from: https://doi.org/10.1016/j.gloenvcha.2019.01.007

[17] BPS-Statistics Indonesia. Statistical Yearbook of Indonesia [Internet]. Sub-directorate of Statistical Compilation and Publication, editor. Jakarta: BPS - Statistics Indonesia; 2019. 719 p. Available from: https://www.bps.go.id/publication/2018/07/03/5a963c1ea9b0fed6497d0845/statistik-indonesia-2018.html

[18] Santoso AB. Pengaruh Perubahan Iklim terhadap Produksi Tanaman Pangan di Provinsi Maluku. J Penelit Pertan Tanam Pangan. 2016;35(1):29.

[19] Hosang PR, Tatuh J, Rogi JEX. Analisis Dampak Perubahan Iklim Terhadap Produksi Beras Provinsi Sulawesi Utara Tahun 2013-2030. Eugenia. 2012;18(3).

[20] Ruminta R, Handoko H, Nurmala T. Indikasi perubahan iklim dan dampaknya terhadap produksi padi di Indonesia (Studi kasus : Sumatera Selatan dan Malang Raya). J Agro. 2018;5(1):48-60.

[21] Dewi LWL, Suamba IK. Manfaat Asuransi Usahatani Padi dalam Menanggulangi Risiko Usahatani Krama Subak di Kecamatan Penebel Kabupaten Tabanan. J Agribisnis dan Agrowisata. 2020;9(2):221-31.

[22] Saragih IR, Chalil D, Ayu SF. ANALISIS RISIKO PRODUKSI PADI DALAM PENGEMBANGAN ASURANSI USAHATANI PADI (AUTP) (Desa Panca Arga, Kecamatan Rawang Panca Arga, Kabupaten Asahan). J AGRISEP Kaji Masal Sos Ekon Pertan dan Agribisnis. 2018;17(2):187-96.

[23] Makarim AK, Ikhwani I. INOVASI DAN STRATEGI UNTUK MENGURANGI PENGARUH BANJIR PADA USAHATANI PADI. J Ilmu Tanah dan Lingkung [Internet]. 2011;13(1):35-41. Available from: http://jurnal.ipb.ac.id/index.php/jtanah/article/view/11538/9029

[24] Nuraisah G, Budi Kusumo RA. Dampak Perubahan Iklim Terhadap Usahatani Padi Di Desa Wanguk Kecamatan Anjatan Kabupaten Indramayu. Mimb AGRIBISNIS J Pemikir Masy Ilm Berwawasan Agribisnis. 2019;5(1):60.

[25] Fauzi NF. SOSIALISASI DAN PENDATAAN PESERTA PROGRAM ASURANSI USAHATANI PADI (AUTP) SOSIALISATION AND DATA COLLECTION OF RICE FARMING INSURANCE PARTICIPANTS Nurul Fathiyah Fauzi Program Studi Agribisnis Fakultas Pertanian Universitas Muhammadiyah Jember PENDAHULUA. J Pengabdi Masy Ipteks. 2018;4(1):1-8.

[26] Hardiana J, Elwamendri, Nurchaini DS. FAKTOR-FAKTOR YANG MEMPENGARUHI KEPUTUSAN PETANI DALAM

MENGIKUTI PROGRAM ASURANSI USAHATANI PADI (AUTP) DI KABUPATEN TANJUNG JABUNG TIMUR. J Agribus Local Wisdom. 2019;2(2).

[27] Yanuarti R, Aji JMM, Rondhi M. Risk aversion level influence on farmer's decision to participate in crop insurance: A review<br />. Agric Econ (Zemědělská Ekon [Internet]. 2019 Oct 22;65(No. 10):481-9. Available from: https://www.agriculturejournals.cz/web/agricecon.htm?type=article&id=93_2019-AGRICECON

[28] Kementerian Lingkungan Hidup. Kajian Risiko dan Adaptasi Perubahan Iklim Tarakan Sumatera Selatan Malang Raya Ringkasan untuk Pembuat Kebijakan. 2012.

[29] BAPPENAS. National Action Plan For Climate Change Adaptation (RAN-API) [Internet]. Jakarta; 2012. Available from: https://www.bappenas.go.id/files/2913/4985/2794/national-action-plan-for-climate-change-adaptation-ran-api-synthesis-report__20121226163242__0.pdf

[30] Rondhi M, Fatikhul Khasan A, Mori Y, Kondo T. Assessing the Role of the Perceived Impact of Climate Change on National Adaptation Policy: The Case of Rice Farming in Indonesia. Land [Internet]. 2019 May 10;8(5):81. Available from: https://www.mdpi.com/2073-445X/8/5/81

[31] Ali A, Erenstein O. Assessing farmer use of climate change adaptation practices and impacts on food security and poverty in Pakistan. Clim Risk Manag [Internet]. 2017;16:183-94. Available from: http://dx.doi.org/10.1016/j.crm.2016.12.001

[32] Deressa TT, Hassan RM, Ringler C, Alemu T, Yesuf M. Determinants of farmers' choice of adaptation methods to climate change in the Nile Basin of Ethiopia. Glob Environ Chang. 2009;19(2):248-55.

[33] Rondhi M, Imelda S, Setyawan H, Aji JMM, Hariyati Y, M M, et al. Asymmetric Information And Farmer's Participation In Tobacco Contract Farming. JEJAK [Internet]. 2020 Mar 17;13(1):84-102. Available from: https://journal.unnes.ac.id/nju/index.php/jejak/article/view/17413

[34] Rokhani R, Rondhi M, Kuntadi EB, Aji JMM, Suwandari A, Supriono A, et al. Assessing Determinants of Farmer's Participation in Sugarcane Contract Farming in Indonesia. Agrar J Agribus Rural Dev Res [Internet]. 2020;6(1):12-23. Available from: https://journal.umy.ac.id/index.php/ag/article/view/6478

[35] Rondhi M, Aji JMM, Khasan AF, Putri ATR, Yanuarti R. Risk Aversion, Risk Preference and Farmers' Decision to Participate in Broiler Contract Farming: A Case Study in Jember, Indonesia. Caraka Tani J Sustain Agric [Internet]. 2020 Mar 13;35(1):98. Available from: https://jurnal.uns.ac.id/carakatani/article/view/37964

[36] Rondhi M, Aji JMM, Khasan AF, Yanuarti R. Factors Affecting Farmers' Participation in Contract Farming: The Case of Broiler Sector in Indonesia. Trop Anim Sci J. 2020;43(2):183-90.

[37] Suwandari A, Hariyati Y, Agustina T, Kusmiati A, Hapsari TD, Khasan AF, et al. The Impacts of Certified Seed Plant Adoption on the Productivity and Efficiency of Smallholder Sugarcane Farmers in Indonesia. Sugar Tech [Internet]. 2020 Apr 8;22(3). Available from: https://doi.org/10.1007/s12355-020-00821-2

[38] Ajuang CO, Abuom PO, Bosire EK, Dida GO, Anyona DN. Determinants of climate change awareness level in upper Nyakach Division, Kisumu County, Kenya. Springerplus. 2016;5(1).

[39] Fosu-Mensah BY, Vlek PLG, MacCarthy DS. Farmers' perception and adaptation to climate change: A case study of Sekyedumase district in Ghana. Environ Dev Sustain. 2012;14(4):495-505.

[40] Runsten L, Tapio-Biström M-L. Land Tenure, Climate Change Mitigation and Agriculture [Internet]. 2011. Available from: http://www.fao.org/climatechange/30353-0c11859e8b0cac7aabe39520498b2df22.pdf

[41] Rondhi M, Hariyanto Adi A. Pengaruh Pola Pemilikan Lahan Terhadap Produksi, Alokasi Tenaga Kerja, dan Efisiensi Usahatani Padi. Agrar J Agribus Rural Dev Res [Internet]. 2018;4(2):101-9. Available from: http://journal.umy.ac.id/index.php/ag/article/view/5033

[42] Rondhi M, Khasan AF, Mori Y, Kondo T. Absence of legislation and the quest for an effective mode of governance in agricultural water management: An insight from an irrigation district in central Java, indonesia *. Irrig Drain [Internet]. 2020 Apr 23;(August 2018):ird.2450. Available from: https://onlinelibrary.wiley.com/doi/abs/10.1002/ird.2450

[43] Lebel L, Käkönen M, Dany V, Lebel P, Thuon T, Voladet S. The framing and governance of climate change adaptation projects in Lao PDR and Cambodia. Int Environ Agreements Polit Law Econ [Internet]. 2018;18(3):429-46. Available from: https://doi.org/10.1007/s10784-018-9397-x

[44] Lebel L, Sreymom S, Sokhem P, Channimol K. Empirical and Theoretical Review of Climate Change and Water Governance to Enable Resilient Local Social-Ecological Systems. In: Sam S, Pech S, editors. Climate change and water governance in Cambodia: Challenges and perspectives for water security and climate change in selected catchments, Cambodia. Phnom Penh: CDRI; 2015.

[45] Turral H, Burke J, Faures J-M. Climate change, water and food security. Rome; 2011.

[46] Rondhi M, Pratiwi P, Handini V, Sunartomo A, Budiman S. Agricultural Land Conversion, Land Economic Value, and Sustainable Agriculture: A Case Study in East Java, Indonesia. Land. 2018;7(4):148.

[47] Fatikhul A, Rondhi M, Mori Y. Geolocation data of irrigation network in water user association' s operation area under community-based and provider-based network governance. Data Br [Internet]. 2020;32:106168. Available from: https://doi.org/10.1016/j.dib.2020.106168

[48] Dayamba DS, Ky-Dembele C, Bayala J, Dorward P, Clarkson G, Sanogo D, et al. Assessment of the use of Participatory Integrated Climate Services for Agriculture (PICSA) approach by farmers to manage climate risk in Mali and Senegal. Clim Serv [Internet]. 2018;12(June):27-35. Available from: https://doi.org/10.1016/j.cliser.2018.07.003

[49] Visscher K, Stegmaier P, Damm A, Hamaker-Taylor R, Harjanne A, Giordano R. Matching supply and demand: A typology of climate services. Clim Serv [Internet]. 2020;17(July 2019):100136. Available from: https://doi.org/10.1016/j.cliser.2019.100136